AF823180

Optical Technologies for Aerospace Sensing

Critical Reviews Series

Applications of Electronic Imaging (v.1082, J. C. Urbach)
Applications of High-Power Lasers (v.527, R. R. Jacobs)
The Aviation Security Problem and Related Technologies (CR42, W. H. Makky)
Characterization of Very High Speed Semiconductor Devices & Integrated Circuits (v.795, R. Jain)
Design of Optical Systems Incorporating Low-Power Lasers (v.741, D. C. O'Shea)
Diagnostic Imaging Applications (v.516, E. S. Beckenbach)
Digital Image Processing (v.528, A. G. Tescher)
Digital Optical Computing (CR35, R. Athale)
Electro-Optical Imaging Systems Integration (v.762, R. Wight)
Fiber Optic Communication Technology (v.512, C. W. Kleekamp)
Fiber Optic Sensors (CR44, E. Udd)
Flashlamp-Pumped Laser Technology (v.609, F. Schuda)
Free Electron Lasers (v.728 B. E. Newnam)
Geometrical Optics (v.531, R. E. Fischer, W. H. Price, W. J. Smith)
Gradient-Index Optics and Miniature Optics (v.935, J. D. Rees, D. C. Leiner)
Highly Parallel Signal Processing Architectures (v.614, K. Bromley)
Holographic Nondestructive Testing: Status and Comparison with Conventional Methods (v.604, C. M. Vest)
Holography (v.532, L. Huff)
Image Pattern Recognition: Algorithm Implementation, Techniques, and Technology (v.755, F. J. Corbett)
Industrial Optical Sensing (v. 961, J. Carney, E. Stelzer)
Infrared Detectors (v. 443, W. L. Wolfe)
Infrared Detectors and Arrays (v.930, E. L. Dereniak)
Infrared Optical Design and Fabrication (CR38, R. Hartmann, W. J. Smith)
Infrared Thin Films (CR39, R. P. Shimshock)
Integrated Optics and Optoelectronics (CR45, K.-K. Wong, M. Razeghi)
Interferometric Metrology (v.816, N. A. Massie)
Laser Processing of Semiconductors and Hybrids (v.611, E. J. Swenson)
Laser Wavefront Control (v.1000, J. F. Reintjes)
Lens Design (CR141, W. J. Smith)
Machine Vision Systems Integration (CR36, B. G. Batchelor, F. M. Waltz)
Modeling of the Atmosphere (v.928, L. S. Rothman)
Optical Component Specifications for Laser-Based Systems and Other Modern Optical Systems (v.607, R. E. Fischer, W. J. Smith)
Optical Computing, (v.456, J. A. Neff)
Optical Pattern Recognition (CR40, J. L. Horner, B. Javidi)
Optical Specifications: Components and Systems (v.406, W. J. Smith, R. E. Fischer)
Optical Technologies for Aerospace Sensing (CR47, J. E. Pearson)
Optomechanical and Electro-Optical Design of Industrial Systems (v.959, R. J. Bieringer)
Optomechanical Design (CR43, P. R. Yoder, Jr.)
Photovoltaics (v.543, S. K. Deb)
Radiation Effects on Optical Materials (v.541, P. W. Levy)
Real-Time Signal Processing for Industrial Applications (v.960, B. Javidi)
Remote Sensing (v.476, P. N. Slater)
Robotics and Robot Sensing Systems (v.442, D. Casasent, E. L. Hall)
Spatial Light Modulators and Applications III (v.1150, U. Efron)
Standards for Electronic Imaging Systems (CR37, M. Nier, M. E. Courtot)
Technology of Stratified Media (v.387, R. F. Potter)

U.S. AIR FORCE

Critical Reviews
of Optical Science
and Technology

Volume CR47

Optical Technologies for Aerospace Sensing

James E. Pearson
Editor

Proceedings
of a conference held
16–17 November 1992
Boston, Massachusetts

Sponsored by
SPIE—The International Society for Optical Engineering

SPIE OPTICAL ENGINEERING PRESS

A Publication of SPIE—The International Society for Optical Engineering
Bellingham, Washington USA

Library of Congress Cataloging-in-Publication Data

Optical technologies for aerospace sensing : proceedings of a conference held 16–17 November 1992, Boston, Massachusetts / James E. Pearson, editor ; sponsored by SPIE—The International Society for Optical Engineering.
p. cm. — (Critical reviews of optical science and technology ; v. CR47)
ISBN 0-8194-1042-X (hardcover) — ISBN 0-8194-1041-1 (softcover)
1. Optical detectors—Congresses. 2. Remote sensing—Congresses. I. Pearson, James E. II. Society of Photo-Optical Instrumentation Engineers.
III. Series
TA 1505.06753 1993
621.36—dc20
93-18557 CIP

Published by
SPIE—The International Society for Optical Engineering
P.O. Box 10, Bellingham, Washington 98227-0010 USA
Telephone 206/676-3290 (Pacific Time) • Fax 206/647-1445

Copyright ©1993, The Society of Photo-Optical Instrumentation Engineers.

Copying of material in this book for internal or personal use, or for the internal or personal use of specific clients, beyond the fair use provisions granted by the U.S. Copyright Law is authorized by SPIE subject to payment of copying fees. The Transactional Reporting Service base fee for this volume is $4.00 per article (or portion thereof), which should be paid directly to the Copyright Clearance Center (CCC), 27 Congress Street, Salem, MA 01970. Other copying for republication, resale, advertising or promotion, or any form of systematic or multiple reproduction of any material in this book is prohibited except with permission in writing from the publisher. The CCC fee code is 0-8194-1041-1/93/$4.00.

Printed in the United States of America.

Cover illustration: MODIS imaging spectroradiometer. From W. L. Barnes and V. V. Salomonson, "MODIS: a global imaging spectroradiometer for the Earth Observing System," pp. 285–307.

Contents

vii Conference Committee
ix Preface

SENSOR TECHNOLOGIES

3 **Recent developments in infrared detectors and focal plane arrays**
R. E. Sampson, Environmental Research Institute of Michigan

19 **Optical components and coatings for aerospace sensing**
A. F. Slomba, United Technologies Optical Systems

45 **Lasers for airborne applications**
S. Guch, Jr., D. P. Stanley, Litton Systems, Inc.

64 **Imaging IR sensors: future directions for test and evaluation**
G. D. Boreman, CREOL/Univ. of Central Florida

SENSOR SIGNAL PROCESSING

81 **Optical processing and computing**
R. A. Athale, George Mason Univ.

94 **Optical data storage and retrieval: research directions for the '90s**
S. E. Esener, J. E. Ford, S. Hunter, Univ. of California/San Diego

131 **Optical pattern recognition**
D. P. Casasent, Carnegie Mellon Univ.

AEROSPACE SENSORS

153 **Star sensors for autonomous attitude control and navigation**
R. W. H. van Bezooijen, Lockheed Palo Alto Research Lab.

181 **Laser remote sensing in atmospheric sciences, aviation safety, aeronautical research, and experiments from space platforms and high-flying aircraft**
F. Allario, P. Brockman, R. R. Antcliff, W. B. Grant, NASA Langley Research Ctr.

234 **Photonics for aerospace sensors**
J. M. Pellegrino, E. Adler, A. N. Filipov, L. J. Harrison, J. van der Gracht, D. J. Smith, T. Tayag, E. A. Viveiros, Army Research Lab.

SENSING APPLICATIONS

273 **Optical instruments for remote sensing from space**
M. P. Chrisp, Jet Propulsion Lab.

285 **MODIS: a global imaging spectroradiometer for the Earth Observing System**
W. L. Barnes, V. V. Salomonson, NASA Goddard Space Flight Ctr.

308 **Laser radar systems and applications**
M. J. Halmos, J. H. S. Wang, Hughes Aircraft Co.

Conference Committee

Conference Chair

James E. Pearson, United Technologies Research Center

Session Chairs

Session 1—Sensor Technologies
James E. Pearson, United Technologies Research Center

Session 2—Sensor Signal Processing
Brian M. Hendrickson, Rome Laboratory

Session 3—Aerospace Sensors
Edward R. Washwell, Lockheed Missiles & Space Company, Inc.

Session 4—Sensing Applications
James B. Breckinridge, Jet Propulsion Laboratory

Session 5—Applications of Fiber Optic Sensors
William B. Spillman, Jr. University of Vermont

Preface

Optical technologies for sensing in aerospace applications have matured to the point where they are in widespread use, rapidly replacing many types of mechanical, electronic, and microwave-based sensors. Optical sensing techniques are now common in such diverse applications as navigation and guidance; image and data capture, storage, and retrieval; velocimetry; remote species detection; object identification and ranging; earth observation; and high-speed imaging. Photon-based sensing techniques have always been the technology of choice when the application required high information collection rate, remote location of the sensor, high precision, immunity to electromagnetic interference, or some combination of these features. Passive imaging—ground-based, airborne, or space-based—is perhaps the most obvious optical sensor application.

Early aerospace optical sensing was limited, however, by the resolution and speed of recording and transmission media, and by the long times required to process the large amount of data. In addition, the power, brightness, and spectral purity of optical sources limited the precision and remote sensing distance in many applications. The advantages of optical sensing thus also were the source of many of their limitations. Low application volume often kept costs high, further limiting application.

The advent of high-speed signal processing, megapixel detector arrays, and the laser has removed or greatly reduced many of the limitations of optical sensing. In addition, low cost, small size, low power consumption, and high reliability are features of photon-based sensors now available to the sensor designer and user.

This critical review was designed to survey the state of the art in optical sensing and sensor-related technologies, and to forecast future directions for the technologies as well as their applications. Selected applications are described to illustrate current use and technical challenges. The papers were originally presented as part of the Boston OE/Technology symposium in a two-day conference, 16–17 November 1992.

Session 1, Sensor Technologies, deals with the components of sensors: detectors (Robert Sampson), optics (Albert Slomba), and lasers (Steve

Guch and Doug Stanley). Infrared scene projection is described by Glenn Boreman to illustrate the current state of the art and future developments in selected components.

Session 2, Sensor Signal Processing, addresses the use of optical technologies for signal processing, which offers the prospect of even higher processing speeds than that of microelectronic computers. Ravindra Athale discusses a unified view of different approaches to optical information processing and computing. The paper by Sadik Esener, Joseph Ford, and Susan Hunter reviews the exploitation of three-dimensional optical storage to meet the needs for high-performance information storage and retrieval. David Casasent discusses optical pattern recognition using optical correlator techniques.

Session 3, Aerospace Sensors, looks at selected applications of optical sensors in aerospace applications. The paper by Roelof W. H. van Bezooijen describes how optical star sensors have improved the accuracy of autonomous attitude control and navigation. Laser remote sensing for a variety of purposes is reviewed in a paper by Frank Allario, Philip Brockman, R. R. Antcliff, and William Grant. John Pellegrino *et al.* address fiber optic links and optical signal processing, particularly their use in synthetic aperture radar (SAR) and automatic target recognition (ATR) applications.

Session 4, Sensing Applications, focuses on remote sensing and laser radar applications. Michael Chrisp reviews the design of passive remote sensing instruments for use on space platforms. William Barnes and Vincent Salomonson describe the MODIS instrument, one of the several optical sensors planned for NASA's Earth Observing System (EOS). Laser radar (LADAR) techniques, technologies, and applications are reviewed in the paper by Maurice Halmos and John Wang.

James E. Pearson

Sensor Technologies

Chair
James E. Pearson
United Technologies Research Center

Recent developments in infrared detectors and focal plane arrays

Robert E. Sampson

Environmental Research Institute of Michigan
Ann Arbor, Michigan 48113-4001

ABSTRACT

Infrared radiation is detected and converted to electrical signals by either thermal detectors or photon detectors. Development of focal plane arrays based on these processes continues with progress achieved in increased uniformity for photon detectors and increased sensitivity for thermal detectors. The current state of development for several materials for infrared focal plane arrays is discussed and recent developments reviewed. Applications and requirements for infrared focal plane arrays are described.

1. INTRODUCTION

Progress continues in several areas of development for infrared focal plane arrays. In this paper, the status of development for several materials is discussed and the current state of the art described. We will discuss the state of developments of photon detector materials (HgCdTe, PtSi, InSb, PbS, PbSe) and extrinsic silicon as well as room temperature thermal detectors (Barium Strontium Titanate, Lithium Tantalate, Lithium Niobate and Vanadium Oxide). The more mature detector technologies such as InSb and PtSi are still evolving as applications for larger array configurations with smaller pixel sizes and improved uniformity continue to drive the technology. Other mature technologies such as PbS, PbSe, and photoconductive HgCdTe have been significantly enhanced in their performance in the past 5 to 10 years in response to demands for high responsivity, lower noise, better uniformity, and greater producibility. There have been significant developments in pyroelectric and bolometric thermal detectors (BST, $LiNbO_3$, $LiTaO_3$, VO) with resolvable temperature differences of 0.1° C demonstrated.

Development continues in photovoltaic photon detectors with two to three inch wafer size InSb focal plane arrays reported. While these arrays are more mature and still exhibit superior material uniformity than HgCdTe, progress continues in the development of HgCdTe with arrays up to one inch reported. PtSi Schottky-Barrier arrays represent the most advanced technology for large-area high density focal plane arrays with up to 1040 x 1040 element arrays reported. Increasing attention is being devoted to focal plane array (FPA) reliability, stability, performance limitations by point defects, advanced readouts with signal processing on the focal plane, and two-color arrays. Although little data is available showing reliability problems in very large areas FPAs, several groups are addressing approaches for adapting to the thermal expansion differences between substrates and readouts. Approaches have been reported for two-color IR sensors including stacked MWIR/LWIR diodes for staring arrays and doped silicon scanning FPAs.

High performance LWIR detectors have been fabricated using molecular beam epitaxy (MBE). This implies that the technology is becoming sufficiently mature to support advanced detector structures such as two-color and superlattice devices.

MWIR 640 x 480 FPAs have been successfully demonstrated in both HgCdTe and InSb. The mean NEΔT of the HgCdTe array was 0.013° C and the residual nonuniformity after two-point correction was less than 0.01%. Arrays with 640 x 480 detectors have been reported in LWIR material on CdZnTe substrates. Progress has also been reported in uncooled FPAs where sensors sensitivities of 0.1° C reported using both microbolometer and pyroelectric approaches. The status of development of focal plane arrays made of these detector materials are discussed in detail in the following paragraphs.

2. PHOTON DETECTORS

Photon detectors convert radiation into an electrical signal based on the number or mobility of charge carriers (electrons and/or holes) that are caused by changes in the number of incident infrared photons. Most current infrared focal plane arrays are based on photovoltaic (voltage change), photoconductive (resistance change), or photoemissive (measured as voltage change) effects.

2.1 Platinum Silicide (PtSi)

Platinum Silicide detectors based on internal photoemissive effects offer the largest IR image sensor formats currently available. PtSi represents the most established Schottky-barrier detector technology for applications in the short wave and mid wave infrared bands. The basic construction and operation of Platinum Silicide infrared detectors with a silicon CCD readout is illustrated in Figure 1. The infrared radiation with photon energy less than the bandgap of silicon is transmitted through the substrate. The absorption of the infrared radiation in the platinum silicide layer results in the excitation of photocurrent across the Schottky barrier by internal photoemission. This causes a net negative charge to accumulate on the silicide electrode which is transferred into a CCD MOS readout structure. Progress in the development of platinum silicide technology is summarized in Table 1. Line sensing arrays with up to 4096 x 4 elements and staring arrays with up to 1040 x 1040 elements are available. PtSi Schottky-barrier detectors are fabricated by a well established silicon VLSI process that assures low production cost for large volume, good response uniformity, and provides at this time, the most promising IR technology for very high-resolution and large-area IR imagers. Good thermal imaging near 300K-background has been demonstrated with noise equivalent temperature (NEΔT) of 0.04K to 0.15K for staring PtSi FPAs with 320 x 244 elements to 640 x 486 elements for operation at a temperature of 77K and 30 frames per second.

2.2 Mercury Cadmium Telluride (HgCdTe)

HgCdTe is one of the most important materials for IR detection due to a number of uniquely suitable physical, electronic and thermal properties. High quantum efficiency is achieved with this material for obtaining maximum detection range and temperature resolution in the longwavelength (8-12 μm) infrared region. HgCdTe IR detectors may be made for operation either in the photoconductive

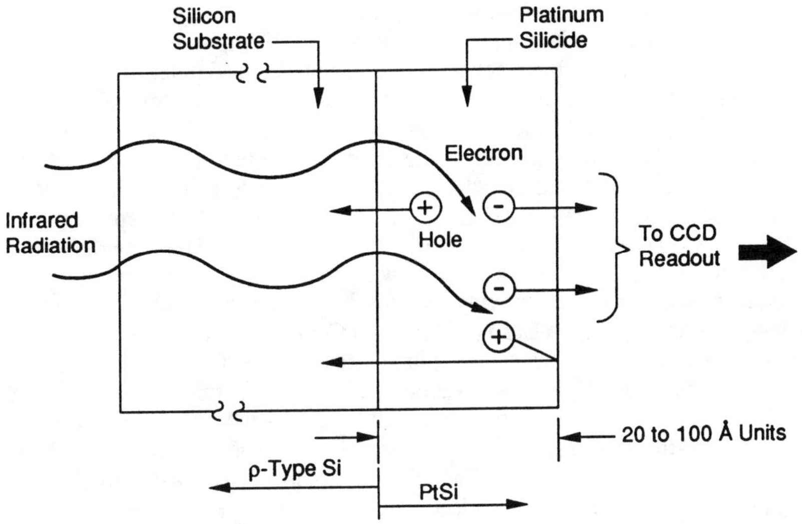

Figure 1. Operation of Scottky-Barrier Detector

Table 1. Schottky-Barrier PtSi IR Focal Plane Arrays

Type of FPA	Pixel Size H x V (μm)^2	Fill Factor (%)	Year	Company
128 x 128 CTD-MOD (1T)	60 x 60	42	1987	Reticon
4096 x 4-CCD	10 x 10	100	1987	Mitsubishi
320 x 244 IT-CCD	40 x 40	43	1988	Sarnoff
324 x 487 IT-CCD	42 x 21	42	1988	NEC
190 x 244 IT-CCD	60 x 36	20	1988	Loral Fairchild
320 x 244 IT-CCD	40 x 40	44	1989	Sarnoff
512 x 488 IT-CCD	31 x 25	36	1989	Loral Fairchild
512 x 512 LACA	30 x 30	54	1989	Reticon
2 ϕ non-interlaced 2048 x 16 TDI-CCD	30 x 30	43	1990	Kodak
256 x 256 IT-CCD	40 x 40	52	1990	Ford
640 x 480 MOS (2T)	24 x 24	38	1990	Sarnoff
648 x 487 IT-CCD	21 x 21	40	1991	NEC
640 x 486 IT-CCD	25 x 25	54	1991	Kodak
2 ϕ non-interlaced 640 x 488 Hybrid	20 x 20	80	1991	Hughes
1040 x 1040 CSD	17 x 17	53	1991	Mitsubishi
512 x 512 CSD	26 x 20	71	1992	Mitsubishi

(photoconductor) or photovoltaic (photodiode) mode. Photoconductors have already been mass produced and used in numerous applications. Compared with photoconductors, photodiodes have several advantages. First, the detectivity of photodiodes can be higher than that of photoconductors by a factor of square root of two at the BLIP (background limited infrared photodetector) condition. Secondly, its response time is limited by transit time, while for photoconductors, it is related to the life-time of carriers as well as the detectivity. HgCdTe detectors are available to cover the spectral range from 1 to 25 μm. Figure 2 illustrates representative spectral response from HgCdTe photovoltaic devices. The versatility of HgCdTe detector material is directly related to being able to grow a broad range of alloy compositions to optimize the response at a particular wavelength. In the last three years, Hughes Aircraft Company and the Santa Barbara Research Center have demonstrated that LWIR photovoltaic HgCdTe staring focal plane arrays can be fabricated reproducibly with high performance and high yield. Record yields for detectors and readouts, in excess of 50%, have been achieved on 128 x 128 arrays with 40 x 40 μm pixels on 40 μm centers. Key to the successes are the development of p-on-n photovoltaic HgCdTe detector technology with high density CMOS readout circuitry. Larger HgCdTe focal plane arrays (640 x 480) are under development at Rockwell. Based on 256 x 256 results to date, detectors have been produced with mean quantum efficiency of greater than 30% and RMS responsivity nonuniformity of less than 8% across a typical array.

2.3 Indium Antimonide (InSb)

Indium Antimonide is a well developed and well behaved detector material for midwave infrared focal plane array applications. Improvements continue in both fabrication and readout technology. Recent work at Westinghouse, Santa Barbara and Cincinnati Electronics as well as at Technicon - IIT, Israel, National Taiwan University in Taiwan and Luoyamg Optic - Electronic Institute in P.R. China has demonstrated improved performance, especially sensitivity and spatial resolution for InSb infrared focal plane arrays.

Photovoltaic InSb remains a popular detector with well understood characteristics for the MWIR spectral band at 80 K. The detectors spectral response at 80 K is shown in Figure 3. InSb material is highly uniform and combined by a planar implanted process in which the device geometry is precisely controlled. The resulting detector array responsivity uniformity is excellent, as shown in Figure 4. Staring arrays of direct hybrid InSb detectors in 58 x 62, 128 x 128, 200 x 200, 256 x 256, and 640 x 480 formats are available with readouts suitable for both high background operation and for low background astronomy applications. There continues to be research devoted to improving readout techniques and focal plane processing for systems of this material.

2.4 Lead Sulfide and Lead Selenide (PbS and PbSe)

Lead sulfide and lead selenide are widely used for high temperature short wavelength infrared applications. Both materials represent mature detector technologies with well understood performance behavior. PbS and PbSe detector materials are made by chemically depositing polycrystalline thin films of these materials on insulating substrates. Both behave as photoconductors and can operate at any temperature between 300 and 77 K. Figure 5 illustrates typical spectral response curves for PbS and PbSe. The responsivity uniformity of PbS and PbSe is good,

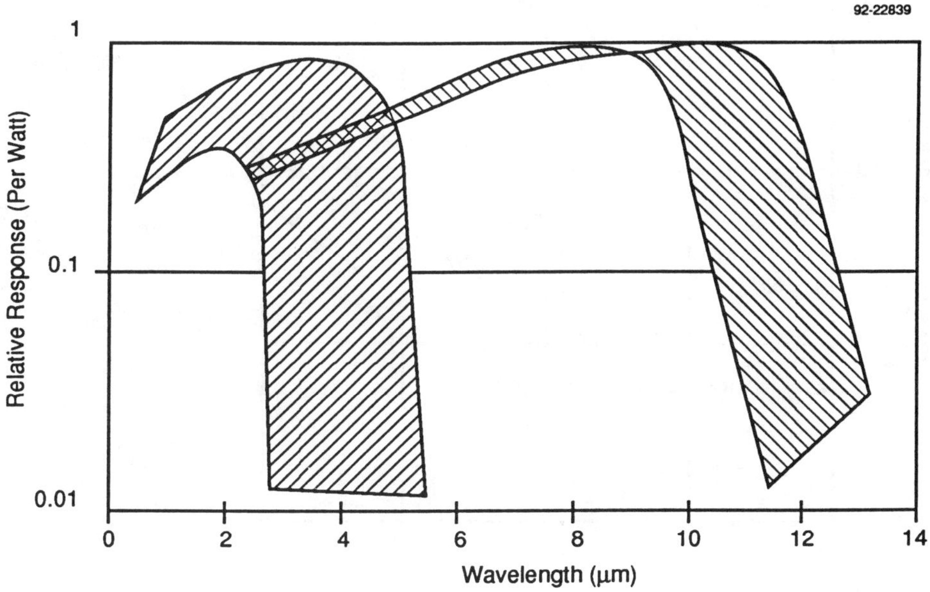

Figure 2. Representative Spectral Response Data for Photovoltaic HgCdTe Detectors. Spectral cutoff can be tailored by adjusting the HgCdTe alloy composition.

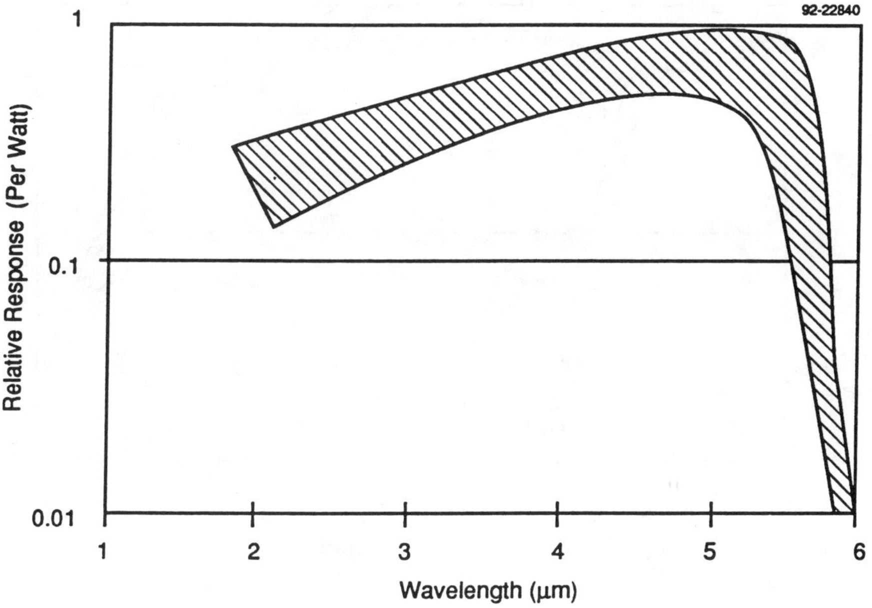

Figure 3. Range of Spectral Response of InSb at 80° K

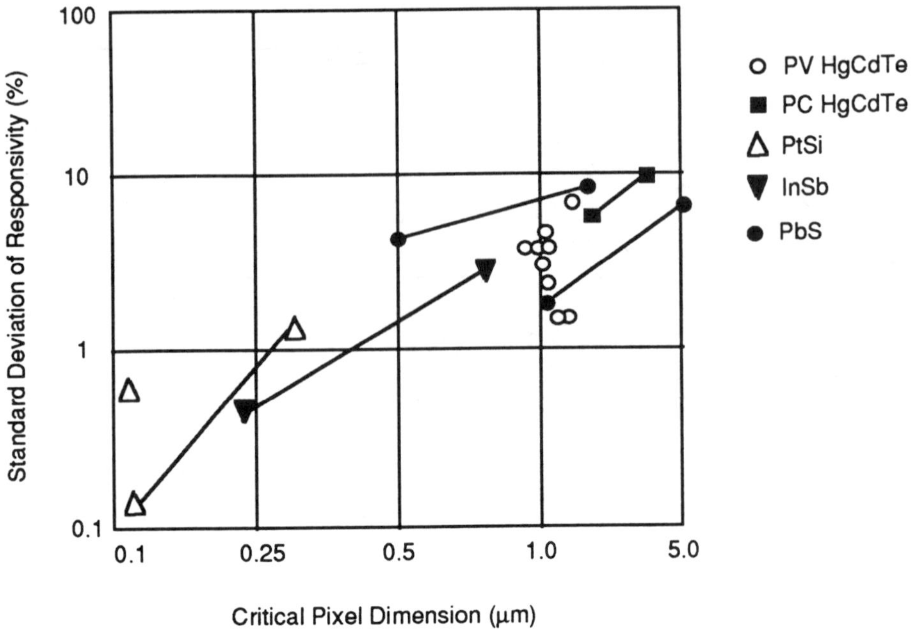

Figure 4. Responsivity Uniformity (one sigma standard deviation/average) for a Variety of IR Sensor Arrays. Comparison is shown with uniformity limits imposed by pixel size (fixed critical dimensions).

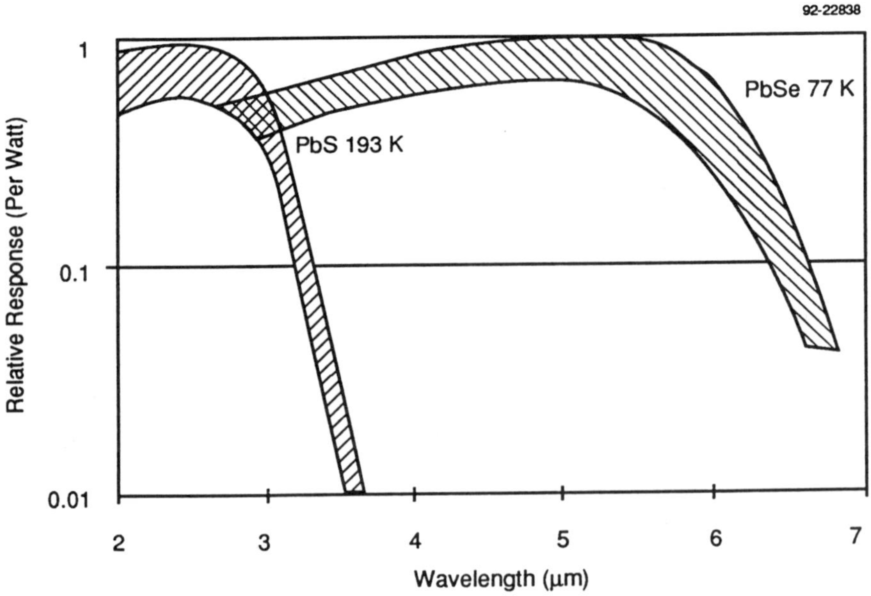

Figure 5. Spectral Response Range of PbS Detectors at 193 K and PbSe Detectors at 77 K. The spectral response for these materials varies with temperature.

generally on the order of 3 to 10%. When sufficiently cooled to eliminate thermal noise, the D* performance of PbS and PbSe at high background flux levels comes within about a factor of 2 of the background limit, implying a quantum efficiency of about 30%. Quantum efficiency is limited by incomplete absorption of the incident flux in the relatively thin (1 to 2 μm) detector material deposited by the chemical process.

PbS and PbSe arrays have been made in a variety of linear array formats for use in focal planes not having cooled readouts. Operability of 99 to 100% is readily achieved for these arrays having 100 or fewer elements. Large arrays of between 1000 and 2000 elements on a single substrate have also been produced with operability exceeding 98%.

2.5 Extrinsic Silicon

Extrinsic silicon detectors rely upon photoexcitation of impurity levels within the bandgap of silicon. The spectral response of the detector depends on the energy level of the particular impurity state and the density of states as a function of energy in the band to which the bound charge carrier is excited. Table 2 lists some of the common impurity levels and the corresponding long wavelength cutoff of the extrinsic silicon detector based on them. The exact long wavelength spectral cutoff is a function of the impurity doping density, with higher densities giving slightly longer spectral response. Figure 6 illustrates the spectral response for several extrinsic silicon detectors. The longer wavelengths of the Arsenic doped detectors is caused by impurity band conduction. The longer spectral response of the impurity band conduction Si:As detector compared with the bulk Si:As detector is due to the higher doping level in the former that reduces the binding energy for an electron.

The performance of extrinsic silicon detectors is generally background limited with a quantum efficiency that varies with the specific dopant and dopant concentration, wavelength, and device thickness. Typical quantum efficiencies are in the range of 10 to 50% at the response peak.

Extrinsic silicon detectors operate either as high-impedance photoconductors or as impurity band conduction devices. Both differ from intrinsic photoconductive devices. Conventional extrinsic photoconductors have only one type of mobile charge carrier, unlike intrinsic devices in which both electrons and holes can move.

3. THERMAL DETECTORS

Thermal detectors absorb radiation and change temperature with a resulting change in electrical characteristics. Currently infrared focal plane arrays rely on either bolometric processes (changes in electrical conductivity with temperature) or pyroelectric processes (change in an observable electric field due to change in dipole moment with temperature). Thermal detectors operate at ambient (or near ambient) temperature and have been known for many years. All of these detectors are sensitive in the infrared region; they operate by first converting the incoming radiation into heat and then sensing the resulting temperature change. Different types of detectors use different methods of sensing the temperature change: the bolometer (temperature

Table 2. Common Impurity Levels Used in Extrinsic Si Infrared Detectors. Operating Temperature Depends Upon Background Flux Level.

Impurity	Energy	Cutoff (μm)	Temp (K)
Indium	155	8	40-60
Bismuth	69	18	20-30
Gallium	65	19	20-30
Arsenic	54	23	13
Antimony	39	32	10

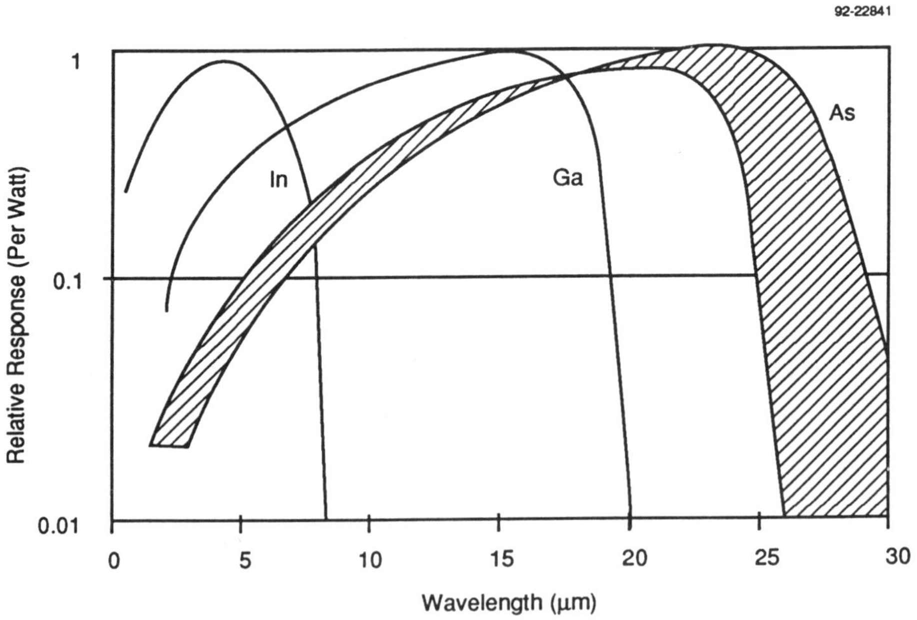

Figure 6. Examples of Extrinsic Silicon Detector Spectral Response. Shown are Si:In, Si:Ga, Si:As bulk detectors.

coefficient of resistance), the capacitance bolometer (temperature coefficient of dielectric constant), the thermocouple (thermoelectric effect), the Golay cell (thermal expansion of a gas), and the pyroelectric detector.

Three programs currently underway at Loral, Honeywell and Texas Instruments have recently reported significant progress in the development of room temperature focal plane arrays. These rely on one of two basic thermal detection processes to accomplish infrared radiation detection: bolometric or pyroelectric. The bolometric effect is a change in the electrical resistance of the responsive element due to temperature changes produced by absorbed incident infrared radiation. The pyroelectric effect is a change in the surface charge of the responsive element due to a temperature change produced by absorbed incident infrared radiation. The Loral and Texas Instrument systems are based on the pyroelectric effect whereas the Honeywell system is based on the bolometric effect. These systems are discussed in more detail in the following paragraphs.

The long wavelength infrared imaging system under development at Loral features a solid state pyroelectric focal plane array sensitive to 8-14 μm radiation operating at ambient temperature. The focal plane array is fabricated as a hybrid structure with a fully reticulated lithium tantalate detector array bump mounted to a CMOS multiplexer. The prototype imaging system uses a focal plane array of 192 x 128 pixels on 50 μm centers. The next version will use a focal plane with approximately 330 x 240 pixels for standard television compatibility. The predicted noise equivalent temperature difference (NETD) for these systems is less than 0.1°K at a 30 Hz display rate.

Progress in uncooled technology and the short range thermal sight program, lead NVEOD and DARPA to launch the High-Density Array Development (HIDAD) program in early 1987. Honeywell and Texas Instruments were funded under the HIDAD program, which called for the development of uncooled focal plane arrays continuing approximately 80,000 detector elements on 2 x 2-mil centers, with a required noise-equivalent temperature difference (NETD) of not more than 0.3°K.

The Texas Instruments approach for the HIDAD program uses an array of detectors whose polarization and dielectric constant change with temperature, resulting in a change in charge on a capacitor dependent on the target scene radiance. The detectors are fabricated from a ceramic material-barium strontium titanate (BST)-and then bump-bonded to a silicon readout circuit. Barium strontium titanate is a feroelectric insulator that has a lattice phase transformation near room temperature. At this particular temperature (the Curie temperature) the polarization and dielectric properties of this material vary strongly with changes in temperature.

A low-power, single-stage thermoelectric cooler is used to stabilize the focal plane array at room temperature. In order for the detector to sense a change in temperature between the target scene and some background reference temperature, it is necessary to have the detectors look alternately at some fixed background and then at the target scene. The incoming infrared scene is AC-coupled via a rotating mechanical chopper, the only moving part in the system. Without a chopper in the system, the surface charge would dissipate and cause the scene contrast to decrease to zero, resulting in fadeout of the scene, except for any moving objects in the scene. This type of system may be desirable for some applications such as security where only moving objects are of interest. The elements of the array are read out in sequence by applying appropriate voltages to the rows and columns of a two-dimensional switch matrix. The signals from the columns are then amplified as they pass through an amplifier at the end of each column of the readout circuit. The analog signal is then transferred off the focal plane and multiplexed through an analog/digital converter (ADC) prior to further signal processing. Gain and offset correction are then applied to remove the fixed-pattern noise due to the nonuniformity of response of 80,000 detectors. Dead-pixel correction is also applied to substitute information from an adjacent pixel to replace the signal from detectors that are unresponsive. Further processing including additional amplification, formatting for TV-compatible video output, and digital-to-analog conversion (DAC) is then performed.

The Honeywell approach uses a two-dimensional array of bolometers, small thermal mass detectors whose resistance changes approximately 2% per degree Celsius change in temperature caused by heating from the incident radiation. The equivalent circuit is shown in Figure 7. The temperature-sensitive resistor material is vanadium oxide, which is suspended on a bridge of silicon nitride which results in very low thermal conductance to the substrate containing the readout electronics. This bridge structure forms a very efficient absorber with absorption efficiencies of over 70% reported. The focal plane temperature is stabilized via a low-power, single-stage thermoelectric cooler. The transistors in each unit cell act as an array of gate-controlled switches. As a voltage is applied to each row of gates in sequence, a bias current flows through all of the detectors in the row, and they are read out in parallel via the column leads to a high-seed multiplexer at the base of the columns. This row of information is then read out through the high-speed multiplexer before the information from the next row is transferred. Because the Honeywell approach does not require any AC coupling of the incoming radiation, no mechanical chopper is required. The only moving part of the Honeywell uncooled system is a shutter that is periodically closed momentarily for system calibration. This lack of a requirement for chopping the signal increases the sensitivity of the system by a factor of two. The increase in sensitivity is due partly to the increased integration time for the signal and partly to the elimination of time-dependent noise.

After the signals are multiplexed off the focal plane, they are converted from analog to digital via an ADC, and then gain- and offset-correction are added to remove any nonuniformities among the 80,000 detectors. Dead-cell correction is applied, and the digital signal is converted back to analog and formatted for RS-170 video output.

4. APPLICATIONS

The temperature sensitivity and resolution required for infrared focal plane arrays depends on the particular application. Civilian applications such as night vision, navigation, and security requires only nominal temperature sensitivities (1°C) and resolution (2 milliradians). Industrial applications and some military applications requiring temperature monitoring have more stringent requirements (0.1°C and 1 milliradian). Threat and target detection requires discernible signals at the maximum possible range which means high sensitivity (.01°C) and high resolution (0.1 milliradian).

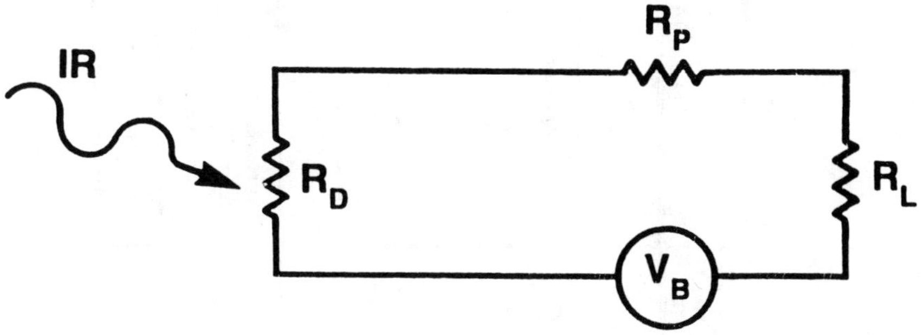

Figure 7. Equivalent Circuit for Honeywell Bolometer

Table 3.

INFRARED IMAGING FOCAL PLANE ARRAYS

Manuf.	Architect.	Size	Pixel Size (µm)	Pitch (µm)	Detector Material	Spectral Range (µm)	Oper Temp (K)	Comments/Readout
Aerojet Electro-Systems (818) 812-2875	Hybrid Hybrid	20x64 8x192	100 120	100 120	Si:As Si:As	8-28 8-28	10-15 10-15	Parallel output Parallel output
Amber (805) 683-6621	Hybrid Hybrid Hybrid Hybrid	128x128 256x256 128x128 128x128	40 43 40	50 38 50 50	InSb InSb HgCdTe Si:Ga	1-5.5 1-5.5 8-10 3-18	77 77 77 11	CMOS multiplexer CMOS mux CMOS, Photovoltaic Photoconductor
Cincinnati Electronics (513) 573-6100	Hybrid	64x64	87	100	InSb	1-5.5	8-77	CMOS, Photovoltaic
Eastman - Kodak (716) 722-4385	Monolithic	486x640	25	25	PtSi	1-5	77	Schottky barrier
EG&G Reticon (714) 583-2250	Monolithic	128x128	60		PtSi	1-5.5	77	Schottky barrier diode
GE (315) 456-0514		1x64 1x128			HgCdTe HgCdTe	> 11 > 11	77 or lower	Custom applications, low noise CMOS mux
Loral (Boston) (617) 863-3082	Hybrid	128x128	30		HgCdTe	8-13	77	Custom applications
Mitsubishi (714) 220-2500	Monolithic IT	512x512	26x20		PtSi	3-5	77	Schottky barrier

Table 3. (continued)

INFRARED IMAGING FOCAL PLANE ARRAYS

Manuf.	Architect.	Size	Pixel Size (μm)	Pitch (μm)	Detector Material	Spectral Range (μm)	Oper Temp (K)	Comments/Readout
Rockwell (714) 762-5570	Hybrid	64x64	100		Si:Ga	1-17	20	Photoconductor
	Hybrid	64x64	100		Si:In	1-8	42	Photoconductor
	Hybrid	128x128	40	40	HgCdTe	1-3	77	Photovoltaic
	Hybrid	256x256	40	40	HgCdTe	7.5-10.5	77	Photovoltaic
SBRC (805) 968-3511	Hybrid	256x256	30	30	InSb	1-5.5 0.3-5.5 w/coating	10-77	Non-destructive readout, quasi random access
	Hybrid	58x62	76	76	Si:Ga	3-17	12-15	Random access
	Hybrid	58x62	76	76	Si:As	8-24	8-10	Random access
	Hybrid	256x256			HgCdTe	2.5-10.5	77	Only on contract basis
SRI (609) 734-3056	Monolithic IT	320x244	40		PtSi	1.1-5	77	Schottky barrier
Texas Instruments (214) 995-3333	Diode Technology	64x64	61	61	HgCdTe	8-12	77	CMOS optimized for 77K
	Hybrid	328x245	50		Pyro	2-40	Rm	

Detectors are used for these various applications based on how well the characteristics achievable by the system can match the requirements of the applications. Most available infrared focal plane arrays today require operation at low temperature (77°K) and thus cryogenic cooling (see Table 3). These are virtually all photon detectors and represent a significant portion of the infrared focal plane arrays that are commercially available.

Photon detectors have the high sensitivity and resolution desired for threat and target systems whereas thermal detectors are capable of meeting most industrial and civilian requirements. Thermal detectors also have the potential for not requiring coolers which reduces initial system cost and complexity. In addition to the lower initial cost for uncooled systems, there are advantages in life cycle costs. Uncooled systems do not require a cryogenic cooler or a high vacuum . The detector material is simpler, with no requirements for carrier concentration, lifetime, or mobility. Only one mechanical component (the chopper) is required. Uncooled systems are lightweight, have low power requirements, and have an instantaneous turn on capability.

Thermal detector focal plane arrays make low cost compact systems possible. It is expected that this technology will very soon make infrared systems viable for many applications for which cooled infrared systems are too expensive.

REFERENCES

Infrared Detectors and Focal Plane Arrays, E.L. Dereniak and R.E. Sampson, Volume 1308, SPIE, Orlando, 1990.

Infrared Detectors and Focal Plane Arrays II, E.L. Dereniak and R.E. Sampson, Volume 1685, SPIE, Orlando, 1992.

J.M. Edwards, J.L. Gates, A. Thompson, R.H. Wyles, et al., "Large Hybrid Schottky Focal Plane Arrays (U)," 1988 Meeting of the IRIS Specialty Group on Infrared Detectors, Vol. 1, pp. 347-368, February 1989.

J.A. Finch, S.H. Black, A.J. Lewis, K.A. Kormos, et al., "Performance and Producibility of Photovoltaic LWIR HgCdTe Focal Plane Arrays (U)," 1989 Meeting of the IRIS Specialty Group on Infrared Detectors, Vol. 3, pp. 211-228, October 1990.

Robert E. Flannery and James E. Miller, "Status of uncooled infrared imagers," SPIE Vol. 1689 Infrared Imaging Systems (1992), pp. 379-395.

A. Hoffman and D. Randall, "High-Performance 256x256 INSB FPA for Astronomy and Space Surveillance Applications (U)," 1991 Meeting of the IRIS Specialty Group on Passive Sensors, pp. 79-86, May 1991.

S.B. Horn, R.E. Flannery, J.E. Miller and D.A. Ferrett, "Low-Cost Uncooled Focal Plane Arrays (U)," Proceedings of the Infrared Information Symposia, Vol. 35, No. 2, pp. 147-159, January 1991.

James Janesick and Tom Elliott, "History and Advancements of Large Area Array Scientific CCD Imagers," pre-print, Astronomical Society of Pacific Conference Series '91, Tucson Arizona.

J.P. Karins, S.C.H. Wang, J.M. Swab, L.J. Smith et al, "Dualband Focal Planes for Infrared Systems (U)," 1989 Meeting of the IRIS Specialty Group on Infrared Detectors, Vol. 1, pp. 273-287, September 1989.

C. Kauffman, T. Fischer, M. Grenn, A. Gray, et al., "Performance and Producibility of Monolithic 488x512 Platinum Silicide Focal Plane Arrays (U)," 1990 Meeting of the IRIS Specialty Group on Infrared Detectors, Vol. 2, pp. 229-248, December 1990.

L.J. Kozlowski, D. Lo, R.B. Bailey, K. Vural, et al., "640x480 Pace HgCdTe FPA Development for High Performance FLIRS (U)," 1991 Meeting of the IRIS Specialty Group on Passive Sensors, Vol. 1, pp. 55-65, May 1991.

L.J. Kozlowski, K. Vural, W.E. Tennant, J.T. Viola, et al., "Reproducible, High Performance LWIR HgCdTe Staring Focal Plane Arrays (U)," Proceedings of the Infrared Information Symposia, Vol. 35, No. 2, pp. 203-214, January 1991.

J.J. Lange and R.R. Sandys, "Phenomenology/Sensitivity Analysis of Platinum Silicide, Indium Antimonide, and Mercury-Cadmium-Telluride Detector Performance in the 3 to 5 Micrometer Infrared Band for Air-to-Ground Applications (U)," 1990 Meeting of the IRIS Specialty Group on Passive Sensors, Vol. 1, pp. 281-303, April 1990.

J.W. Meehan, "Advances in HgCdTe Focal Planes (U)," Proceedings of the Infrared Information Symposia, Vol. 31, No. 1, pp. 241-248, September 1987.

M.L. Noble, A.F. Milton, J.P. Karins, A.J. Manoni, et al., "Focal Plane Array/Dewar Technology for Wide-FOV Dual Band Air Defense IRST Systems (U)," Proceedings of the Infrared Information Symposia, Vol. 35, No. 1, pp. 171-186, January 1991.

Paul R. Norton, "Infrared Image Sensors," Optical Engineering, Vol. 30, No. 11, pp. 1649-1663, November 1991.

P. Norton, J. Myrosznyk, T. Wise, R. Grannenman, et al., "Microlens Arrays Offer Enhanced IR Detector Performance (U)," 1990 Meeting of the IRIS Specialty Group on Infrared Detectors, Vol. 3, pp. 87-91, February 1991.

J.H. Pollard, "Report for IRIS Specialty Group on Infrared Materials (U)," Proceedings of the Infrared Information Symposia, Vol. 35, No. 1, pp. 55-75, January 1991.

W. Rogatto, "IRIS Detector Specialty Group (U), Proceedings of the Infrared Information Symposia, Vol. 34, No. 1, pp. 25-33, October 1990

W. Rogatta, "IRIS Detector Specialty Group (U)," Proceedings of the Infrared Information Symposia, Vol. 35, No. 1, January 1991.

B. Seymour, G. Garrett, S. Strom, J. Elkind, et al., "Advances in LWIR 64x64 Element and 128x128 Element Staring VIMIS Focal Plane Arrays (U)," 1990 Meeting of the IRIS Specialty Group on Infrared Detectors, Vol. 3, pp. 275-288, February 1991.

F.D. Shepherd, "An Overview of Schottky Barrier Sensors (U)," 1989 Meeting of the IRIS Specialty Group on Passive Sensors, Vol. 1, pp 133-140, April 1990.

Optical Components and Coatings for Aerospace Sensing

Albert F. Slomba
United Technologies Optical Systems
Post Office Box 109660, West Palm Beach, Florida 33410-9660

1.0 INTRODUCTION

The application of advances in a variety of disciplines, has made significant impact on the area of optical sensing technology. This impact has not evolved so much from new optical designs, component prescriptions, or revolutionary coating designs, although there have been substantial innovations in certain areas. Instead, progress has resulted from the combined mix of a variety of ideas including the areas of advanced materials and material forming processes, coatings that do their job more efficiently, the application of computer control to manufacturing, wavefront correction, and alignment, as well as the collection, integration, and display of relevant data in a format that makes possible interpretation at-a-glance.

2.0 SUMMARY

The entire topic of Components and Coatings for Aerospace Sensing covers an enormous area. A technology review is shaped by a variety of influencing factors, not the least of which is the writer's view of the world, and first hand experience. References are given throughout the text, and are listed at the end. There is always the danger of omission, but it is hoped that the bibliography contained in each of the referenced papers will cover most cases. For the current survey, the discussion will be limited to several of those areas which have recently seen a number of important and intriguing changes - changes that hold the promise of further advances in critical applications. Some of the sensors which will not be included in this discussion are the earth resources satellites, or the airborne systems used for forestry and land management, drug interdiction, or forest fire control. Instead, emphasis will be given to specific examples from three general categories. These categories are Tactical and Reconnaissance Systems, Space Based Sensors, and Ground Based Systems. As an integral part of each topic, it has been necessary to give consideration to the impact of new materials, along with new methods of producing the raw optical substrates from them. Likewise, developments in optical finishing technology have been included where these have a direct bearing on the success of the application. Optical Coating developments in key areas will be treated under each of the primary topics. Finally, an assessment is given of the probable direction these innovations will take in the future, as well as some perceived problem areas that need to be addressed.

3.0 TACTICAL AND RECONNAISSANCE SYSTEMS

A variety of tactical and reconnaissance systems have been developed which rely on advanced technology in the areas of material science, mechanical fabrication, optical finishing and coating. These systems are deployed on tanks, missiles, helicopters, and both manned and unmanned aircraft. Some are hand carried onto the battlefield. Examples are PAVE TACK, ATARS[1,2] LOROPS, M1-A2 fire control systems, AMDAS, plus numerous rangefinders and designator systems.

In general, the trend is toward unobscured optical systems of ever increasing aperture, lighter weight, and rugged construction for operation under severe environmental condition. These are designed to survive even greater extremes during storage and transport. Helicopters, target acquisition and designation systems, pilot night vision systems and various navigation devices have fostered the need for sophisticated optical components and coatings. These cover the visible, near infrared, and infrared spectral regions. Examples are windows, lightweight scanning mirrors, and compact broadband afocal telescopes feeding a variety of focal plane instruments such as infrared arrays, and video cameras.

3.1 Telescope components

3.1.1 Fast focal ratios

In order to satisfy both the compact and large unobscured aperture requirements, the optical designer is driven toward the use of extremely fast prescriptions which are applied far off-axis. As an example, an unobscured aluminum 4-X afocal telescope was recently built under a feasibility study, that had an aperture of 30 cm. The primary was designed to be used nearly 35 cm off-axis. As shown in Figure 1, the parent f-number was faster than 0.4. The parent diameter in this case was just under one meter! Fabrication of this optic presented a number of challenges. Economical quantity production indicated single point diamond turning (SPDT) to near net shape, nickel plating, and final turning to the required optical tolerance. As the parent diameter increases, the number of shops equipped to do the SPDT diminishes rapidly. Any larger and they can be counted on one hand - at least those shops that have a reasonable production capability. Another problem arises due to the high slope with respect to the turning axis for such low f-number surfaces. The diffraction grating-like or groovy surfaces produced by the SPDT process is greatly aggravated. This in turn requires a good deal of very skillful post-polishing in order to reduce the diffraction pattern at the image plane to acceptable levels. For glass or ceramic substrates, precision grinding operations replace SPDT. The idea is to precision surface generate (PSG)[3] the substrate to near net optical shape while producing zero to, at most, a few microns of subsurface damage. It makes no sense to generate a surface to extremely high tolerances and then have to free abrasive grind (FAG) away nearly half a millimeter, to remove damage caused by traditional generation processes. The surface figure is guaranteed to be ruined! It takes as good a machine to PSG a glass surface as it does to SPDT a metal one. In addition, extreme care must be given to balancing and dressing the grinding wheel.[4] The successful process then allows the optical surface to be generated to near net optical shape, eliminating the FAG step, and polishing the optic to final specification. In the course of polishing, a few microns of material is almost always removed, along with any residual subsurface damage from the PSG operation. After figuring, smoothing is sometimes required to obtain microroughness values below about 8 Angstroms, a value that can be achieved directly with some types of computer controlled (CCP) machines. Post-polishing to eliminate the diffraction from SPDT surfaces, and smoothing the surface further, in order to achieve low microroughness, or scatter is difficult to do without degrading the surface figure. Fortunately, CCP techniques have been developed[5,6,7] which address a portion of the problem by removing the grooves, while maintaining optical figure. For the other part of the problem, the design and use of special smoothing laps has been fairly successful in reducing surface microroughness. These laps are elastically tailored to have different compliances in orthogonal directions, or over different portions in contact with the work.[3,8] The laps can be driven by the CCP machines, or conventional spindle polishing machines can be rigged either singly or ganged to produce the correct polishing action. Exact techniques are usually regarded as proprietary, and seldom disclosed in detail in the literature.

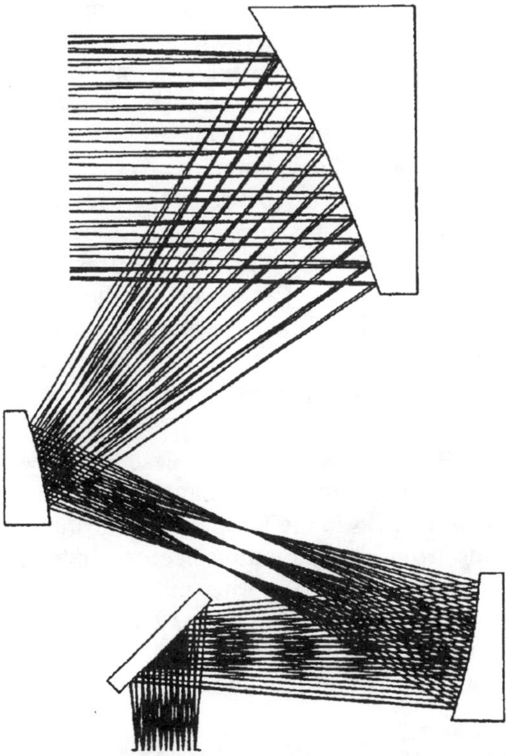

OFF-AXIS AFOCAL TELESCOPE

Fig. 1 Fast Focal Ratio Aspheric Telescope

3.1.2 Asymmetrical aspherics

In order to achieve a large field of view the classic Schmidt design offers simplicity in a compact package. But to achieve an extremely broad spectral range, without the added complexity and weight of various achromatization schemes, a reflective corrector plate has frequently been suggested. A small unobscured 5cm instrument has recently been demonstrated with a tilted, off-axis, aspheric reflective corrector. Again, CCP methods, this time using a roller lap polishing head, successfully polished both the symmetric Zernike terms (with respect to the parent axis) plus the asvmmetric terms occasioned by the tilt condition.

3.2 Scanning systems

3.2.1 Lightweight flats

Another class of advanced optical components which is often used in these systems are scan flats. The criteria for these elements are light weight and low figure distortion under dynamic conditions, combined with a high tolerance to vibration and shock. Recent advances in this area have been made possible through the development of advanced materials and substrate forming processes. Scan flats can be used in a single or two axis gimbal system which forms part of the system, or they can be used in various diagnostic/test sets. In the latter application, such instrumentation is sometimes hung off of the system under test. Therefore, weight must be held to an absolute minimum so as not to distort the assembly under test, thus avoiding generation of spurious test data. A development prototype of a helicopter scan mirror is shown in Figure 2. This mirror combines the use of silicon carbide (SiC), a material that has a very high modulus, but not the lowest density available, with a forming process which allows extremely thin web features to be achieved in a closed back substrate design (trade marked "Ceraform®")[9] shown in Figure 3. This combination then produces an optimum "specific section stiffness". A further discussion of trades and design applications will be given later under the section Space Based Systems.

Figure 4 shows an all beryllium open back design. Beryllium has the highest modulus, lowest density of all mirror grade materials. Prototype mirrors intended for use as a replacement for glass in the Leopard I tank fire control unit[10] have also been successfully made from an aluminum/silicon carbide (Al/SiC) metal matrix composite (MMC), trade named SXA. Other matrix combinations have also been used successfully, including beryllium/aluminum. The relevant mechanical and thermal constants of the various materials, thermal and mechanical figures of merit, and an extensive discussion on design trades will be given in Section 4.0.

3.3 Infrared Windows

Requirements on infrared windows have become increasingly more demanding, both in terms of spectral performance and durability. As aircraft speeds increase, and systems are used and serviced in harsh operational environments, new or greatly improved window materials and more durable multi-function coatings need to be developed to keep pace. Many modern windows need to transmit in the visible as well an in the infrared, have an electrical defog/defrost capability, provide low observable (LO) characteristics, and have rain/sand erosion antireflection coatings. These must all be deposited on the same substrate! For long wave (8-12μm) ir windows that experience no significant heating, i.e., for subsonic operation, an electrically conductive germanium substrate with a diamond-like carbon (DLC)[11,12] coating offers good performance and excellent durability. However, for high speed applications, or multispectral use over the 3-5 and 8-12μm regions, ZnS is the best current material. ZnSe is best for windows that must operate over an even broader range, for example throughout the 0.55μm to near ir, 3-5, and 8-12μm (or longer) wavelength regions. For added durability, ZnSe substrates with a thin layer of ZnS

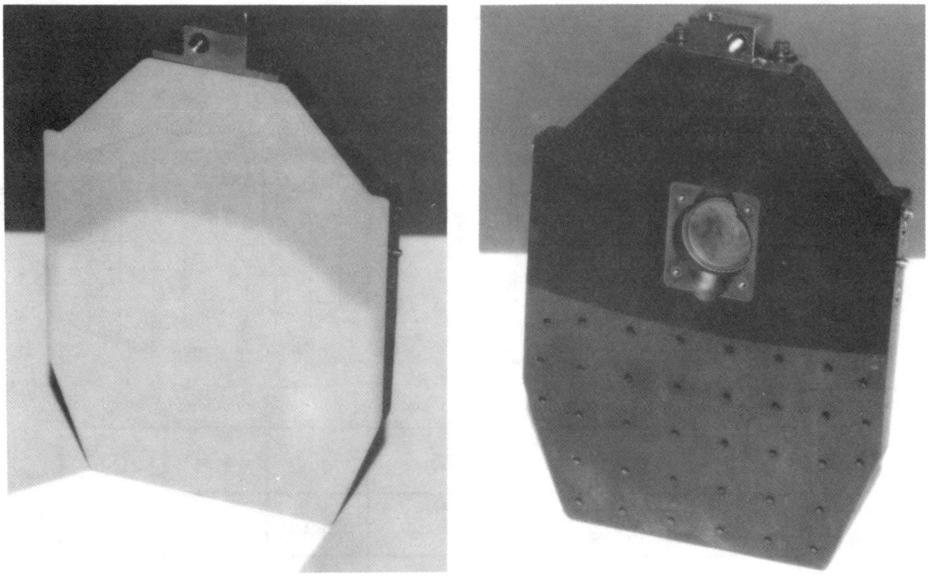

Fig. 2 Lightweight Infrared SiC Scan Flat

92–11–15–2

Fig. 3 Internal Detail of Web Features in a SiC Light-Weight Mirror

92–11–15–3

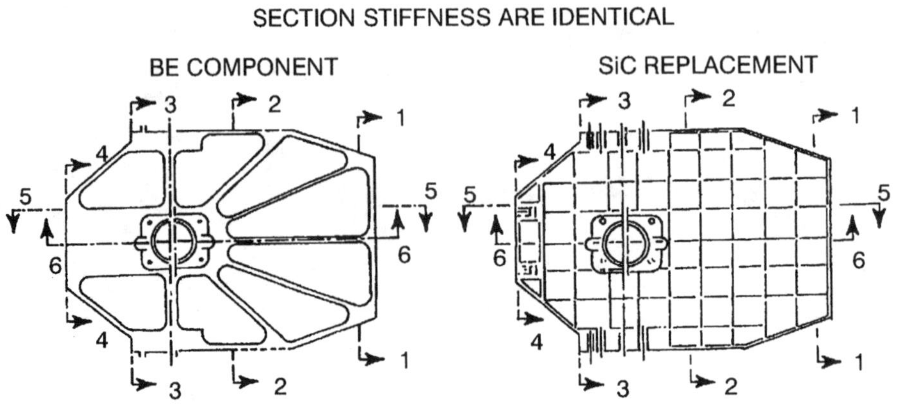

SECTION	1-1	2-2	3-3	4-4	5-5	6-6
SiC	0.219	0.97	2.96	4.47	3.35	2.85
Be	0.125	0.79	2.35	1.14	1.69	2.81

Fig. 4 Design Comparison Between Beryllium and Lighter SiC Replacement Mirror

92–11–15–4

may be used with some reduction in short wavelength transmission. Unfortunately, DLC coatings do not adhere well to these materials, and tend to scatter excessively at the shorter wavelengths. One way to improve coating properties is to use higher energy deposition methods The use of Ion Assisted Deposition (IAD), or Ion Beam Sputtering (IBS) as well as ion plating and other advanced sputtering techniques have led to coatings with improved microstructure. Therefore, they are more durable, and show much smaller or no spectral shifts due to changes in ambient humidity. Very narrow reflectors are now achievable due to the continued development of Rugate technology, where the refractive index of the coating is made to vary in a sinusoidal fashion, or in a digitized way termed a "flip-flop" coating technique. While no ideal coating solution is currently available, considerable progress has been made in understanding the physics of rain and particulate damage at high velocities, and developing broadband windows that can better stand up to harsh environments and rough handling in the field. Recent results on erosion damage mechanisms of coatings have been reported which show good correlation between substrate and the thin film mechanical properties and rain erosion damage resistance.[13,14] Other studies have modeled the aerodynamic behavior of rain drops at various velocities to determine drop size distribution and impact effects at various angles of incidence.[15] Figure 5 shows a typical window assembly that has a defog/defrost and durable antireflection coating. This is a PAVE TACK window that incorporates two smaller windows as part of the assembly.

3.4 Binary Optics

Binary optics is a term which applies to the application of a grating structure to a surface to produce an optical element. A familiar example is the classical zone plate.

Fig. 5 Infrared Testing of a PAVE-TACK Window Assembly. Perkin-Elmer Corp.

92-11-15-5

With the advent of fairly large scale microlithographic processes, ion beam etching techniques, and SPDT machines, a wide range of possibilities has opened. Hundreds of papers have recently been published[16,17,18] describing various techniques and applications. SPDT methods are well suited to infrared systems. It is sometimes possible to greatly reduce the number of lens elements by forming binary elements on the surfaces of spherical or aspheric lenses. On simpler systems, the binary surface can be designed to turn a singlet into an achromat. Although the SPDT process increases the effective surface microroughness, the increased scatter is not objectionable, particularly for the mid and long wave infrared spectral regions. Many new applications are emerging for these elements, since they can offer better performance, lighter weight, lower cost, and can be more robust than the more complex assemblies they replace.[19]

4.0 SPACE BASED SENSORS

Space-based optical sensors that will be discussed cover the spectrum from x-rays to the far infrared. Here too, advances in materials, optical processing methods, and coatings have made possible the realization of instruments that have literally changed the way we perceive our own planet and the universe. And here, too, a long list of examples from various categories may be cited. There have been x-ray telescopes for astronomy like the Einstein Observatory,[20] ROSAT,[21] and AXAF[22] (under construction), and the SXT[23,24] for solar imaging. In the extreme ultraviolet there are the UVCS and SUMER instruments for solar coronal and solar emission studies, plus FUSE for spectroscopic stellar observations. Moving toward longer wavelengths, Hubble was designed to span the ultraviolet from about 1200Å through the far infrared.[25] IRAS, an all beryllium telescope operating at superfluid helium temperatures, performed a detailed all-sky survey at high sensitivity in the infrared at wavelengths out to about 120μm.[26] SIRTF, is planned as a five year mission, and will provide imagery from 1.8μm out to 1200μm, at a sensitivity level of from 100 to 10,000 times higher than previous instruments.[27] New instruments like LAWS,[28] trained earthward, will map wind velocities at high spatial resolution up through the atmosphere on a global scale. The SPOT series of satellites provides mapping, earth resource, and imaging information at moderate resolution to commercial customers. A few interesting examples have been chosen for discussion which illustrate application of enabling technologies that allow those instruments to fulfill their missions.

4.1 Lightweight components

For space applications, weight is a performance and cost driver. The trend is always toward optical systems that are lighter. A number of factors enter the trade of performance, weight and cost. For uncooled systems, most of the interest is centered around the primary mirror, since it is usually the largest optical element. Its design influences the rest of the system. Segmented mirrors aside, size is limited to the physical dimensions of the available launch vehicle (about 4-meters in diameter for the Space Shuttle). Weight impacts choice of vehicle also, as well as orbital height, which determines payload lifetime. All of this translates into cost. For a given material and limits on substrate construction processes (i.e., achievable facesheets an rib or cell wall thickness), weight can be reduced at the expense of decreased rigidity. As the mirror blank becomes more flimsy, there comes a point at which the costs associated with handling, the need for more sophisticated and expensive metrology mounts for testing under 1-G, and the extra time required to verify that everything is working properly, begins to have an appreciable impact on overall costs. For a given flexibility and threshold of pain in regard to handling and testing, only three things can be done to improve the situation. One may look for or develop mirror materials with better properties, develop construction methods that produce thinner facesheets and webs, or provide an active on-orbit mirror figure and alignment capability. Some of these options can have a negative cost impact, and the overall process of design trades can become very complicated. But ultimately, for any given set of ground rules for the primary mirror - one piece passive or active, segmented passive or active, or an assemblage of many active or passive segments from multiple payload launches - the optimum solution depends upon a combination of material properties and section geometry that can be reasonably fabricated, polished, and tested. Table 1 lists the relevant mechanical and thermal properties of a variety of candidate mirror materials. Table 2 lists some figures of merit. The choice of material depends on the application, and thermal environment. But these tables tell only part of the story. Not all materials can be formed to equivalent geometries, shapes and section thicknesses. There is no single material/process which is best for every situation. Substrate designs must be optimized to determine the best choice for any specific application.

Table 1. Relevant Mechanical Properties for Various Optical Materials						
Properties	**Be**	**SiC(*)**	**Al**	**ULE™**	**Fused Silica**	**SXA™**
Physical						
Density (g cm^{-3})	1.85	2.92	2.73	2.21	2.20	2.96
Mechanical						
Young's Modulus (GPa)	303	311	71	68	70	145
Microyield Stress (MPa)	35	(**)	98	24	11	124
Thermal						
CTE (10^{-6} K^{-1})	11.4	2.57	24.0	0.03	0.55	10.8
Diffusivity (cm^2 s^{-1})	0.52	0.80	0.96	0.008	0.007	0.47
Specific Heat (J kg^{-1} K^{-1})	1,880	670	900	766	790	800
Conductivity (W m^{-1} K^{-1})	180	156	237	1.3	1.3	111

(*) Values shown are for CERAFORM®
(**) None detected prior to fracture

Table 2. Figures of Merit for Various Optical Materials						
Figures of Merit	**Be**	**SiC(*)**	**Al**	**ULE™**	**Fused Silica**	**SXA™**
Structural (higher is better)						
Specific Stiffness (E/ϱ)	***164***	107	26	31	31	49
Thermal (lower is better)						
1st Order Distortion (CTE/k)	0.063	***0.017***	0.101	0.023	0.042	0.090
Thermal Distortion (CTE/D)	22.2	***3.26***	25.0	3.92	78.6	23.0
Dim. Change (CTE•ϱ/k)	118	***48.1***	276	51.0	931	288

(*) Values shown are for CERAFORM®

To speak of mirror weight by itself is meaningless. What counts is weight for a given stiffness. Competing designs must be compared on an equivalent basis. How does one quantify mirror stiffness? When is a mirror considered too flimsy? This can be somewhat subjective. A mirror is too flimsy when it deforms beyond the ability to deal with these deformations while trying to polish or measure surface figure. This will vary, depending upon the time and money available for fixturing, mounts etc. Fortunately, stiffness can be precisely quantified. The free-free natural frequency of a mirror structure can be calculated. For any given frequency, one may calculate the peak-to-peak (p-p) and rms self-weight deflection for a specified support condition in a gravity field-say a 3-point support and 1-G. For example, a 200 Hz mirror will deform 0.3 μm p-p on a simple 9-point mount. If two lightweight mirrors of equal size have been optimized and have been designed to have the same natural frequency, and supported identically in the same G-field-they will have identical self weight distortions-independent of the material of which they are made, and independent of their internal feature or geometry sizes. They may have different web heights or thickness, different cell dimensions or shapes, and different faceplate and backplate thicknesses. The better material/forming process combination will weigh less. That is, its area density (areal density) will be lower. The two parameters of importance then, are areal density, (A_d), and natural frequency (f_o). In addition to mechanical properties, the best mirror must also possess the required thermal characteristics. Will the mirror experience a thermal gradient across its diameter, or from front to back? Will it see gradual or rapid temperature changes? Will it need to be cooled to cryo temperatures? The structural geometry of the mirror, its coating, and mounting configuration all play a role in determining its thermal response.

Figure 6 shows areal density plotted against frequency for Ceraform® SiC mirror designs of various diameters. These are all closed back designs. A closed back design is always superior to open back provided the fabrication process can produce features thin enough to take advantage of the material properties. Except as indicated, all of the designs are for two-tier construction as shown in Figure 7. An additional imposed constraint was that the deflection over any cell had to be less than 0.06μm under a polishing pressure of 0.5 psi. As an exercise, two separately optimized 200 Hz, 4-meter diameter mirrors were compared. One was made of ULE®, and the other of Ceraform® SiC. Based upon realistic web and feature thickness constraints for each material, the SiC weighed about one third of the ULE® mirror.

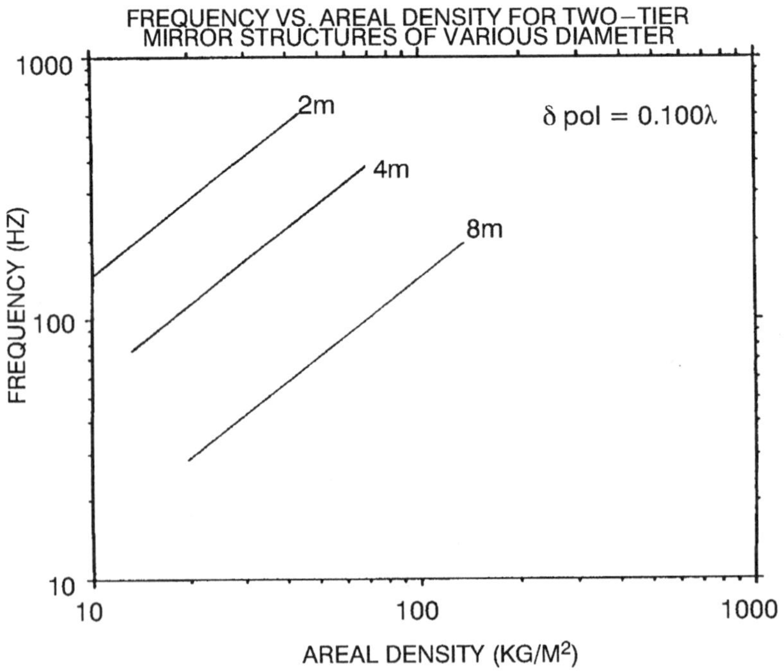

Fig. 6 Results of Scaled Optimization Study for Ceraform SiC Lightweight Mirror Design 92–11–15–6

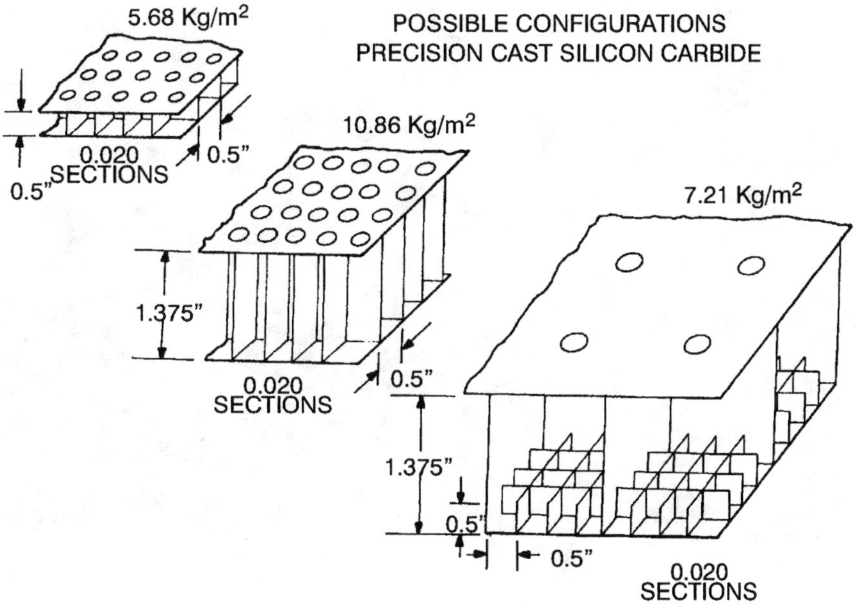

Fig. 7 Illustration of Single and Two-Tier Substrate Designs

92–11–15–7

Figure 8 shows a 44-cm scan flat with an areal density of 12kg/m^2 weighing 1.3kg (2.9lbs) and with a natural frequency of 2472Hz. On the right, the facesheet has been ground away to reveal the internal structure; all cast in a single step, simultaneously with the face and back sheet. Figure 9 shows a 0.5-m hexagonal shaped portion of a 1.5-m design. A_d is about 14kg/m^2 and f_o is 2250Hz for the mirror, and 250Hz for the 1.5-m design.

4.2 X-ray optical systems

Considerable progress in x-ray optics has been made, especially over the past dozen years. Advances have come not only to the fabrication and coating technology, but also in the basic understanding of the physical optics aspect of image formation, and better optical designs.

Up through the time of the development and flight of the HEAO-B (High Energy Astronomical Observatory-B), subsequently christened the Einstein Observatory, x-ray image formation in the 2 to 30 Å range was thought of and described in geometric ray tracing terms. For the grazing incidence Wolter telescopes, no consideration was given to diffraction. Later, while studying the figure, slope error and microroughness tolerances proposed for the AXAF/TMA (Advanced X-Ray Astrophysical Facility/Technology Mirror Assembly), the author noted an inconsistency among these specifications. It was soon realized that physical optics effects must be taken into account even at so short a wavelength. Subsequently, surface figure requirements were thought of in terms of a "power spectral distribution" of figure error

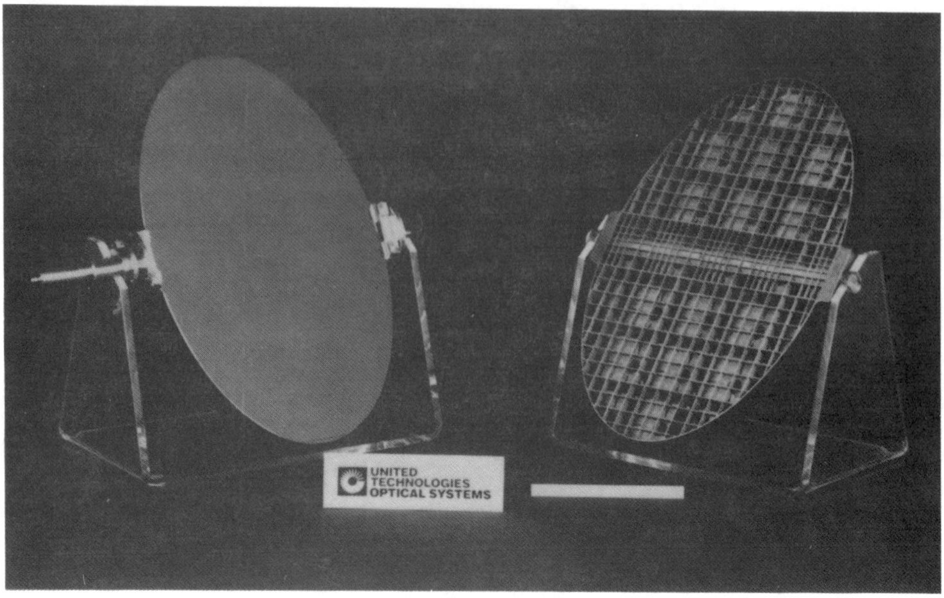

Fig. 8 44 cm Ultralightweight SiC Scan Mirror Weighing 1.3 kgm (2.9 lbs) with a 2472 Hz natural frequency (left side). Cutaway photo on right shows fine interval structural detail.

92–11–15–8

60cm PROTOTYPES

Fig. 9 Ultralightweight 60cm Spherical SiC Mirror. This is a Portion of a 1.5 Meter Design

92–11–15–9

amplitude versus the spatial periods of the errors.[29] Focal plane performance, particularly encircled energy versus spot size can be accurately modeled using methods similar to those employed for conventional systems used near normal angles of incidence.[30,31] It turns out, however, that the surface figure requirements on spatial surface wavelengths of from 5-mm to about 1-mm are very tight, at about 50 to 10Å rms respectively. This presents a real measurement challenge. The errors in this critical mid-spatial frequency regime lie between the long period, or aberration errors which are measured in terms of several cycles per aperture, and the high spatial frequency errors of 1-mm to 1-μm that are associated with microroughness. Encircled energy performance is critically dependent on these mid-spatial frequency errors as well as low microroughness in order to obtain maximum image contrast. This realization has fostered the development of better polishing methods and more sensitive metrology instrumentation.

Another area of great current interest, with application beyond astronomy is the development of multilayer reflectors for use at normal incidence.[32,33,34,35] Although these coatings have a rather narrow bandwidth, they possess several advantages. Since grazing incidence telescopes have very highly obscured apertures, a much smaller primary used at normal incidence provides the same collection efficiency. This means that a dozen or more much smaller telescopes, each covering a different spectral region, can be launched for an equivalent weight.[36] In addition, multispectral data can be recorded simultaneously, rather than in series, as is the case with a single large grazing incidence telescope. At the same time, this does mean that a multiplicity of detectors and data preprocessing/transmission equipment is required.

Going toward longer wavelengths, finding materials to be used bare or as coatings to provide reasonable reflectance, remains a problem up to the vacuum ultraviolet. While SiC should provide good reflectance down to 500Å, only recently have polishable chemical vapor deposited (CVD) claddings been developed which achieve near theoretical reflectance. Both the SUMER and UVCS instruments make use of this material (and the supersmooth polishing techniques that allow rms finishes of 2 to 5 Å), to help accomplish their missions.

4.2.1 SXT

Figure 10 shows the Solar-A X-ray Telescope (SXT), which is currently transmitting high contrast, high resolution images of the sun and its corona. Because of the severe launch and operational environment, both optical elements were generated on the same Zerodur® blank using PSG techniques, figured with the CCRP, and then smoothed using special anisotropically compliant laps.[3,8] In order to reduce field curvature and substantially improve off-axis imagery, this telescope uses two hyperbolas (the Nariai configuration) instead of the parabola-hyperbola design for a Wolter telescope. Figure 11 shows a solar image obtained. Figures 12 and 13 show the effect of trading microroughness for mid-frequency error in order to achieve various values of encircled energy at the desired spot size for different wavelengths.

4.2.2 ROSAT

The Roentgen Satellite (ROSAT), which is currently in orbit, is the largest and highest performance x-ray telescope used to date for high resolution astronomy.[22]

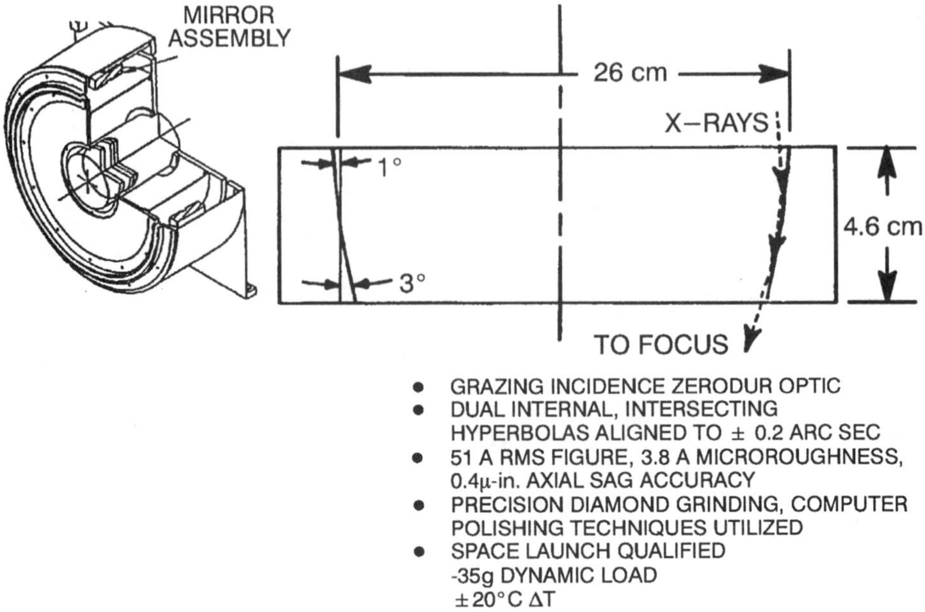

Fig. 10 Solar-A X-Ray Telescope (SXT) in use on the Yohkoh Spacecraft of the National Astronomical Observatory of Japan. Surfaces were Ground and Polished on a Single Substrate. 92–11–15–10

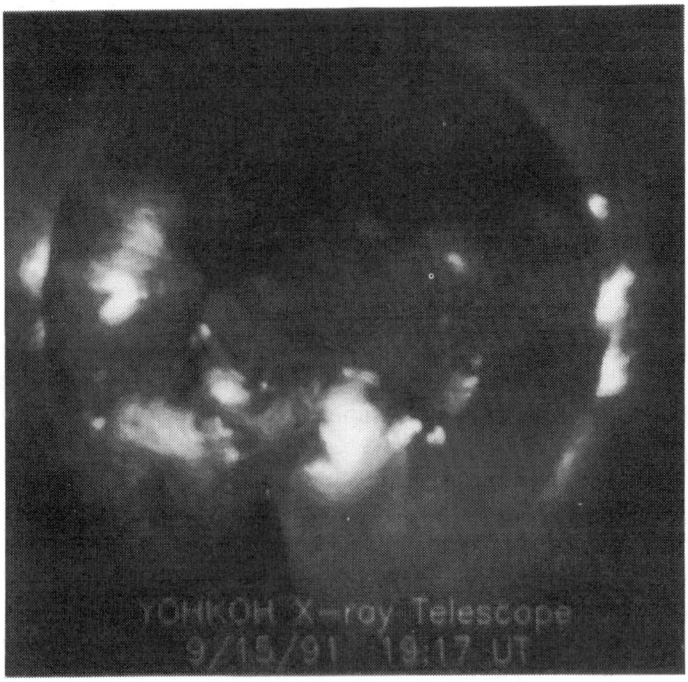

Fig. 11 X-Ray Image of the Sun Taken by Yohkoh Observatory. Lockheed Polo Slto Research Laboratory 92–11–15–11

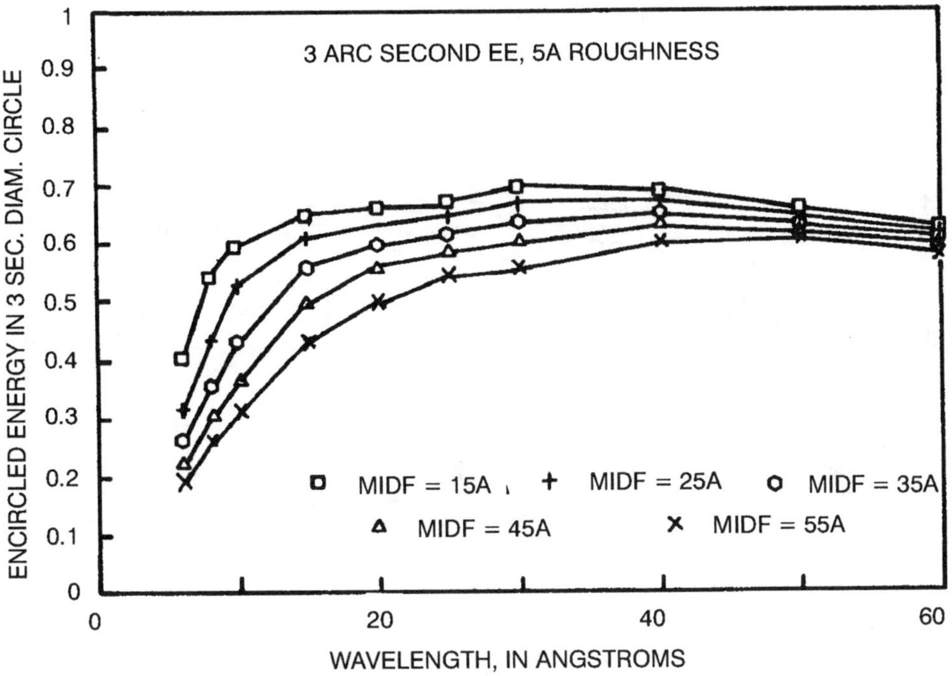

Fig. 12 Encircled Energy Predictions Within a 3 Arc Second Image Diameter, for Various mid Frequency Error Values, and 10 Angstroms RMS Micro Roughness

92–11–15–12

The mission for this instrument is to initially perform an all-sky survey, and then return to study interesting objects in much greater detail. It consists of a coaxially nested array of four telescopes, ranging in diameter from 47 to 83 cm. It was designed to have about 3 times the energy collection efficiency of the Einstein, which was also a four telescope array, but of 60cm maximum diameter. Surface figure was controlled to a strict power spectrum error budget, and a microroughness of 3Å rms was achieved in order to obtain high encircled energy.[37] As with modern instruments, ROSAT was tested and calibrated in a specially constructed x-ray facility under 1-G and, necessarily, at finite conjugates to verify performance before launch.

5.0 GROUND BASED SYSTEMS

Fifteen or twenty years ago, there was the five-meter Mt. Palomar, then the six-meter Soviet telescope. These are massive instruments. Had the practical limits been reached for aperture size? Wouldn't the thermal time constant of larger telescopes grow longer than the diurnal temperature variations? Greater light gathering power would be nice, but how could imagery be improved, since atmospheric turbulence limits resolution to about 0.3 arc second, i.e., equivalent to that of a 45-cm aperture. How could one hope to afford instruments even as large as the Mt. Palomar device today? Because of truly significant advances in several areas of technology, from today's perspective, apertures much larger than these, and a tenfold improvement in image resolution is entirely feasible. New materials, construction, and manufacturing processes, advances in compensated imaging, and active control

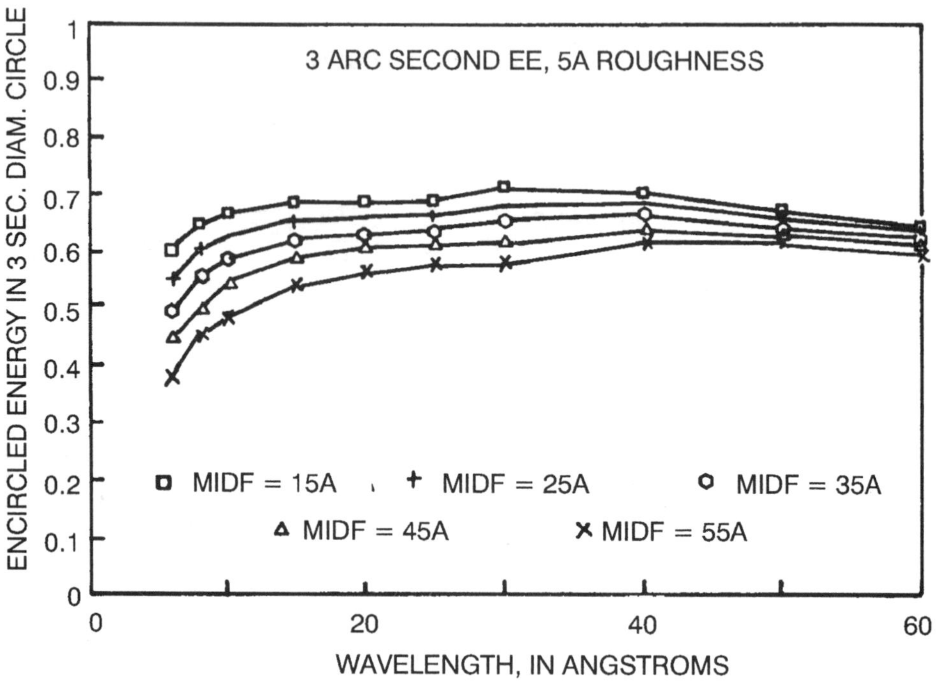

Fig. 13 Encircled Energy Predictions Within a 3 Arc Second Image Diameer, for Various mid Frequency Error Values, and 5 Anstroms RMS Micro Roughness

92–11–15–13

of large structures, coupled with completely different design philosophies for both instrument and observatory, have rendered such projects practical as well. The last ten years has witnessed the burgeoning of a large number of new program starts. Examples include the Keck 10-meter telescope shown in Figure 14 plus its twin, the dual 8-meter GEMENI, the ESO VLT array with four 8-meter telescopes (Figure 15), the 6.5-meter MMT conversion (Figure 16), the 8.3-meter SUBURA telescope, as well as several other instruments in the 3.5-meter range.

5.1 Compensated imaging

Probably the most exciting development in the technology of building large ground based optical systems has been the rapid progress made in the area of compensated imaging.[38] That is, realtime sensing of the spatial variation of optical path over the entrance pupil due to atmospheric turbulence or boiling, and correction of these wavefront errors by creating the inverse distortion on an optical element such as a deformable mirror located at the pupil conjugate. The widespread use of advanced computer technology has very effectively been applied to wavefront sensing and correction. Because the scale of atmospheric distortions, i.e., the isoplanatic patch size is fairly constant, the implication is that the number of mirror actuators and data channels increase approximately as the square of the primary diameter. Sensing distortion at an increasingly larger number of points over the aperture, and computing the required amplitude and timing of the appropriate actuator responses are the greatest challenges to be met for increasing instrument size. Better, more

Fig. 14 View of the Keck Telescope with 36-Segment, 10-Meter Diameter Primary Mirror 92–11–15–14

Fig. 15 ESO Very Large Telescope Array which Includes Adaptive Optics and Active/Chopping Secondary Mirror Systems 92–11–15–15

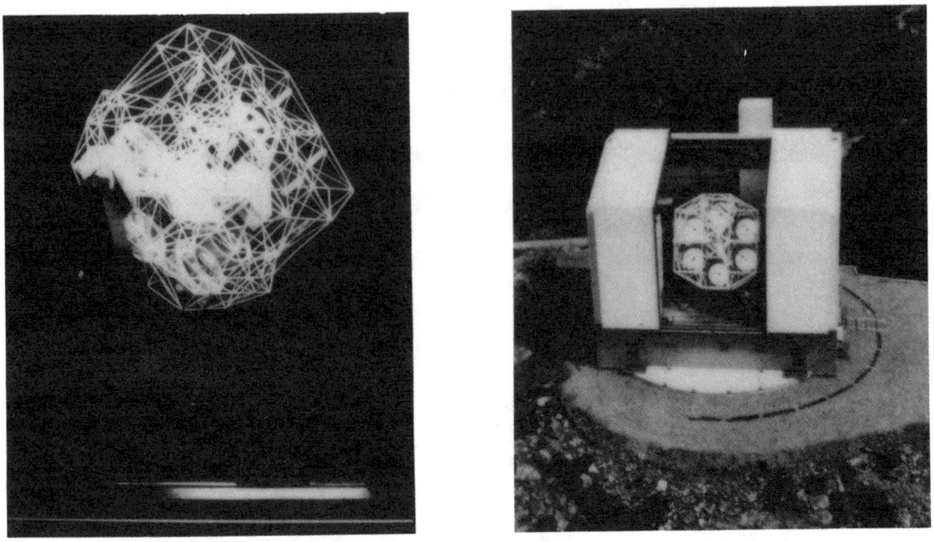

Fig. 16 Multiple Mirror Telescope (MMT) Scheduled to Receive a Single 6.5 Meter Diameter Lightweight Primary Mirror

92–11–15–16

reliable actuators with reduced hysteresis, and faster computers, coupled with refined predictive algorithms for the instantaneous atmospheric transfer function, hold great promise for achieving near diffraction limited performance for imaging through the atmosphere. One of the best aspects of these developments is that they can potentially be designed for a given application and instrument and then retrofitted to many of the telescopes which already exist!

There have been such a large number of works published on all aspects of compensated imaging, that it is difficult to summarize. Topics include comparisons of optimization for directed energy versus imaging,[39] descriptions of various wavefront sensing methods,[40] choices of beacons,[41] both natural reference stars or artificial references, various designs for deformable mirrors, the improvements possible implementing only piston, tip and tilt corrections versus additional Zernike terms, computational architecture for wavefront sensing and actuator command,[42] discussions of optimal models for the atmosphere, experimental data showing improvements in star images actually obtained in the visible and infrared,[43] and descriptions for future work. Figures 17 and 18 show examples of active optics.

5.2 Fabrication technology

5.2.1 Advanced polishing methods

Several developments in manufacturing, overlapping only slightly in time, have combined to contribute to such advances in large ground based optical systems. In

241 ACTUATOR MOLYBDENUM
DEFORMABLE MIRROR

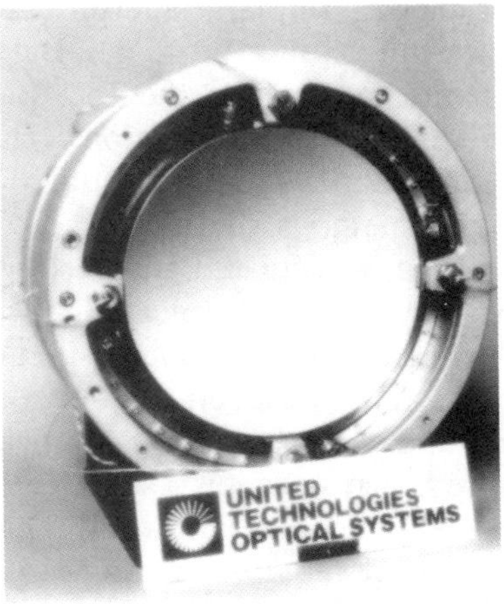

24cm, 1KHz
FAST STEERING MIRROR

Fig. 17 Examples of Active Optical Elements. Perforated Cover on DM Allows Testing of Each Actuator While Protecting Mirror Surface

92–11–15–17

Fig. 18 Deformable Mirror with 221 Actuators. Itek Corp.

92–11–15–18

order to build truly large apertures, it seems obvious that a transition to a segmented primary mirror must be made. But this implies the necessity of producing off axis segments. For example, the Keck 10-meter telescope uses 36 segments comprised of 5 different off-axis prescriptions. The feasibility to fabricate such segments in a cost effective manner was demonstrated by Nelson et al.[44] in 1979 using stressed mirror polishing methods. In this method, moments are introduced into the substrate in a carefully pre-calculated manner Figure 19, while a sphere is ground and polished into its surface. Upon release of these stresses, the substrate elastically springs back to assume the shape of the desired off-axis parabolic segment. Over the following years, methods were refined for fixturing, calculating the required stress distribution fields, and demonstrating the principle on a large glass meniscus. The outstanding feature is that since it is a spherical surface that is polished, the resultant asphere is extremely smooth. Further, the process converges rapidly, so that relatively few iterations are required to produce the final surface. When applied to the Keck project, this process was carried out on 2-meter diameter circular Zerodur® blanks which were then trimmed to final hexagonal shape. Much was learned during the program. After being under stress for long periods of time, for example, the mirrors did not immediately spring back to their proper levels, even though the strain was below the elastic limit for the material. Instead, a delayed relaxation was observed, and had to be accounted for. Even though the material was of the very highest quality, with internal stress levels well below 2μm/cm, some edge distortion was observed after trimming to the required hexagonal shape, despite having carefully avoided any surface stress due to subsurface damage associated with earlier processing steps.

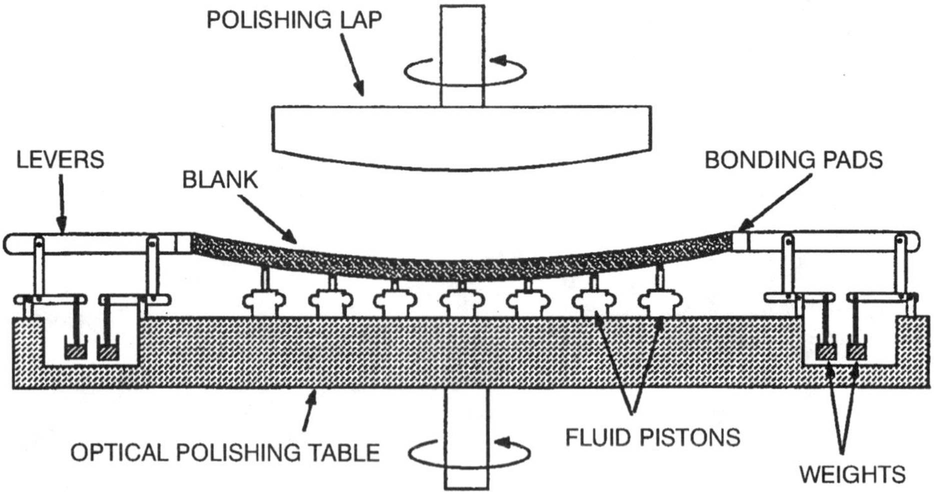

Fig. 19 Schematic of Stressed Mirror Polishing Process to Produce Off-Axis Aspheric Mirror Segments

92–11–15–19

This was corrected for at Eastman-Kodak by using computer controlled ion polishing methods after cutting to shape. This method has been around since the mid-1960's, but met with limited acceptance, because material removal rates were impractically low using ion sources available at that time. Focused beams, usually of argon ions running at about 15kv, and 150mA were used at 60°, and scanned under computer control to remove surface errors. Recent improvements as reported by

Allen and Keim[45] have dramatically increased removal rates and at much lower beam energies.

Computer control of ordinary pitch polishing methods has been used successfully by several manufactures, each with some variation designed to produce specific results, according to some particular objective or fabrication philosophy. Figures 20, and 21 show a 4-meter and a 0.5-meter capacity machine respectively. Control of surface figure has been demonstrated down to less than 1.0-μinch (0.04wave) p-p , at surface location accuracies of 0.1mm, and producing microroughness levels of less than 6Å rms.[3]

Fig. 20 Four Meter Capacity Computer Controlled Polishing Machine. Itek

92–11–15–20

5.2.2 Substrate Formation

In terms of new substrate manufacturing technology, mention has already been made of advanced SiC forming methods where ultralightweight mirror blanks are made of this very high modulus material by means of a slip casting process. The remarkable feature of this process is that complex closed back hollow blanks can be made complete with extremely thin multi-tiered internal features by a one step forming process, fired at high temperature to reaction-bond the slip, and then infiltrated with silicon to produce a fully dense product. Once the tooling has been designed and built, the advantage is that additional copies can be produced very economically, making it practical to carry along spares in the event of an accident.

Another innovative technique described by Angel[46] has been the introduction of spin casting to form lightweight borosilicate glass substrates. The glass is melted in a

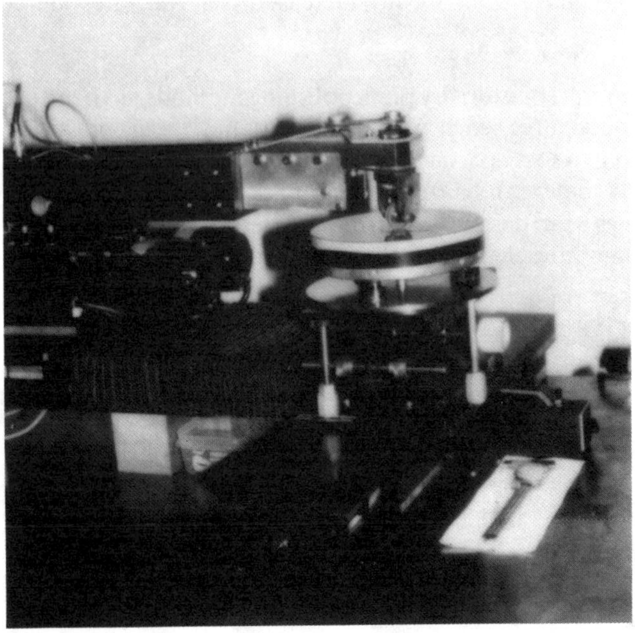

Fig. 21 Computer Controlled Roller Polisher (CCRP) Half–Meter Capacity. UTOS

rotating oven. Forms are arranged to produce hollow cells. During the casting process, the molten mirror assumes a natural parabolic shape. Mirrors of successively larger diameter have been cast. The most recent is the 6.5-meter diameter $f/1.25$ honeycomb sandwich replacement mirror for the MMT.

Some workers at Schott have produced monolithic meniscus blanks up to 8.6-meters in diameter of Zerodur® for the ESO Very Large Telescope (VLT), using the spincasting technique[47] seen in Figure 22. Others are constructing an 8.3-meter ULE® mirror using the fusion hex-seal process[48] for the Subura (JNLT) Telescope.

Recently, high pressure abrasive water jet cutting techniques have been successfully applied at Eastman-Kodak to cut cores for very lightweight mirror blanks.[49] These cores can be made of fused silica, ULE®, or Zerodur®. Thin facesheets can be fused or frit-bonded on to form the complete mirror blank. Operating under robot control, the technique is applicable to curved or flat mirrors. Use of these techniques in conjunction with the 2.5 meter generator, built by Cranfield Precision Engineering[50] should reduce the overall time and cost to produce large lightweight mirrors.

6.0 CONCLUSIONS

There have been a number of advances in a broad area of technologies which have enhanced the capabilities of many types of sensor systems based on the ground, on aerial platforms, and in space. Advances have come not so much from revolutionary new optical forms, but from substantial improvements in materials, techniques for forming these materials into more efficient structures, ultra-durable, more effi-

Fig. 22 8.6 meter VLT Blank. Schott Corp.

cient optical coatings, the application of more deterministic optical fabrication methods using computer control, and the use of active, deformable mirrors to provide real time wavefront correction. As these technologies mature, we can look forward to truly spectacular achievements in astronomy like long baseline coherent imaging; the detection of planets outside of our solar system; earth observations which more accurately measure over a larger spectral range for more efficient management of our natural resources; and understanding the heat engine which drives climatic changes, by recording the dynamics of global wind patterns, and ocean currents.

7.0 REFERENCES

1. "TAC will shift to Electro-optical real-time reconnaissance by 1993," Aviation Week & Space Technology, August 29,1988.

2. D.S. Pickard, "Low/medium altitude sensor," *SPIE Proc.*, Vol. 1763, 1992.

3. G. Hull-Allen, "Precision optical fabrication and test methods applied to the SXT grazing incidence mirror assembly," *SPIE Proc.,* Vol. 1160, 1989.

4. P.N. Blake, R.O. Scattergood, "Chip topography of diamond turned ductile metals," *SPIE Proc.,* Vol. 676, 1986.

5. R.A. Jones, R.L. Plante, "Rapid fabrication of large aspheric optics," Precision Eng. 9(2), 1987.

6. R.A. Jones, "Optimization of computer controlled polishing," *Appl. Opt.,* Vol. 16 (1), 1977.

7. G. Hull-Allen, "A computer controlled polisher for advanced fabrication," *Photonics Spectra,* March 1990.

8. R.A. Jones, W.K. Rupp, "Rapid optical fabrication with computer-controlled optical surfacing," *Optical Engineering*, December 1991.

9. J.R. Bolch, R. J. Drake, "Feature: Military Optics," *Laser Focus World,* August 1989.

10. E. Ulph, "Fabrication of a metal matrix composite mirror," *SPIE Proc.,* Vol. 966, 1988.

11. K.A. Snail, "Status of CVD diamond as an optical material," *Int. Journal of Opt. Materials,* March 1991.

12. A. Feldman, S. Holly, ed., Diamond Optics V, *SPIE Proc.,* Vol. 1759, 1992.

13. A.F. Slomba et al, "Testing of erosion resistant infrared coatings by nanoindentation techniques," Fifth Topical Meeting, *OSA,* Tucson Arizona, 1992.

14. J.A. Detrio "Rain impact testing of durable coatings on water clear ZnS," *SPIE* Proc., Vol. 1760, 1992.

15. W.F. Adler, "Infrared-transmitting window survivability in hydrometer environments," *SPIE Proc.,* Vol. 1760, 1992.

16. M.J. Riedl and J.T. McCann, "Analysis and performance limits of diamond turned diffractive lenses for the 3-5 and 8-12 micrometer regions," *SPIE Critical Review,* Vol. CR 38, 1991.

17. G. J. Swanson, "Binary optics technology: Theoretical limits on the diffraction efficiency of multilevel diffractive optical elements," *Lincoln Laboratory Tech Rept.,* Vol. 914, 1991.

18. T.J. McHugh, D.A. Zweig, "Recent advances in binary optics," *SPIE Proc.* , Vol. 105, 1989.

19. M.J. Riedl, "Predesign of diamond turned refractive/diffractive elements for ir objectives," *SPIE Critical Review,* Vol. CR 41 08, 1992.

20. R. Giacconi et al, "The Einstein (HEAO-2) x-ray observatory," *The Astrophysical Journal,* Vol. 230, 1979.

21. J. Trumper, "The ROSAT mission," *Advances in Space Research,* Vol. 2, 1983.

22. R.F Schilling, "Design options for the Advanced X-ray Astrophysics Facility (AXAF)," *SPIE Proc.*, Vol. 597, 1985.

23. Y. Ogawara, "The Solar-A mission," *Solar Phys.,* Vol. 113, No 1/2, 1987.

24. L. Acton et al, "The Solar-A Soft X-Ray Telescope experiment," *Adv. Space Res.,* Vol. 8, No 11, 1988.

25. D. Ghitelman, "The space telescope," *Gallery Books,* New York, N.Y., 1987.

26. R. Kerrod, "The illustrated history of NASA, anniversary edition," *Gallery Books*, New York, N.Y. p. 194, 1988.

27. R.J. Spehalski, M.W. Werner, "Objectives for the Space Infrared Telescope Facility," *SPIE Proc.* Vol. 1540, 1991.

28. J. W. Bilbro et al., "Airborne Doppler lidar wind field measurements," *Bull. Amer. Meteor. Soc.,* Vol. 65, 1984.

29. P. Glenn, A.F. Slomba "Derivation of requirements for surface quality and metrology instrumentation for AXAF/TMA," *SPIE Proc.,* Vol. 597, 1985.

30. G. Hull-Allen et al, "Optical performance vs. prediction analysis for the SXT mirror assembly at x-ray wavelengths," *SPIE Proc.,* Vol. 1160, 1989.

31. D.J. Schroeder Astronomical Optics, Academic Press, New York, N.Y. p. 213, 1987.

32. S. Seki et al, "W/C multilayers deposited of plastic films," *SPIE Proc.,* Vol. 1742, 1992.

33. P. Boher, "Ni/C and Rh/C multilayers for soft x-ray optics: influence of the deposition conditions on the microstructure," SPIE Proc. 1742, 1992.

34. J.M. Slaughter, et al, "Non-specular x-ray scattering from Mo/Si multilayers," *SPIE Proc.,* Vol. 1742, 1992.

35. G. Gutman et al, "Method for fabrication of multilayer uniform/graded coatings on curved substrates," *SPIE Proc.,* Vol. 1742, 1992.

36. A.B.C. Walker, R.B. Hoover, T.W. Barbee "Multispectral solar telescope array: initial results and future plans," *SPIE Proc.,* Vol. 1742, 1992.

37. K. Beckstette, E. Heynacher, "Contour line measurements on ROSAT x-ray mirrors," *SPIE Proc.*, Vol. 429, 1983.

38. J.W. Hardy, "Adaptive optics-a progress review," *SPIE Proc.,* Vol. 1542, 1991.

39. R.K. Tyson, "Measuring phase errors of an array or segmented mirror with a single far-field intensity distribution," *SPIE Proc.,* Vol. 1542, 1991.

40. L. Dame et al., "ASSI: An optimized fringe tracking stellar interferometer," *SPIE Proc.,* Vol. 1114, 1989.

41. L.A. Thompson, C.S. Gardner, "Experiments on laser guide stars at Mauna Kea Observatory for adaptive optics in astronomy," *Nature,* Vol. 328, 1987.

42. A.P. Doel et al., "MARTINI: System operation and astronomical performance," *SPIE Proc.,* Vol. 1542, 1991.

43. F. Merkle et al., "Adaptive optics systems tests at the ESO 3.6-m telescope," SPIE Proc. 1542, 1991.

44. J. Lubliner, J. Nelson, "Stressed mirror polishing. 1: A technique for producing nonaxisymmetric mirrors," *Applied Optics*, Vol. 19, 1980.

45. L.N. Allen, "Precision figuring of optics by ion machining processes," *Conference on Lasers and Electro-Optics*, Vol. 11, 1989.

46. R. Angel, "Casting of the 6.5-meter borosilicate mirror for the MMT conversion," ed. M.H. Uhlrich Progress in Telescope and Instrument Technology, *ESO Conf. Proc.*, 1992.

47. Schott press release, "Hot start for supertelescope: 8-meter mirror successfully cast," *No. 4/91*, March 27, 1991.

48. J. Spangenberg-Jolley, C.L. McGill, "Large mirror blank fabrication," OSA Optical Fabrication & Test Workshop, November 1992.

49. M. Baumler, "Abrasive water jet cutting of mirror cores," *OSA Optical Fabrication & Testing Workshop,* November 1992.

50. P.B. Leadbeater, et al., "A unique machine for grinding large, off-axis optical components: the OAGM 2500," Precision Eng. 11(4), October 1989.

Lasers for airborne applications

S. Guch, Jr. and D. P. Stanley

Litton Systems, Inc.
Laser Systems Division
Apopka, Florida

1. INTRODUCTION & SUMMARY

The use of lasers in sensing systems carried by aircraft has been a topic of active research and development since the invention of the first coherent optical sources in the 1960's. The added-value provided by airborne laser sensors lies in four unique capabilities:

- Rapid deployment to remote sites or regions
- Access to large instantaneous fields of regard
- High area coverage rates
- Long operating ranges

These capabilities are possible because of the intrinsic mobility, speed, and altitude of airborne platforms. Achievement of these benefits requires that system performance needs be met simultaneously with platform-driven requirements. Some of these needs and requirements are:

System Performance Needs

- wavelength
- bandwidth
- pulse repetition rate
- pulse width
- beam divergence
- pointing accuracy
- pointing stability

Platform Driven Requirements

- vibration resistance
- shock resistance
- low electromagnetic signature and susceptibility
- minimum weight and volume
- electrical power requirements
- heat rejection compatible with aircraft capabilities

The progress of airborne laser system development has been driven by the need to strike a balance of the following factors:

- Laser optical performance
- Platform and laser compatibility
- Cost

The first factor has encouraged developers to explore the capabilities of airborne laser systems. Unfortunately, the realities associated with the latter two factors represent hurdles that have restricted the practical applications of lasers to a very small number. Of the broad variety of air, ground, and ocean sensing systems that have been proposed,

FL2827
30SPW/XPOT TECHNICAL LIBRARY
BLDG. 7015
806 13th ST., SUITE A
VANDENBERG AFB, CA 93437-5223

the only ones to be fielded in quantities more than 30 or so of a given design have been electrooptical targeting systems for precision weapons delivery.

This is not to say that no other worthwhile applications have been identified. Numerous airborne applications with both military and commercial interest have been suggested and in some cases demonstrated:

- Aircraft systems for air velocity measurements have potential for improving the operational economy and safety of commercial aircraft operations
- Environmental monitoring systems can provide rapid surveys of airborne contaminants around potential polluters or accident sites
- Geophysical surveys, from either aircraft or spacecraft, can provide accurate mapping of surface topography features and, in some cases, shallow underwater bottom features

Additional uniquely military applications remain largely untapped:

- Measurement of deep underwater bottom profiles and objects, and acoustic environment monitoring for sonar optimization
- Aircraft wire avoidance systems
- Advanced targeting and guidance systems for precision munitions guidance

Many of the technical capabilities necessary to implement laser systems that accomplish these missions have been demonstrated. However, high cost has frustrated the development of mature system designs.

In this paper, we will first summarize the characteristics of the few currently operational lasers used in airborne sensors, and will then present brief looks at some of the systems that are being developed or might be developed in the future. Perhaps not surprisingly, it appears likely that the emergence of breakthroughs in laser technology, in areas such as diode pumping of solid state lasers, high efficiency spectral conversion, and high brightness resonator design, may not dramatically broaden the proliferation of airborne laser systems. The relatively modest cost impact of the lasers used within mature systems, typically less than 15% of the total system, means that system cost drivers must be aggressively attacked if the promise of airborne laser sensors is to be realized.

2. OVERVIEW OF OPERATIONAL EQUIPMENT

Despite extensive research, development, and engineering, only a few airborne laser systems have achieved operational status. By operational status, we mean being installed for permanent use in performing a dedicated mission, as opposed to verifying or validating the capability of the laser system for a function. The most important of these are the following:

- Precision targeting for military weapon guidance
- Geophysical mapping

2.1 Precision weapon delivery

The only airborne laser sensor systems proliferated in significant numbers are with the precision guidance of munitions for the military. The key characteristics of such systems include some or all of the following:

- Night vision observation, acquisition, and tracking using forward looking infrared (FLIR) systems
- Day vision capabilities using direct-view, television, or intensified-video cameras
- Automatic tracking of targets and stabilization of system line-of-sight
- Laser target rangefinding, for use in developing ballistic fire control solutions
- Laser target marking and designation, for guidance of sensor-equipped bombs and rockets
- High speed computers for system control and, potentially, target recognition
- Thermal management subsystems

Despite the extensive publicity associated with the precision guided weapons used in recent military conflicts, the number and proliferation of such systems has been relatively modest. While numbers and locations of deployed systems are still closely held, it is probably safe to estimate that of order 3000 may be operational in Western nations. The level of technology involved in development of compact systems required for use on tactical platforms is high. Thus it is not surprising that the western nations primarily involved in developing these systems are the United States, Great Britain, and France. While sales to other nations are progressing, the developing countries have retained the technology and most of the installed-based of equipment to date. Figure 1 provides a summary of some of the characteristics of existing laser-based targeting systems. Figure 2 shows photographs of several of these systems.

The lasers used in airborne targeting systems exhibit very similar performance characteristics. Detectors used to generate ranging information and/or produce guidance command data also have nearly identical performance characteristics from system to system. Commonality of the weapons systems delivered by airborne targeting systems is the primary reason for the similarity.

The key parameters around which laser designers optimize their devices include the following:

- Pulse energy
- Pulse rate
- Beam divergence
- Beam uniformity
- Pulse width
- Wavelength
- Boresight alignment
- Boresight stability

Platform compatibility issues that represent major design drivers include the following:

- Temperature environment
- Vibration and shock environment
- Electrical power availability
- Heat removal capability
- Electromagnetic interference control to minimize platform to payload conflicts
- Available volume and system mass

Many of the characteristics of the laser output of particular targeting systems are classified. Unclassified, typical performance and environmental levels at the level of the laser transmitter/receiver unit in such systems are listed below:

• Pulse Energy	100-150 mJ
• Pulse Rate	10-20 Hz
• Beam Divergence	1-2 milliradian
• Transmitter Boresight Accuracy	0.1 X Divergence
• Transmitter Boresight Stability	0.1 X Divergence
• Ranging Probability of Detection	99%
• Ranging False Alarm	1%
• Weight	15-20 pounds
• Electrical Power Input	200-350 watts
• Operating Temperatures	
• Minimum	-54 to -10 °C
• Maximum	+50 to +71 °C
• Nominal	+10 to +40 °C
• Storage Temperatures	
• Minimum	-65 to -20 °C
• Maximum	+50 to +85 °C
• Nominal	25 °C
• Operational Vibration	3 to 12 g_{rms}
• Survival Vibration	6 to 18 g_{rms}
• Humidity	100%
• Salt Fog	No degradation after long-term exposure

The evolution of airborne laser systems for rangefinding and target marking or designation has generally proceeded in the direction of reducing weight while maintaining optical and environmental performance. The lack of dramatic optical performance improvements is largely the result of the requirement for transmitter compatibility with the large inventory of seekers that have been fielded. The detectors and filters in these systems set the wavelength and energy needs of the weapons systems. Targets and standoff ranges establish the operational scenarios. Given that all of these factors have remained, and are likely to remain, essentially unchanged over the past twenty years, it is not surprising that laser development has been evolutionary rather than revolutionary.

The most significant changes that have appeared in operational units include the following:

- Weight reduction: Flashlamp pumped laser rangefinder/designators have been reduced in weight. Over the past twenty years, airborne laser rangefinder/designator weight has fallen from nearly 50 pounds to under 15 pounds. This improvement has been achieved chiefly through improved design processes, in which optimized placement of structural members provides environmental stability without the need for massive metal enclosures.
- Eyesafe operation: Raman conversion of 1.06 μm Nd:YAG laser output to 1.54 μm in methane dramatically increases the allowable intra-ocular exposure. Operational efficiencies in the 25-35% range are achieved. Switchable systems allow training or operational ranging above friendly forces in the eyesafe mode. The system retains full tactical performance at 1.06 μm when required.

Interestingly, a number of potentially revolutionary changes in laser technology have not impacted precision targeting systems. Some of the more significant technology breakthroughs that have not yet impacted the installed base of airborne rangefinders and marker/designators include the following:

- New materials (in particular Nd/Cr:GSGG): Even with potential for increasing laser efficiency by more than a factor of two, new materials have not been adopted for military systems. This is primarily because their wavelengths are sufficiently removed from that of Nd:YAG to preclude their use with the existing sensor inventory. Additional factors, such as thermal limitations of the new materials, represent secondary factors that could have been overcome.
- Diode-pumping: The potential for achieving a two to three fold improvement in laser efficiency has not been realized in systems to date chiefly because of diode array costs. Thermal management of the diode arrays has also posed challenges. However, this limitation appears on the verge of solution by virtue of the recent availability of high temperature laser diodes. Diode costs appear to be rapidly approaching a level which is consistent with laser system cost requirements. If this turns out to be true, future airborne laser systems will likely be diode pumped.
- Unstable resonators: Improvement of beam quality by factors of two to three, to achieve between two and three times diffraction limited performance, has the potential for reducing the output beam divergence of airborne laser sensors. This could either reduce the size of the projected spot on the target to improve system accuracy or drive down the diameter of the host system beam collimating telescope, thus saving weight of the telescope and the tracking gimbal that supports it. With the exception of one system that is just beginning to go into production, this benefit has not been realized. Most transceiver systems are driven as much by the need for a large receiver collecting aperture as by output beam divergence considerations.

In looking ahead at future laser targeting systems, it appears likely that technology evolution will continue, but that the overall performance improvements are unlikely to be dramatic. The areas in which this evolution is likely to occur include the following:

- Weight: As high temperature diode pumping becomes affordable in the latter half of the 1990's, the efficiency benefits achieved will allow reduction in power supply, pulse drive electronics, and cooling system sizes. Full-performance systems are likely to become routinely available at weights below 10 pounds.
- Reliability: The advent of long-life diode pumping and newly emerging optical compartment contamination control techniques will eliminate planned maintenance requirements. Improved beam quality, deriving from the combination of diode pumping and unstable resonators, will provide improved margins to optical damage and dramatically reduce in-service failures. It is probable that no laser maintenance actions will be required over a typical 20 year system life.
- Electrical power and heat removal: In the same manner as weight reductions will be achieved, diode pumping efficiency benefits will drive down electrical power and thermal management needs by 30 - 40% from current levels associated with flashlamp pumped systems.

While underpinning technologies, power supplies, energy storage, spectral conversion, may change, it is unlikely that they will provide major impacts on overall system capabilities or cost. It appears more likely that laser rangefinders and designators will simply become ubiquitous in future military aircraft fire control suites, as technological solutions resolve the issues that posed significant problems in the development of current airborne systems.

2.2 Geophysical mapping

Laser altimeters have traditionally been used in relatively modest numbers to provide precision height measurements for use in mapping studies. Traditionally, microwave altimeters used at altitudes of order 1,000 m were effective but projected sufficiently large beams on the ground to obscure some features of interest and/or had difficulty in achieving accuracy and resolution under 1 m.

Laser diode transmitters have been employed with success in aerial surveying at modest altitudes, achieving small spot sizes and good resolution with high PRF, sub-microsecond devices using AlGaAs and GaAs laser diodes. Prototypical of systems likely to be used more broadly in the future, however, is a diode pumped Nd:YAG laser currently in use in the Antarctic to map ice flows. Figure 3 shows a photograph of this system. Key performance characteristics of the laser transmitter are summarized below:

- Pulse energy: 20 micro Joules
- Pulse rate: 4 kHz
- Pulse duration: 15 nsec
- Range resolution: 10 cm
- Max. altitude capability: > 1.7 km

- Weight: 16 kg
- Size: 16 x 22 x 38 cm (Laser/Receiver)
  13 x 18 x 28 cm (Power Supply)
- Electrical input: 120 watts

The key advantage of this system over diode laser transmitters is higher peak power while retaining high pulse rate capability. The higher peak power provides enhanced range and, due to the higher receiver signal-to-noise ratios and reduced pulse rates, better resolution and accuracy. Of order 30 of these systems are operational worldwide, providing accurate topographical mapping surveying at approximately 1% of traditional surveying. While the capability of such systems is considerable, and represents a major improvement over prior microwave units, it clearly represents a niche application for which broad proliferation is unlikely.

3. DEVELOPMENTAL SYSTEMS & LASERS

One of the hallmarks of laser system developers has been optimism regarding the potential capabilities, cost effectiveness, and user acceptance of their handiwork. Optimism should be tempered considerably as indicated by the relatively small penetration of laser systems in operational environments. While research is ongoing on numerous fronts, only the following potential applications are considered in detail in this discussion based on their potential for ultimate operational applications:

- Underwater sensing systems
- Advanced weapon targeting and guidance
- Aircraft wire avoidance systems
- Airborne atmospheric sensors
- Airborne identification: Friend Or Foe

3.1 Underwater sensing systems

Underwater sensing systems are constrained both by the nature of the aqueous beam propagation environment and laser technology. The transmission characteristics of water are such that only the blue-green wavelengths (roughly 400 - 540 nm) offer sufficient optical transmission to make laser sensors feasible. With maximum practical laser energies in this band ranging up to several joules, typical linear attenuation values in water limit sensing systems to 30 to 100 meter depths. Achievement of significantly greater depth capability would require order-of-magnitude laser and energy power scaling that will clearly be impractical in the foreseeable future.

The most mature underwater sensing technology lies in the area of depth sounding. Countries, such as Australia, with large continental shelf areas have a clear requirement for accurate and economical charting of relatively shallow littoral regions. Similarly, requirements for accurate navigational data in support of military amphibious operations may include the ability to rapidly and accurately determine ocean bottom profiles over large, remote regions. Typical of the transmitters used in bathymetric systems is a unit developed by AWADI and routinely used for shallow water charting. The laser is a 5 mJ, 5 to 7 nanosecond, frequency doubled, flashlamp-pumped Nd:YAG unit operating at 150 to 200 Hz pulse rate. The green output is scanned with approximately 10 m spacing, while residual 1.06 μm radiation is transmitted vertically from the aircraft to provide a surface/attitude reference. This system provides depth measurement capability from approximately 2 to 50 m, with a 2 sigma accuracy of approximately 1 m, validated through comparison of data with ground truth established by conventional marine surveys. An area coverage rate of 5 square kilometers per hour is achieved by operating with a 250 m wide beam scan pattern as the aircraft proceeds forward at 70 m/sec. While other systems optimize performance slightly differently than the Australian one, by increasing pulse rate and/or tightening the scan pattern to improve resolution the general features of all are reasonably represented by the above example. In all cases, the lasers that have been used in developmental testing to date have been hardened sufficient to tolerate the airborne environment but not fully packaged for use in tactical platforms. An example of this is the system-level requirement for 9 kW of electrical power for the Australian system, which exhibits a total mass of approximately 1600 kg. Of this, the laser head contributes approximately 30 Kg of the mass and 2 Kw of the power consumption.

Airborne sensors have also been evaluated for measurement of deeper bottom profiles, location of the deep sea floor underwater objects, and for measurement of underwater temperature profiles. The reflection of optical energy from submerged objects is, like shallow water bathymetry, a straightforward extension of laser radar principles. As in the case of bathymetry, details of system design and water attenuation determine the maximum depth and search rate that are attainable. However, radiometric considerations, i.e., attenuation coefficient of order .1 m^{-1} and blue-green laser energy of order 1 J/pulse, will limit operating depths to under 100 meters. Measurement of underwater temperature profiles represents another area in which development activities have taken place. Two alternatives have been considered:

- Raman shift: Using a broad-bandwidth transmitter, exploit the temperature dependence of the large wavelength Raman shift of sea water.
- Brillouin shift: Using a narrow-bandwidth transmitter, exploit the temperature dependence of the small wavelength Brillouin shift of sea water.

Preliminary analytical studies indicated that incoherent Raman systems may be capable of providing measurements accurate to several degrees Celsius at depths up to 100 m with realizable transmitters. Incoherent Brillouin systems appear to provide signal strengths that are too low to provide useful depth resolution capabilities for most applications. However, systems exploiting coherent stimulated Brillouin scattering may achieve sub-Celsius resolution with depth capability of 50 to 100 m. To provide useful data rates and performance, all of these systems are likely to require typical laser design features and characteristics listed below:

Laser type	Frequency-doubled Nd:YAG, Q-switched. Diode pumped, using high temperature diodes. Injection-seeded, single frequency (Brillouin sensor system only)
• Pulse energy	0.3 to 1 J
• Pulse rate	20 to 200 Hz
• Wavelength	400 to 550 nm

• Beam Quality	Not critical
• Laser Efficiency	1% min., 3% desired (optical/PFN)
• Lifetime	> 100 hours between maintenance
• Spectral Bandwidth	Single axial mode, referenced to local oscillator (Brillouin system only)
• Weight	Under 50 Kg

Open-literature sources indicate that some of the potential applications of airborne underwater laser sensor technology include fish school or submarine location, ship or submarine wake detection, and temperature mapping for fishery or underwater acoustic propagation prediction purposes.

3.2 Advanced weapon targeting and guidance

Advances to conventional rangefinders/designators referred to in Section 2 are likely to produce incremental improvements in military targeting system performance. More advanced laser radar systems have been considered that require revolutionary changes in the laser devices and associated receivers. High-resolution miniature systems that provide 3-dimensional imaging of targets for identification in surveillance or final fire control commands during terminal guidance maneuvers may prove to be feasible and economical. While systems and components for this application are early in concept analysis and development, typical parameters may include the following:

• Range:	1 to 2 km, to provide standoff/divert capability
• Accuracy:	1 foot resolution in 3 dimensions
• Pulse Rate:	Consistent with development of video scenes at > 10 Hz
• Weight:	Under 10 kg, including power supplies
• Cost:	Under $30,000

At least four different laser.systems appear to offer promise for meeting most or all of these characteristics in fully productionized systems:

- Waveguide CO_2 lasers: High PRF cavity dumped CO_2 lasers offer operation at 10^5 Hz at average powers of order 10 watts in very small packages, as illustrated in Figure 4. By using heterodyne detection techniques, good performance has been demonstrated in a small package volume, although operation over broad environmental regimes has yet to be demonstrated.
- Diode-pumped Nd:YAG lasers: CW-pumping, preferably via end pumping using multi-fiber coupling, offers prospects for operation at 10's of kHz in Q-switched operation or at 100's of kHz in cavity dumped operation. While these systems operate more naturally in a direct video detection (vice heterodyne) mode, either may be implemented in future systems to achieve the required performance.
- 1.5 to 2 μm diode pumped lasers: Crystalline-host Ho, Tm, and Er lasers have been postulated as sources for this application, owing to the possibility of achieving eyesafe operation for use in training and over civilian or friendly areas. While such systems would be desirable, the relatively low

stimulated emission cross section of known media are likely to render them less attractive than the Nd:YAG system unless eye safety considerations become truly compelling.

- Near-infrared laser diodes: Laser diodes operating in the 800 to 900 nm, 1300 nm, and 1550 nm regimes offer attractive efficiency and packaging characteristics, but are likely to be problematical from the standpoint of providing adequate peak power to achieve the signal-to-noise levels needed to achieve the required depth resolution at the ranges of interest.

As in the case of the laser transmitters used in many applications, it appears that several attractive technologies exist which meet the technical performance requirements of the host systems. However, breakthroughs in component prices, either in diode pump lasers or in processing of CO_2 laser tubes, will be required if any of these laser transmitters are to reach affordable levels.

3.3 Aircraft wire avoidance systems

The requirements for lasers for airborne wire avoidance systems are similar to those of the advanced weapons guidance systems discussed in the above paragraph. The most significant differences are as follows:

- Range: Since most wire avoidance systems are intended to protect helicopters operating at low altitudes, and since helicopters are extremely maneuverable at their modest forward speeds, acceptable operating ranges can be a factor of 2 to 4 lower than required for 3-D imaging sensors.
- Target geometry: Since wires are exceedingly thin but very long, standing out by 10's of m above ground clutter, system design must accommodate unusual scanning and/or "target" recognition algorithms.
- Weight and cost: Operating from a manned airborne platform in a non-disposable package, cost and weight constraints may be relaxed by perhaps 50% from levels required by the imagers.

Given these considerations, all of the laser devices mentioned in Section 3.2 could be used successfully in configuring a wire avoidance system. Depending on the particular system embodiment and detailed system requirements, however, a transmitter employing diode lasers as the primary source, probably as an array of incoherently combined multi-stripe laser diode bars, may have compelling cost advantages.

3.4 Aircraft instrumentation sensors

A broad variety of environmental sensors have been proposed for aircraft instrumentation applications, including the following:

- True airspeed sensors, as replacements for mechanical Pitot tubes, which are subject to physical damage and require periodic maintenance and/or replacement. Because laser systems must displace existing hardware, initial replacement costs must be competitive and life-cycle costs must be compelling.

- Wind shear or microburst sensors, for use during landing operations to avoid hazardous conditions. While inertial and rf systems are being advanced to meet Federal requirements, their marginal performance provides an opening for laser-based systems, but only if they can be implemented at a lower cost.
- Winds aloft sensors, to sense direction and velocity of winds in proximity of an aircraft to allow selection of altitude to optimize fuel efficiency. This application has large potential long-term cost benefits and no competing technologies. However, both the technology and short-term cost benefits of these systems remain to be proven.

The selection of a laser for Pitot tube replacements is probably the simplest, owing to the short ranges involved. The laser sensor must provide Doppler velocity measurements, based on aerosol back reflections from the unperturbed air immediately ahead of the sensor aperture, where required maximum ranges are only a few meters. While a variety of lasers, including CO_2, diode pumped Nd:YAG, and diode pumped 2 μm crystalline lasers, could meet the performance requirements, use of a single-mode diode laser of a few milliwatts output appears compelling. The inherent simplicity of the device and its power source, coupled with inherent compatibility with heterodyne operation, potential for eyesafe operation at 1.55 μm, and low cost represent important factors favoring this approach.

One of the most important factors governing the selection of laser transmitters for the other aircraft instrumentation applications is eye safety. Given that the systems will be checked out and operated in uncontrolled proximity to personnel and other aircraft, operation in an unambiguously eyesafe spectral region is viewed as a mandatory characteristic of transmitters to be used with these systems. Typical performance needs of such systems are listed below:

• Wavelength:	> 1.4 μm (Note: Eyesafety criteria are in flux, and are likely to change dramatically over the next year; the > 1.4 μm criteria has been taken as a guideline for airborne systems, but is not rigorously in conformance with any current or future eyesafety regulations)
• Pulse Energy:	30 to 300 mJ per pulse
• Pulse Rate:	10 to 30 Hz
• Weight:	$<$ 10 kg
• Cost:	$<$ \$30K

Based on these criteria, most of the development work to date has centered on diode pumped Tm or Ho crystalline lasers, using YAG or YLF hosts, operating near 2 μm. While these sources suffer design difficulties owing to their low stimulated emission cross sections, they can meet the basic performance requirements outlined above. Meeting the critical cost criterion is likely to be much more difficult, however, due to the inherent inefficiencies and complexity of the required injection-seeded oscillator/power-oscillator and inherently inefficient energy extraction required to generate millijoule pulse energies at fluence levels well below optical damage thresholds. It appears likely that system-level strategies similar to those below may be needed if truly cost-effective solutions are to be found:

- Multi-pulse signal averaging: Operate at pulse rates in the 3 to 10 kHz regime, at microjoule pulse energies and low single-pulse energy extraction. Employ multi-pulse averaging to achieve the required signal-to-noise performance for the parameters to be sensed, and exploit the long (multi-millisecond) energy storage lifetimes in the laser media to produce high net efficiency despite low single pulse extractions.
- Combine multiple functions within a single sensor: By tailoring a multi-function sensor to provide many capabilities (e.g., airspeed, gust, wind shear, winds aloft) in a single sensor, the cost of a laser and system may be amortized over the several instruments being replaced.

3.5 Airborne identification: friend or foe

The problem of positively determining the identity of a remote platform that has been detected, but not resolved to the extent that it can be recognized, remains a key problem in military airborne systems. Infrared systems can detect, but not resolve, potential airborne targets at long ranges. Ground targets camouflaged and embedded in battlefield smoke and dust are inherently difficult to recognize from the air. Conceptual solutions to these problems are attractive, but require extensive development before they can be brought into routine application:

- Air-to-air IFF: Heterodyne CO_2 laser radars employing quasi-CW waveguide laser transmitters operating at 10's of watts of output power offer the potential for sensing the unique vibration signatures of aircraft at operationally useful standoff ranges.
- Air-to-surface IFF: Modifications of existing airborne designators and rangefinders may be used in conjunction with rf or optical transponders on potential target vehicles to provide low-probability-of-intercept identification. A particularly attractive concept is illustrated in Figure 5, which shows a transponder embodied as an optical retro-reflector that can be modulated. The response is triggered only on receipt of the correct downlink code. The modulator impresses a second code on the passively-but-directionally reflected return signal.

3.6 Other developmental systems

A broad range of additional systems have been proposed for airborne applications, nearly every system conceived for fixed-site use has been considered. Among those investigated to a lesser extent than those outline above are the following:

- Detection of lethal chemical agents or atmospheric pollutants, via real-time backscattering spectroscopy using wavelength scanned pulsed CO_2 lasers
- Crop or resource evaluation, using fluorescence excited by ultra-violet excimer or frequency-converted solid state lasers

- Precision trackers for strategic defense, using short pulse frequency doubled Nd:YAG lasers

The proliferation of laser based airborne sensors will be dependent on improvements both in the performance and economics of the laser transmitters and the systems in which they are embedded. Current trends indicate that the following are likely to be the areas in which the most significant impacts on improving transmitter technology and production processes occur.

- <u>$1/watt diode pumping</u>: The advantages of diode pumping in eliminating coolant and pump cavity maintenance over the life of laser systems are well known, and the implementation of diode pumping for most solid state systems has proved remarkably simple. The only factor limiting the implementation of diode pumping for essentially all solid state lasers is the cost of diodes, currently in the range of $15-20/watt of peak pump power. With 2 to 8 kW of pump required for many applications, diode pumping clearly is not attractive from an initial equipment acquisition standpoint when compared with $300 rare gas discharge lamps. Emergence of a market requiring a total of 3 to 10 MW of peak diode power per year will be required to drive diode pump prices to $1/watt. At this point the cost differential between flashlamp pumping and diode pumping becomes acceptable. While a single application has not yet been identified which will sustain this diode production rate, a number of promising areas are being pursued which should provide this demand by the late '90's. A secondary benefit of diode pumping that may further help push implementation of new applications of airborne laser sensors into reality is the impact of using a diode pumped transmitter on the system itself. Lower power consumption, lower heat dissipation requirements, and lower transmitter weight will all tend to reduce the size, complexity, and cost of the sensor system.
- <u>Damage-resistant Nd:YAG spectral converters using KTP</u>: As aperture and length scaled KTP nonlinear crystals of dimension > 1 cm become available at prices below $2k in the mid-to-late 1990's, proliferation of 0.53 μm frequency doubled lasers and 1.57 μm parametrically converted lasers will become attractive because they can be operated at fluence levels well below optical damage thresholds while maintaining useful conversion efficiencies. While crystal demand and price represent a "chicken and egg" situation in many circumstances, it appears that industrial and medical requirements may drive production of large crystals to the required cost and quantity levels by the late 1990's.
- <u>Narrow linewidth solid-state microlasers</u>: Operating under diode pumping in CW, long-pulse, or Q-switched modes, these devices will provide easy access to narrow, single axial mode operation. Either as injection seeds or in coupled arrays, these devices will form the basis of a new generation of laser systems exploiting heterodyne detection in the mid-to-late 1990's. Technology developments to date indicate that no significant barriers exist to realizing the potential of these devices in future systems incorporating a variety of laser hosts, including Nd, Cr, Tm, and Ho in multiple host crystals.

- Integrated, miniaturized CO_2 laser transmitters: Development of environmentally hardened, fully packaged, rf pumped waveguide transmitters incorporating both multi-watt transmitter and local oscillator structures in a package costing < $10k will provide affordable, long-term operation in the mid-infrared. This goal is most likely to be attained as a by-product of a tactical military hardware program, as yet unidentified, involving high volume production.

5. CONCLUSIONS

Past history indicates that, despite the large number of concepts advanced as applications of laser transmitters to airborne sensors, few systems will pass the tests of functionality and affordability with sufficient margin to compete favorably with existing, often rf, systems. The reasons for this are more often associated with the high cost of acquiring laser systems, driven by the 85% of system costs associated with non-laser equipment including the pointing, data acquisition, and data processing systems. We believe that airborne laser systems will continue in concept evaluation and development, with niche systems (like bathymeters) entering limited production where they can provide capabilities unmatched by competing systems. Broad proliferation of airborne lasers is likely to occur incrementally as technology improvements and multi-function system embodiments are developed to drive cost-per-function to levels well below those achievable today.

	TADS (Target Acquisition/Designator Sight)	Mast-Mounted Sight	Infrared Detection Set	C-NITE (Cobra-Nite)	LTDS (Laser Target Designation Set)	LRMTS (Laser Ranger And Marked Target Seeker)
Country	US, International	US, International	US	US	US, International	UK, International
Length (inches)					4 in. (Wide)	23.9 (Optics)
Diameter (inches)		25**				11.2 (Optics)
Weight (lbs)		140*	415.8			79.2 (Total)
Electrical Power (w)		Mil-Std-1553B	1800			Approx. 550
Data Bus						
LRFD	Yes	Yes	Yes	LRF Only	Yes	LRF + Seeker
FLIR	Yes	Yes	Yes	Yes	No	No
Day TV	Yes	Yes	No	Yes	Day View Optics Only	No
Autotracker	Yes	Yes	Yes		No	Yes
First Article Delivery	1984		1979	1989	Mid-70s	Late 70s
System Contractor	Martin-Marietta	Northrop/ McDonnell-Douglas	Texas Instruments	Hughes	Northrop	GEC
Laser Vendor	Litton	Litton	Texas Instruments	Hughes	Hughes	GEC-Ferranti
Platforms	AH-64	OH-58D	OV-10D	AH-1S	F-5F	Jaguar, Harrier, Tornado

* AV WEEK 1-14-85 ** AV WEEK 9-8-86

Figure 1. Summary of laser targeting systems developed for airborne applications.

	TIALD (Targeting Infrared And Laser	LANTIRN (Low Altitude Navigation & Targeting IR For Night)	F/A-18 FLIR Targeting Pod	PAVE TACK	PAVE SPIKE	ATLIS II (Auto-Tracking Laser Illumination System)	TRAM (Target Recognition & Attack Multi-sensor)
Country	UK	US, International	US, International	US, International	US, International	France	UK, International
Length (inches)	102.4	98.5	72.0		144.1	99.2	
Diameter (inches)	12.0	25	13.0		9.8	12.0	
Weight (lbs)	330		339.9		424.6	374.0	
Electrical Power (w)	2000					2300.0	
Data Bus	Mil-Std-1553B	Mil-Std-1553	Mil-Std-1553A				
LRFD	Yes	Yes	1993	Yes	Yes	Yes	Yes
FLIR	Yes	Yes	Yes	Yes	No	No	Yes
Day TV	Yes	No	No	No	Yes	Yes	No
Autotracker	Yes	Yes	Yes		Yes	Yes	
First Article Delivery	Not In Serial Prod'n	1984	1984	1981	1975	1985	
System Contractor	GEC	Martin-Marietta	Loral	Loral	Westinghouse	Thomson-CSF	Hughes
Laser Vendor	GEC-Ferranti	Litton	Litton	Litton	Westinghouse	CILAS	Hughes
Platforms	GR-1 Tornado	F-15E, F-16C/D	F/A-18	F-111	F-4, Buccaneer	Jaguar, Mirage 2000, F-16	A-6E

Figure 1. (Continued) Summary of laser targeting systems developed for airborne applications.

Pave Tack LRFD Used In F-4/F-111 Aircraft Pods
(Litton)

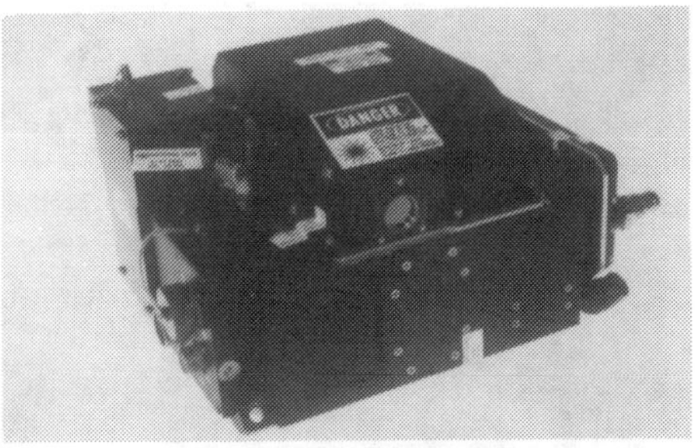

Mast-Mounted Sight LRFD
Used In OH-58D Helicopter (Litton)

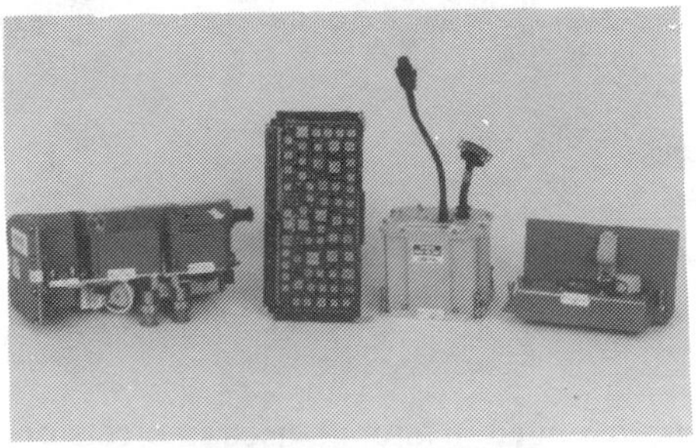

Dual-Wavelength LRFD
Used In F-15/F-16 Aircraft (Litton)

TADS Laser Rangefinder/Designator
Used In AH-64A Apache Helicopters (Litton)

Integrated LRF And Spot Tracker
(GEC-Ferranti)

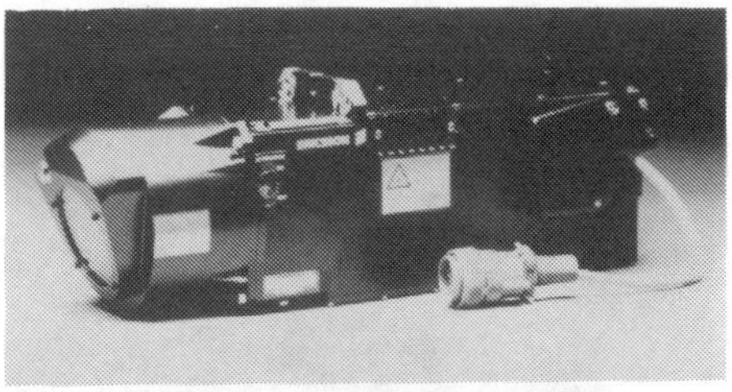

Airborne Fire Control LRF
(Ericsson)

Target Recognition And Marking (TRAM) System
Used On A-6 Aircraft

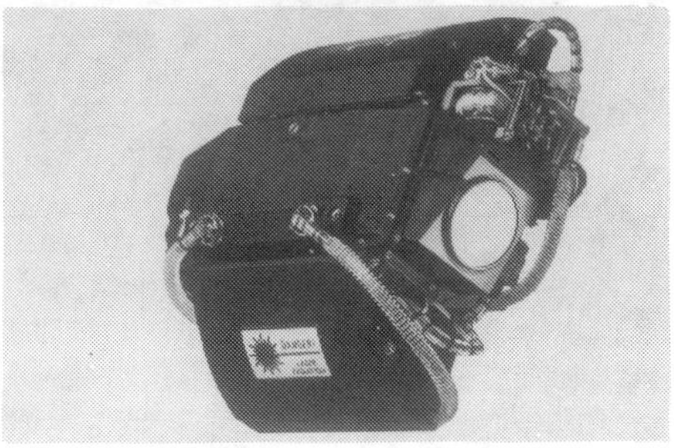

LRFD Miniature Helicopter-Mounted
Targeting Payload (El Op)

Figure 2. Photographs of typical military airborne targeting systems and airborne laser target designator/rangefinders.

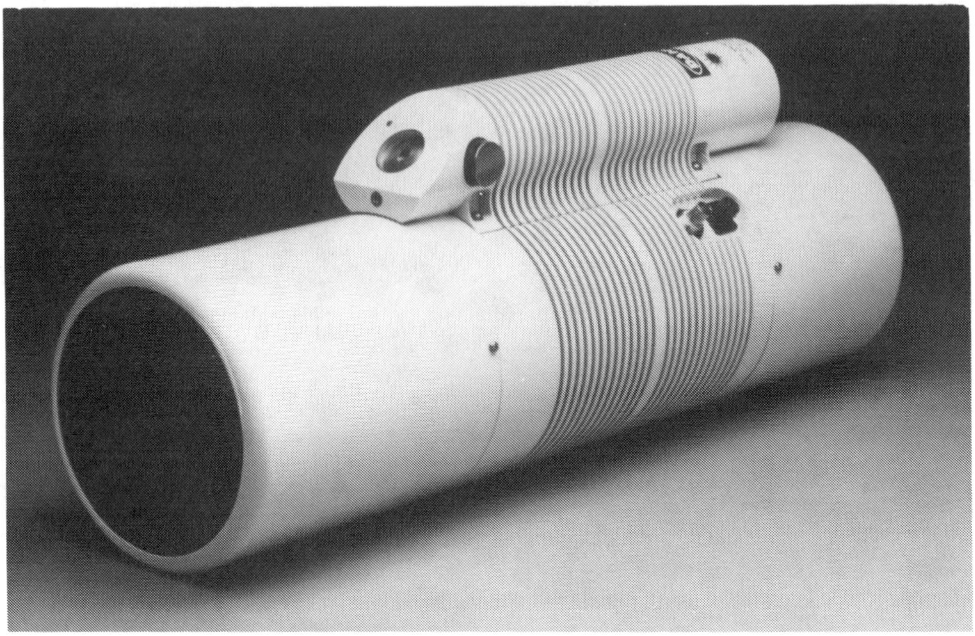

Figure 3. Photograph of laser ranging transmitter mounted in aircraft, used in precision altimetry for geophysical mapping. (Courtesy of Azimuth)

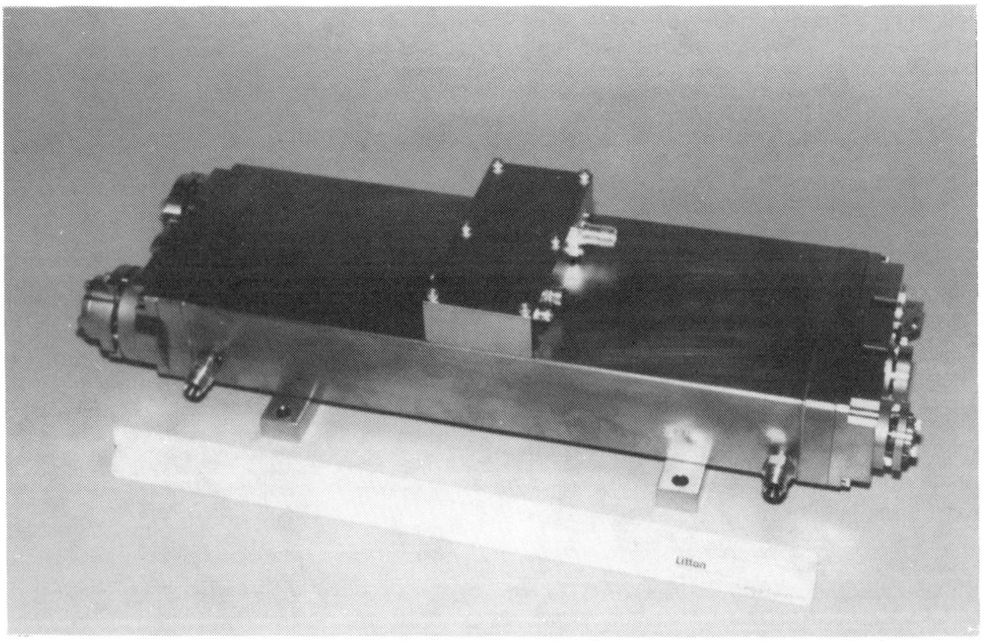

Figure 4. Photograph of RF pumped CO_2 waveguide laser incorporating a built-in local oscillator providing heterodyne detection capabilities. (Courtesy of Litton and McDonnell Douglas)

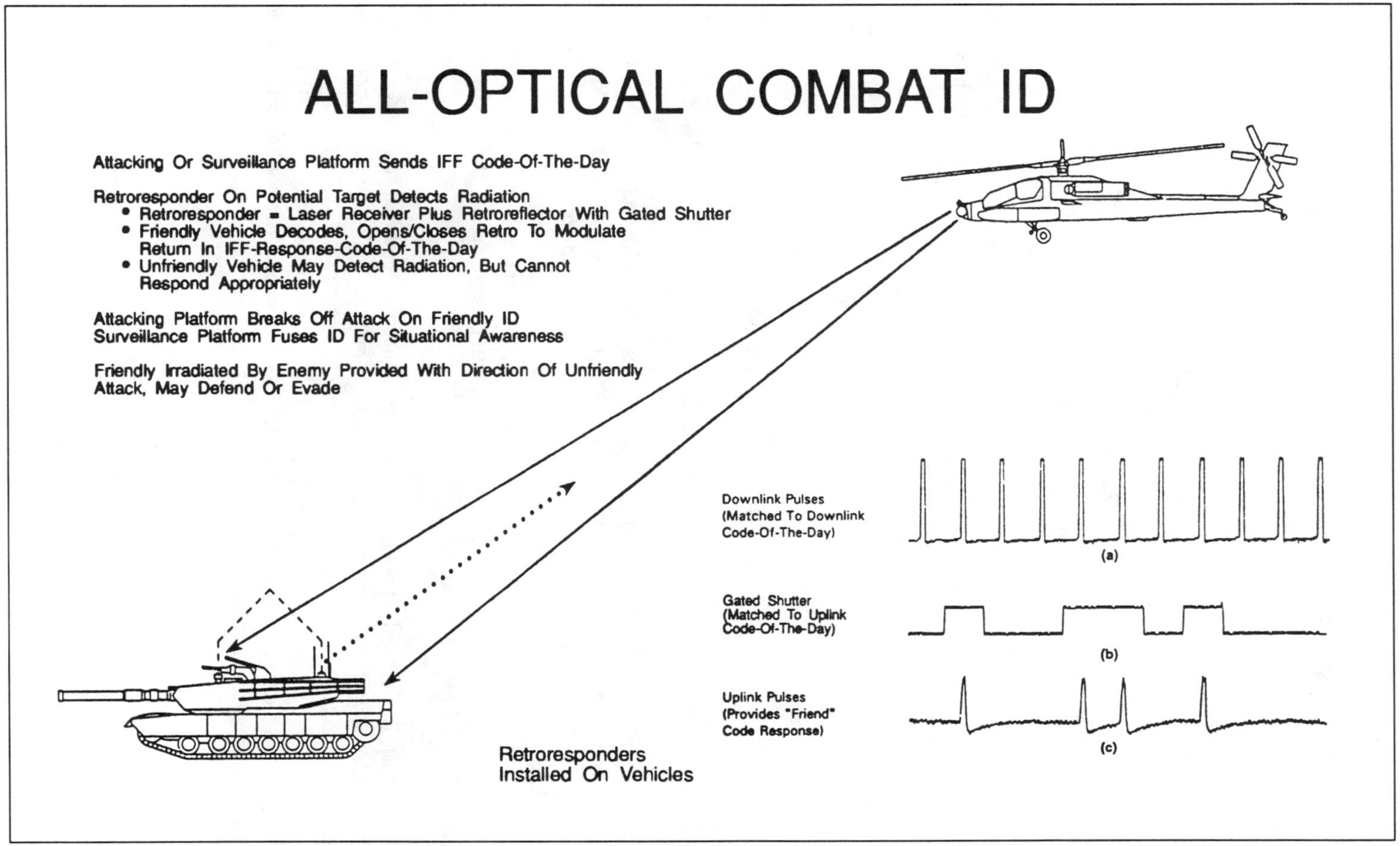

Figure 5. Illustration of All-Optical Battlefield Identification Friend-Or-Foe concept employing a modulatable retro-responder. (Courtesy Litton)

Imaging IR sensors: future directions for test and evaluation

Glenn D. Boreman

University of Central Florida, Department of Electrical Engineering
Center for Research in Electro-Optics & Lasers
Orlando, Florida 32816

ABSTRACT

This paper is a review of present state-of-the-art and future technologies for infrared scene projection, including both target and background simulators. We will address the need to develop appropriate figures of merit for comparison and evaluation of these test methods. Test methods we will review and compare include: raster-scanned laser, liquid-crystal light valve, deformable membrane, reflective deformable-mirror spatial light modulators, thermal emitter arrays, and infrared halftone transparencies.

1. INTRODUCTION

Infrared scene projection (IRSP) is a technology for creating realistic infrared imagery, which is used for hardware-in-the-loop (HWIL) performance testing of thermal imaging systems. Actual field testing of system performance with real IR scenes is time-consuming and expensive. Field testing also lacks a quantitative basis, because natural scenes are ill-defined in terms of the specific characteristics of their radiation signatures. However, the performance of a system in response to real scenes is the final criterion. Methods are desired for the creation of realistic IR scenes that can be precisely described and calibrated, and which are of sufficient quality to measure the performance characteristics of the system under test.

In the past, IR scenes had been created by computer, or recorded from videotape of real IR scenes. Typically, the resulting video was injected behind the detectors in the system (Fig. 1) to exercise the image processor, tracker, and display of the system. This approach did not exercise the optics, detector, and electronics subsystems, and therein lies the need to develop IR scene-projection techniques instead of synthetic IR scene generations for end-to-end system characterization.

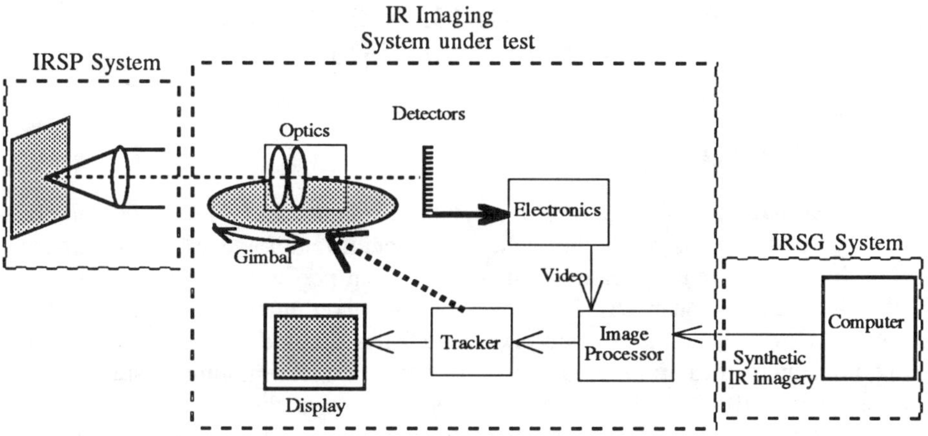

Figure 1. Schematic of IR imaging system under test.

2. REQUIREMENTS FOR INFRARED SCENE PROJECTION

Successful IRSP systems have rather demanding requirements. The scene projector system must be able to access the pertinent performance characteristics for the thermal imager system. Ideally, the scene projector system must have significantly higher performance than the system under test (a factor of ten or so better performance in all aspects). Given the current state of the art in IR imager systems, this requirement is increasingly difficult to meet in an across-the-board manner, which may necessitate the development of test hardware with specific functionality optimized for certain tests. The main technical requirements for IRSP systems fall into the following categories:

Spatial resolution
Dynamic range
Spectral bandwidth
Frame rate
Spatial uniformity
Noise
Flux transfer efficiency.

These requirements parallel the corresponding performance characteristics for IR imaging systems, with some being more important that others for any given application. For example, most current IR imaging systems simply respond to "in-band" flux, but more sophisticated imaging radiometer systems will have spectral discrimination capabilities. The development of an overall figure of merit framework for IRSP systems will necessarily include weighting factors that emphasize the level of system performance being tested for any given parameter.

3. TYPICAL PERFORMANCE RANGES FOR IRSP SYSTEMS

Spatial Resolution

Current IR imaging systems operate at typical resolutions of 512 by 512 pixels, with 1024-by-1024 systems on the horizon. The general requirement of factor-of-ten overkill on IRSP performance versus IR imaging system performance is probably not achievable here in the near term. A one-for-one pixel match is an achievable compromise. However, a one-for-one pixel match has the additional complexity of matching nonsquare-format pixel arrays. If both the IR imaging system pixel arrangement and the IRSP pixel arrangement are square arrays, then a one-to-one match will work, given that the systems are aligned precisely so that the pixels overlay precisely. If the IR imaging system is gimballed where the system field of view moves over a larger field of regard, a strain is placed on the number of pixels required for the IRSP, as well as tight requirements on the pixel alignment. Scanned laser systems can achieve an oversampling to some degree, just in terms of number of resolvable elements, but even those systems have limited data bandwidths. A particularly troublesome problem for scanned laser systems is that they will not work with scanned IR imaging systems without a synchronization of scan. The scan velocity is a hard parameter to vary independently, both for IR imagers and IRSPs. Especially for test equipment that must have flexibility to adapt to different IR imaging systems, a scanned laser IRSP should have a variable scan rate and ability to synchronize to an external source.

Dynamic Range

Testing of imagers intended for air-to-surface, surace-to-surface, and air-to-air engagement scenarios have different dynamic range requirements.

Dynamic range requirements are most often stated as N:1, or as a given number of bits. For IR imaging systems that eventually display to a human observer, the use of 8 bits (256 gray levels) is the usual dynamic range. However, many IR imagers are being built with higher dynamic range requirements, such as 12 bits (4096 gray levels). The main requirement for IRSP systems is to be able to simulate both the hot plume of a missile and the earth-temperature background clutter. Dynamic range has a number of common specifications.

Contrast ratio is the usual measure of dynamic range, though it has no universally accepted definition. Contrast ratio is expressed in terms of the maximum and minimum output levels the IRSP is capable of achieving. It is also commonly expressed as

$$\text{Contrast} = \frac{\text{output}_{max} - \text{output}_{min}}{\text{output}_{max} + \text{output}_{min}} ,$$

or as

$$\text{Contrast} = \frac{\text{output}_{max}}{\text{output}_{min}} .$$

Another way of specifying dynamic range is the range of temperatures that can be simulated. This requires a conversion from flux units to temperature through the Planck radiation equation. A universal means of specifying dynamic range would be the maximum output achievable, divided by the noise-equivalent output (output for S/N=1, or RMS noise expressed in output units).

The dynamic range is limited at the low end by noise in the IRSP system. These noise sources can arise in various ways: temporal noise in the modulation or emission mechanism, spatial noise (pixel-to-pixel variation), stray radiation (crosstalk or leakage), or quantization noise. Dynamic range is limited on the high end by saturation effects in the mechanism that generates the IR energy. For example, the radiance produced in the image plane of a system can not exceed the source radiance being imaged.

Typically the linearity of the IRSP system within its dynamic range is not specified, that being correctable by a look-up-table operation of the desired input data.

Spectral Bandwidth

Laser-based systems provide acceptable testing for systems that operate on in-band flux. To test broadband systems using lasers, several laser channels are often used, taking advantage of the somewhat limited choices one has in IR lasers (e.g., CO, CO_2, and HeNe). Practical problems in testing systems arise primarily in two areas: speckle effects and anti-laser coatings.

Laser speckle is a coherent multipath interference phenomenon that causes the projected image to be received with additional noise. Rotating diffusers are commonly employed to average out the effects of the speckle, with the period of the rotation being much shorter than the integration time of the sensor under test.

Some systems are hardened with interference coatings (optical filters that reject specific wavelengths) against specific laser wavelengths. These coatings effectively block laser radiation that might harm the sensor element in a battlefield scenario, but for convenient in-situ testing of real systems, these coatings also exclude the lasers used by IRSP systems.

In some specific instances, the use of lasers is indicated for IRSP applications, but for the majority of targets upon which the IR imaging systems will be asked to operate, the radiation will be incoherent, and the use of lasers introduces a degree of artificiality into the test, which should be avoided if possible.

Lasers are useful is in the creation of high-brightness regions in the projected image, corresponding to the plume of a missile or other high-intensity target. Often, the background clutter and low brightness targets are projected using a thermal source, while the plume is simulated using the high radiance available from a laser source. The IRSP system is essentially two separate systems with beamsplitter optics used as beam combiners. This effectively increases the dynamic range of the aggregate IRSP system, because the range of temperatures that can be simulated is greatly increased by having a laser take the task of the high-brightness regions.

Frame Rate

While standard video systems operate at 30-Hz frame rates, the requirement for endgame simulation, where the sensor has closed in on the target, is typically in the range of 400-Hz frame rates. This high frame-rate requirement precludes some classes of thermal-emitter IRSPs where the radiation emission mechanism cannot respond sufficiently. Laser-based systems can generally operate at this rate without problems, as can some of the spatial-light-modulator-based systems.

From a system specification point of view, the high-frame-rate requirement must be considered in terms of its effect on all other performance specifications. For example, a system with adequate resolution (MTF) at low frame rate may well have insufficient image contrast when required to operate at these high frequencies.

Also, simply the data bandwidth of the system must be considered in those situations where a high-resolution image needs to be projected at high frame rates. For example, an 8-bit, 512-by-512 image projected at a 400-Hz frame rate requires nearly 1 Gbit/sec data rate. This will generally require parallelism in the drive electronics for the modulators in the IRSP.

For a scanned system with many pixels, the requirement of a certain integrated energy over a frame time can lead to damaging level of power during the short dwell time. This is particularly true for laser-based systems.

Spatial Uniformity

Spatial uniformity is not currently a parameter of major concern, as the current IRSP struggle to meet other more basic requirements. However, as the performance improves in areas of resolution and frame rate, spatial uniformity will become a factor that sets the lower limit of the dynamic range, in much the same way as spatial uniformity sets the lower limit on noise-equivalent temperature difference in staring IR imaging systems today. Uniformity of response of the modulation or emission mechanism is an area that must be addressed as the performance of the IRSP systems eventually improve.

Noise

Similarly, the temporal noise of the projected image is not often specified in current IRSP systems. This is a result of dynamic range usually being specified on the basis of image contrast, rather than as a noise-equivalent flux. All modulation or emission mechanisms have some inherent noise sources, and these, for the most part, remain to be addressed.

Flux-Transfer Efficiency

One issue that is largely ignored in the specification of IRSP systems is the overall flux-transfer efficiency. While specified parameters such as contrast may be reasonably high, if the overall flux level reaching the detectors of the IR imaging system is too low, the dynamic range of the imager will not be adequately tested. Indeed, the contrast levels can often be optimized at the expense of overall flux-transfer efficiency. Again, the system tradeoffs would be clearer if a noise-equivalent flux density were employed, rather than simply contrast ratios.

Development of Global Figures of Merit

The comparison of developmental IRSP systems will be facilitated by more meaningful figures of merit than simply the use of contrast ratios. A signal-to-noise-ratio-based figure of merit will address several of the concerns noted above regarding correct normalization of projected flux values to the noise level of the IRSP.

In addition, meaningful IRSP figures of merit will likely be multivariable combinations of performance parameters, in a functional form that emphasizes their interdependence. A current problem in "specmanship" for IRSP systems is that one performance number is quoted at a time, while the remaining parameters are allowed to be at whatever value optimizes the one performance characteristic being specified. For example: contrast ratios are often quoted for static scenes, or MTFs are quoted for situations where the overall flux transfer efficiency of the system has been compromised.

An approach to specification of IRSP systems may be a signal-to-noise ratio, as a function of both temporal and spatial frequency, along with a specification of the maximum projected flux levels as a function of wavelength. A figure of merit of this sort would account for spectral fidelity, contrast, frame rate, resolution, noise, and flux transfer efficiency.

As the approach shifts from comparison of developmental IRSP systems to the choice of which IRSP system is best for a given application, the framework within which the figure of merit is expressed may be expanded to include weighting factors that emphasize the level of imaging system performance being tested for any given parameter. For instance, for any given imager to be tested, a different importance will be attached to resolution, frame rate, spectral fidelity, etc. The weighting factor applied to a given parameter in the IRSP selection process should be proportional to the performance level of the imaging system to be tested on that parameter.

4. COMPETING APPROACHES

This section discusses the mechanisms for some competing approaches and will compare the advantages, drawbacks, and future directions of each.

Static IRSP Systems

For projection of static scenes, the current practice is quite rudimentary. Outline masks are frequently fabricated as cutouts in metal, and backlit by a blackbody or laser source. If the scale of the target needs to change to simulate endgame scenarios, an IR zoom lens is used to change scale corresponding to target closure. If groups of targets are required to have relative motion, separate cutouts are placed on movable positioner rails.

This class of target can simulate high dynamic ranges, but the drawback is that the cutout is a binary image, either bright or dark.

A recent development in this area that stands to provide more sophistication for static scene projection is the use of infrared-transmitting halftone screens.[1-3] These targets are still constructed of the binary materials characteristic of IR targets, but have sufficient spatial resolution that groups of pixels can be placed within a given resolution element of the IR system to be tested, producing an effective gray-level response. The transmission of any given pixel is determined by the size of the photolithographically defined hole in the opaque material coated on the substrate. As an example of the current capabilities of this technology, a transparency that operates in the 8- to 12-μm band, produced by CI Systems, Inc., is shown in Fig. 2.

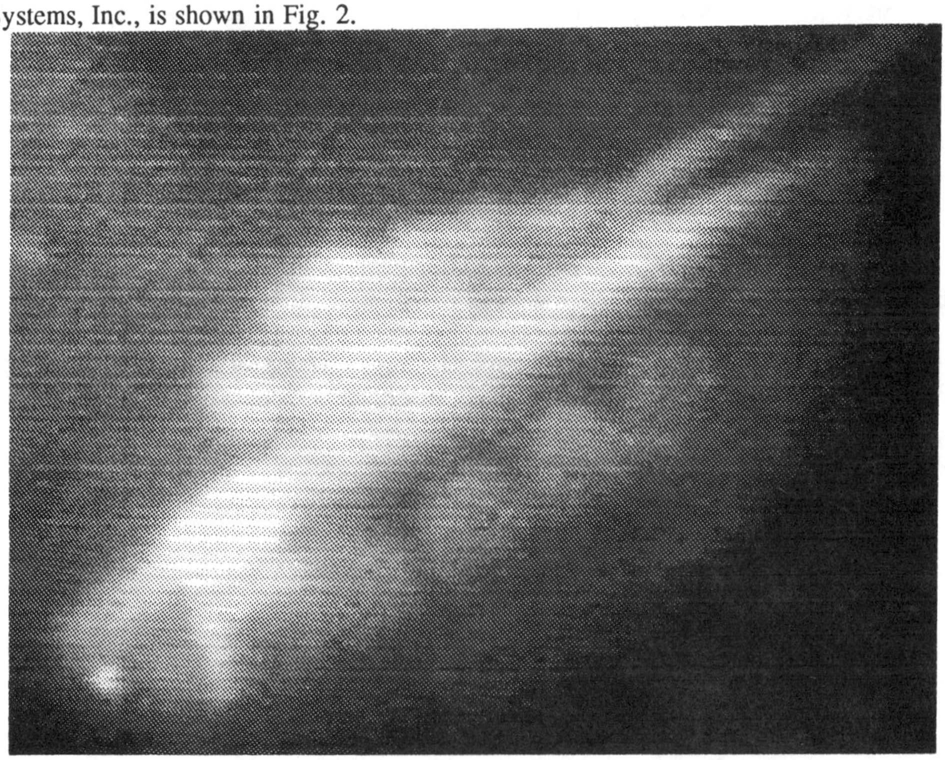

Figure 2. Tank in 8- to 12-μm band. (Courtesy of CI Systems, Inc.)

Dynamic IRSP Systems

The choices for implementation of dynamic IRSP systems are more numerous than the choices for static systems.

Raster-scanned laser. One system approach that has progressed quite far to practical application is the raster-scanned laser, seen schematically in Fig. 3. The laser radiation can be modulated simply as a flying spot scanner, or can be modulated in scophony[4,5] format, which has a larger illuminated aperture in the acoustic cell, and hence a better resolution. Large-scale systems of this type have been built (for example the KHILS projector, built by Aura Systems, El Segundo, CA, installed at Eglin AFB, FL).

That particular system uses four different IR laser channels to cover the long-wave IR, with acousto-optic Bragg cells performing both modulation and scanning functions. Dynamic range, frame rate, and resolution are good, as would be expected of a laser-based system. Effects of laser speckle on the projected image have been noted and are being addressed with a rotating diffuser.

One issue that remains with this type of system is that although this approach is well-suited for testing of staring systems that integrate received power over a frame time, the testing of scanned imagers is hampered by a lack of scan synchronization. Generally, the image from a scanned FLIR system is not useful, consisting of only intermittent bright spots. Future generations of these systems should be designed with programmable scan flexibility in mind, because the IR systems to be tested have many different scan formats.

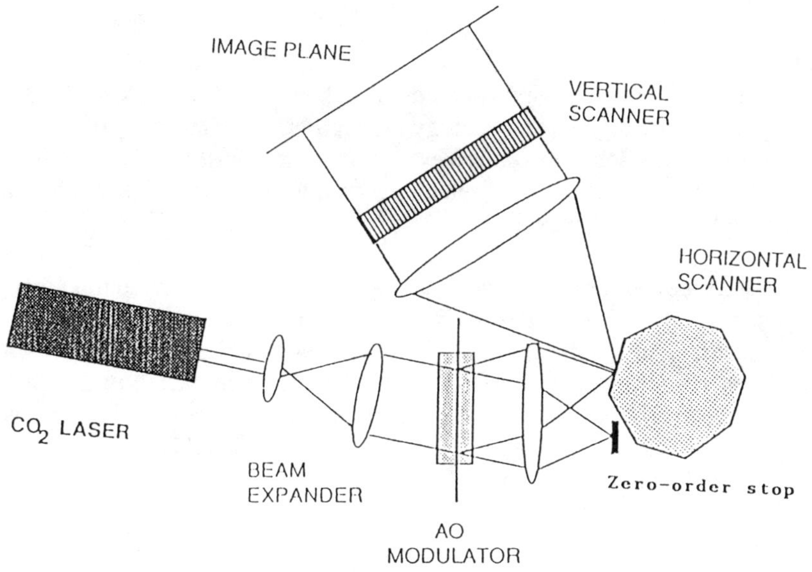

Figure 3. Scophony IRSP system.

Liquid-crystal light valves (LCLV). The technology exists[6] to implement an optical-to-infrared converter using liquid crystal display properties of polarization rotation in the IR. As seen in Fig. 4, the write-side of the LCLV is coupled to a visible CRT display, driven by standard video. The reflection of broadband IR radiation from the read-side of the LCLV is polarization modulated and, after passing through a polarizer, the radiation is intensity modulated. This technology has been demonstrated by Hughes Research Labs in the 8- to 12-μm range. These devices have approximately TV-compatible resolution and frame rate, but the main drawback has been the relatively low contrast ratios produced, in the range of 20:1.

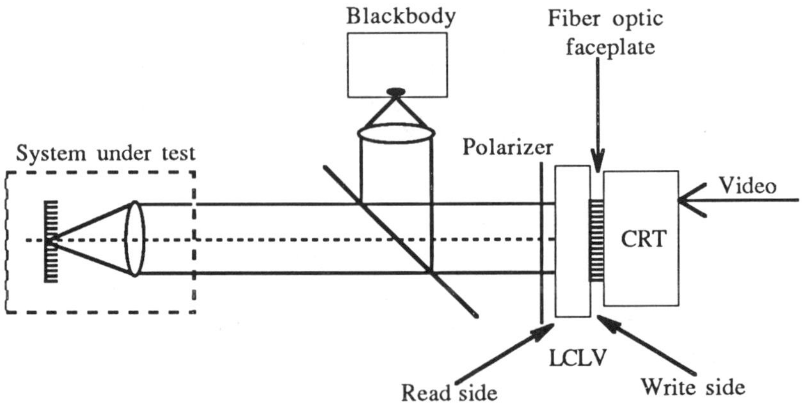

Figure 4. LCLV-based IRSP system.

Thermal-emitter arrays. Substantial activity[7-10] has occurred recently regarding arrays of resistive elements, which would emit their radiation directly in the IR by incandescence. Thermal-emitter technology faces problems in the areas of time response and pixel-to-pixel crosstalk. These issues are addressed by the construction thin-film resistors. The mass of the elements remains small, to minimize their heat capacitance and to improve the time response.

The resistors are fabricated to have small electrical leads, and are commonly operated in a vacuum enclosure. These design features ensure that the main heat loss term is caused by radiation, and minimizes pixel-to-pixel heat flow, but at the cost of a longer response time for cooling. Some designs use a heat sink contacted to the back of the elements, with control of lateral heat transfer and pixel-to-pixel crosstalk by the use of air-filled slots between elements. Some of the systems constructed have extensive cooling systems associated with the heat sink function.

Arrays up to 100 by 100 have been reported, with larger arrays currently under development. The spectral radiance of the emitters is characteristic of a thermal source. Thus, one of the spectral fidelity requirements is met, that of a broad source spectrum.

However, when the finite size of the elements is considered (non-unit fill factor) and the finite flux-transfer efficiency of the projection system is considered, the temperature of the resistive elements must be considerably higher than the temperature of the scene to be simulated, to produce an appropriate amount of in-band flux at the detector plane.

Thus, the radiation spectrum, while proportional to that of a blackbody, does not have the true spectrum corresponding to the temperature of the desired target scene.

Reflective spatial light modulators. Reflective modulators are attractive for IRSP applications because they modulate reflected power rather than emitted power, and thus the issues of frame rate, resolution, and crosstalk are more tractable. In recent years, both membrane-based and movable mirror-based spatial light modulators (SLMs) have been developed, largely for optical processing applications,[13-15] rather than for IRSP applications. The movable mirror modulators have greater capacity for angular tilt in the individual mirror elements, and thus provide higher contrast ratios than the SLMs where the reflective surface is a continuous faceplate. A schematic of a torsion-mirror SLM developed by Texas Instruments is seen in Fig. 5.

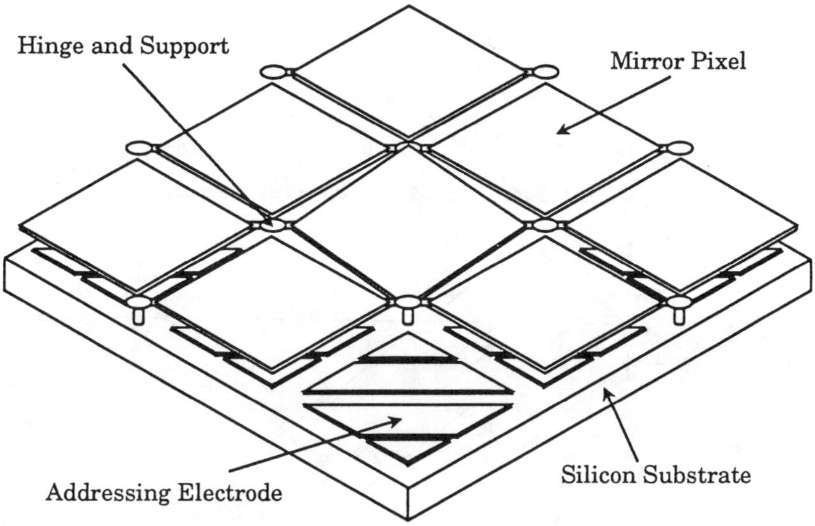

Figure 5. Torsion-mirror SLM.

A schematic of a proposed IRSP based on this SLM is shown in Fig. 6a,b. This IRSP would operate in the binary mode, where an individual pixel is either on or off, with gray levels achievable by temporal pulse-width modulation (PWM) of the mirror tilt within one frame time of the sensor under test. These SLMs are capable of frame rates of 7 kHz, so the PWM approach is capable of providing 8 bits of resolution to a 30-Hz sensor.

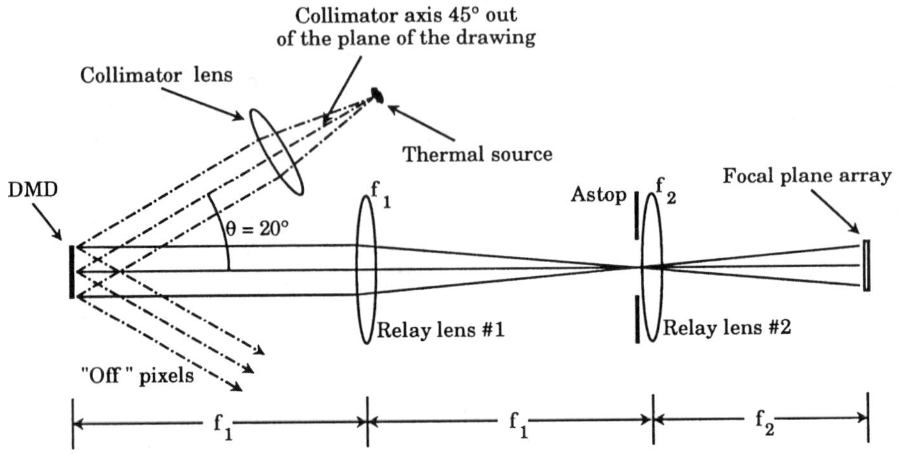

Figure 6a. Layout of proposed IRSP

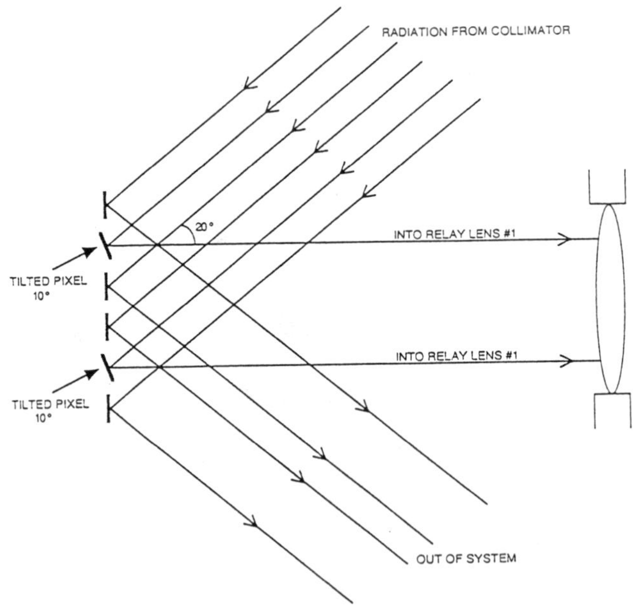

Figure 6b. Close up of modulation mechanism (geometrical optics model).

The main limitation to these devices is the crosstalk caused by diffraction. For IRSP applications, large pixels are desired. Prototypes have been made with 50-μm pixels, but the typical devices made by TI for HDTV applications (1K × 2K pixels) are 17-μm pixels. A diffraction cross-talk model has been developed[16] that predicts that contrast ratios of 256:1 are achievable with 50-μm pixels, and contrast ratios of 2400:1 are achievable with 150-μm pixels, for IRSP applications in the 3- to 5-μm band.

We performed a feasibility demonstration of the basic design seen in Fig. 6a, using a 17-μm pixel device with radiation in the 0.8- to 1-μm band. The actual configuration used is shown in Fig. 7. This should be indicative of performance obtainable in the 3- to 5-μm band with 50-μm pixels.

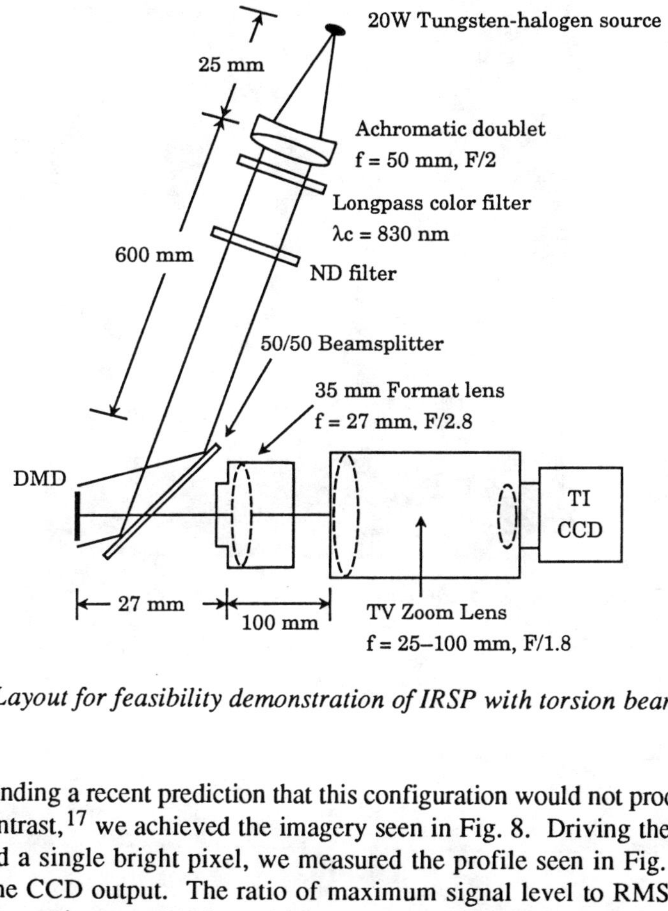

Figure 7. Layout for feasibility demonstration of IRSP with torsion beam DMD.

Notwithstanding a recent prediction that this configuration would not produce images of sufficient contrast,[17] we achieved the imagery seen in Fig. 8. Driving the DMD such that we isolated a single bright pixel, we measured the profile seen in Fig. 9 from one video line of the CCD output. The ratio of maximum signal level to RMS noise level in this measurement is around 100:1, which turned out to be limited by the clock noise of the CCD used to acquire the image. Thus, it appears from the first measurements that our diffraction crosstalk model predicts the correct magnitude of performance levels.

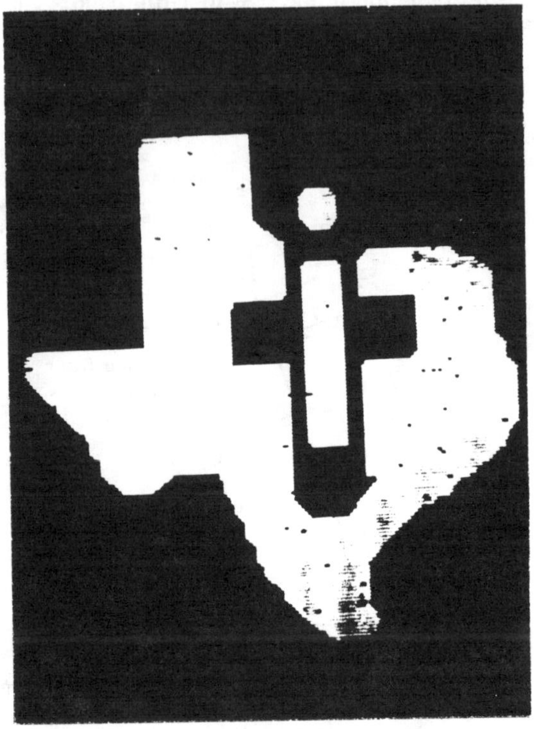

Figure 8. Image obtained using the apparatus shown in Fig. 7.

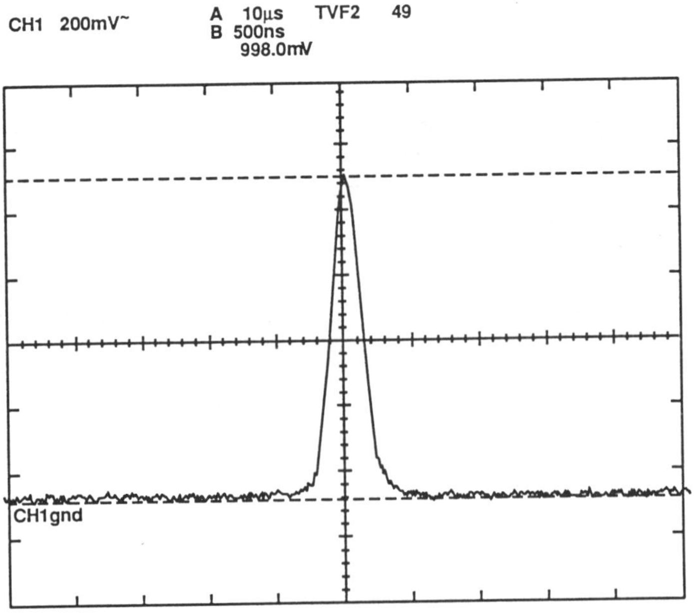

Figure 9. Oscilloscope trace of a selected video line containing a bright pixel.

The main disadvantages of this IRSP approach have been the relative scarcity of the devices, in that TI is presently the only production source of them in high-resolution arrays. However, it seems that the mirror-based SLMs can satisfy most of the requirements for IRSP. A prototype demonstration of IRSP should be done in both the 3- to 5-μm and 8- to 12-μm bands, to ascertain the ultimate performance envelope of this technology.

5. References

1. J. W. Baer, "Infrared dynamic scene generator using halftone reflectance images," Proc. SPIE Vol. 940, 189-193 (1988).
2. S. Ghilai, U. Gera, D. Cabib, A. Lapin, Y. Liran, Infrared transparencies (thermoscenes) for simulation of infrared scenes as a tool to improve FLIR testing," Proc. SPIE Vol. 819, 80-86 (1987).
3. D. Cabib, J. Eliason, S. Ghilai, R. Bracha, "Accurate infrared scene simulation by means of a microlithographically deposited substrate," Proc. SPIE Vol. 1762 (1992).
4. R. V. Johnson, "Scophony light valve," Applied Optics 18, 4030-4038 (1979).
5. E. F. Schildwachter and G. D. Boreman, "Modulation transfer function characterization and modeling of a Scophony infrared scene projector," Optical Engineering 30, 1734-1738 (1991).
6. M. S. Welkowsky, "IR simulation using the liquid crystal light valve," Proc. SPIE Vol. 765, 89-93 (1987).
7. D. Stauffer, B. Haas, and B. Cole, "Performance of a thermal scene generator," Proc. SPIE Vol. 1050 156-164 (1989).
8. A. P. Pritchard and S. P. Lake, "Electrically heated pixel arrays for dynamic infrared scene generation," Proc. SPIE 940, 182-188 (1988).
9. J. L. Hester, "Thermal array target simulation technology," Proc. SPIE Vol. 940, 150-152 (1988).
10. L. Burriesci, D. Keezer, "A dynamic RAM imaging display technology utilizing silicon blackbody emitters," Proc. SPIE Vol. 765, 112-122 (1987).
11. E. R. Schildkraut, "IR image generation by thermoelectric elements," Proc. SPIE Vol. 765, 102-108 (1987).
12. M. Daehler, "Infrared display array," Proc. SPIE Vol. 765, 94-101 (1987).
13. L. J. Hornbeck, "128 × 128 deformable mirror device," IEEE Transaction on Electron Devices, ED30, 539-545 (1983).
14. D. R. Pape, L. J. Hornbeck, "Characteristics of the deformable mirror device for optical information processing," Optical Engineering 22, 675-681 (1983).
15. L. J. Hornbeck, "Deformable-mirror spatial light modulators," Proc. SPIE Vol. 1150, 86-102 (1989).
16. K. J. Barnard, "Crosstalk analysis of a deformable-mirror-based infrared scene projector," PhD dissertation, University of Central Florida, 1992.
17. O. M. Williams, E. J. Bevan, S. B. Mobley, "Optical aspects of infrared projector technologies for hardware-in-the-loop simulation applications," Proc. SPIE Vol. 1687, 49-63 (1992).

Sensor Signal Processing

Chair
Brian M. Hendrickson
Rome Laboratory

Optical Processing and Computing

Ravindra A. Athale

Electrical and Computer Engineering Dept.
George Mason University
Fairfax, Virginia 22030-4444

ABSTRACT

Optical techniques have been investigated for information processing since 1950's. This paper provides a unified view of different approaches to optical information processing and computing. Such a view serves to bring out interrelations (similarities and differences) between seemingly different topics and will thereby identify a common technological infrastructure for this broad discipline. The emerging technology of "Smart Pixels" is also discussed and a framework for organizing different "Smart Pixel" approaches presented.

1. Introduction

The physics and technology for generation, modulation, propagation and detection of optical radiation has unique properties when compared with the electronic technology. Over the past 40 years research has been conducted in exploiting these unique features of optics and optoelectronics for enhancing the performance of systems for information sensing, transmission, storage, processing and display. Collectively, this field has been recently termed "Photonics". Of these different tasks, photonics has been firmly established as providing enhanced performance for sensing, communications, storage and display of information. The main subject of this paper is the area of optical information processing, which lags behind these other applications in terms of practical systems and commercial product development. In this paper I will discuss the major subareas of optical information processing and their interrelation to each other. The hope is that by identifying the common aspects of seemingly different areas of optical processing, a common technological foundation can be identified and further developed. Additionally, this analysis will also help highlight the very close coupling between algorithms, architectures and hardware that exists for optical information processing systems.

Optical information processing has been used as the umbrella term for a number of separate research topics such as optical image processing, optical signal processing, optical interconnects, optical neural nets, optical matrix processing, optical symbolic computing, and digital optical computing. Each of these topic derives its terminology and mathematical foundation from a different discipline, thereby obscuring the underlying similarities. The common theme running between these different topics is that all of them use of optical fields to represent information and use interaction (linear and nonlinear) of these optical fields with active and passive materials and devices to achieve the desired operation. The difference between these topics arises from the different nature of the information (1-D / 2-D signals, logic states, numerical data, symbols) they process, different optical representations used (complex amplitude, intensity, spatial / temporal position, spatial / temporal frequencies, polarization, wavelength), and different operations they try to achieve (linear / nonlinear transformations, numerical computation, logic calculus). In order to find interrelations between these different topics, it is necessary to start with a general framework for analyzing optical information processing systems. In section 2 of this paper I will describe such a framework. A generalized view of an optical information procesing system suggested by this framework is then presented. Section 3 discuses of one of the key technologies for optical information processing systems, "smart pixels". Section 4 will summarize the main conclusions of this paper.

2. Framework for Optical Information Processing

Any computing system can be studied from three different perspectives: (1) algorithmic, (2) architectural, (3) hardware. In general, these three aspects of a computing system are very closely coupled and hence cannot be researched separtely. In electronic computing, however, the algorithmic foundation of binary data representation and stored-program paradigm is so uniform and broadly followed that a discussion of different architectures usually leaves out any references to the algorithms at that fundamental a level. On the other hand, the interaction between hardware and architecture is quite important in electronics, making certain architectures more appropriate for low power consuming CMOS circuits in Si, while certain others are better suited for GaAs technology. Since optical systems are based on very different physical phenomena compared to electronics and since there no single algorithmic paradigm that is dominant in optical information processing, the interaction between these different aspects is all the more crucial.

2.1 Algorithmic Aspet of Optical Information Processing:

Algorithmic aspect of optical information processing deals with the mathematical formulation of the problem solving strategy and, as the term is traditionaly understood, involves a step-by-step description of the procedure. The optical information processing systems can be clasified into three major groups according to the problem solving strategy they employ.

2.1(a) Analog optical computing:

In this group, the *physics* of the problem to be solved is mimicked by the *physics* of the optical system. The problem solution is obtained by setting the initial conditions of the optical system and letting the sytem evolve to the solution state. At that point, measurements are made on the optical fields which correspond to the problem solution. The most celebrated example of this strategy is the original film-based optical processor for processing synthetic aperture radar signals. The synthetic aperture radar imaging was viewed as recording a microwave hologram. The problem of focusing the image is then turned into a holographic image reconstruction operation, which can be accomplished by electromagnetic wave propagation through free space. The optical system to focus the synthetic aperture radar data can be considered to be an analog optical computer, where the desired operation is performed by optical propagation through free space and lenses. Another example of analog optical computers is the Fourier spectrum analyzer for images or 1-D signals. In these systems, the identification between the far field diffraction pattern and the operation of Fourier transform is the basis of the processor. In both cases the computational time required is the time taken by the optical wave to propagate from the input plane to the output plane (hence the celebrated phrase "computing at the speed of light"). Similarly, the computing operation itself does not dissipate any power, since it consists of propagation of an optical wavefront. The drawbacks of such sytems are: (i) lack of programmability, (ii) limited computational accuracy and (iii) input / output bottleneck. The last point effectively negates most of the speed and power dissipation advantages mentioned earlier, since the input and output interface devices are typically slow and power consuming. These analog computers are the descendants of analog mechanical computers employing cams, pulleys and gears as well as analog electronic computers based on operational amplifiers, resistors and capacitors. Since these other systems have been entirely superseded by electronic digital computers, there is a temptation of also considering analog optical computers to be only of "historical interest". Given the much greater diversity and richness of the optical (as compared to mechanical or electronic) phenomena that can be harnessed for complex computational tasks, such a conclusion will indeed be premature.

2.1(b) Optical computers with analog data representation:

The selection of the data representation is one of the first steps in the mathematical formulation of the problem solving strategy. For analog optical computers, the data representation, by definition, is analog, where each of the problem variables is mapped linearly onto an independent variable of the optical wavefront. With the same analog data representation, an alternate strategy can be adopted leading to this second category. Optical modules performing different elementary operations (multiplication, adition, square) are designed and then assembled with appropriate interconnections between them into more complex system for performing the desired operations. A further division within this category occurs depending on the nature of the data - numerical or symbolic.

Examples of processors handling analog numerical data are : optical matrix procesors, optical signal and image processing systems, partial differential equation solvers. In optical image processing systems, optical modules performing the 2-D complex Fourier transforms are combined to create correlators. In optical matrix processing systems, analog multiplication and addition performed over 1-D or 2-D arrays is the primary computation and 1-N fan-out, N-1 fan-in and 1-1 imaging are the primary interconnect topologies tht are employed.

Optical systems for performing the symbolic computing operations of inference and tree searches were briefly studied. These system used analog arrays to represent symbols and attempted to map the desired computing operations into more familiar image correlations and vector-matrix multiplications. The requirement of these systems to only work with rigid and deterministic data structures (such as images, vectors-matrices) represents a major bottleneck when applied to problems in symbolic computation, which cannot ensure such data structures.

Optical processors which use analog data representations suffer from the same lack of accuracy due to error acumulation that analog optical computers suffered from. On the other hand, some degree of programmability is available in these sytems as compared to analog optical computers, since the computational modules that comprise the total system can be appropriately selected to cover a range of problems. The most prominent example of this relative flexibility is the optical vector-marix multiplier since a large number of problems involving discrete data and assumption of linearity can be formulated in term of vector-matrix multiplication. The limited computational accuracy of these systems, however, severely limits the types of algorithms that can be used with them.

2.1(c) Optical computers with digital data representation:

The finite computational accuracy asociated with analog data representation is usually overcome by employing a digital data representation. In digital optical computing, the binary data representation also necessitates nonlinear optical or hybrid devices for implementing the boolean logic operations. The entire algorithmic foundation developed for electronic digital computers can therefore be applied to digital optical computing. However, the unique features of optical technology as compared to electronics (parallelism of interconnects, 3-D topology) present very different constraints and trade-offs. This implies that the algorithmic strategies developed for electronic digital computers may not transfer directly. Indeed some of the major efforts in digital optical computing aim at exploiting such unique features of optics as very large fan-in, ease of implementing regular but global interconnects between 2-D array of logic gates. Digital optical computing can also be divided into two subcategories depending upon whether numerical or symbolic data is to be processed, thus leading to different algorithms based on different computational modules.

2.2 Architectural Aspect of Optical Information Processing:

The architectural aspect of optical processing is concerned with arrangements of physical components. At the very lowest level it involves taking the variables associated with an optical wavefront (amplitude, phase, polarization, wavelength, spatial position, angle, temporal position) and assigning them to independent variables (i.e., used for data multiplexing) or dependent variables (i.e., used for data representation). Some specific examples will help clarify this point. The use of complex light amplitude for data representation and the two spatial variables for data multiplexing leads to coherent optical architectures for frame-by-frame image processing. The wavelength of light can be used both for data multiplexing as well data representation. The spatial and temporal variable, traditionally used for data multiplexing can be used for data representation by employing pulse width or pulse position encoding. This first step of variable assignment limits further architectural choices. These choices can be organized in many different ways. In this paper I organize the optical processing architectures in the following four major categories, each of which leads to numerous individual architectures depending upon further specifications.

2.2(a) Fourier optical processors:

This class of architectures deals with 1-D/2-D spatial array of data encoded by complex amplitude or intensity. The fundamental operation in this class is the measurement or modification of the spatial frequency spectrum of the input 1-D/2-D distribution. This modification can take place directly in the spatial frequency plane or indirectly in the pupil plane of an imaging system. Historically, these architectures are the oldest ones leading to image processing as well as image recognition systems.

2.2(b) Tapped delay line processors:

This class deals with processing temporal signals and explicitly utilize the time variable in computation. As such, these architectures are intimately associated with certain hardware choices (acoustooptic devices) which are naturally modelled as an analog shift register or a tapped delay line with a large number of spatially parallel taps. These architectures are particularly suitable for implementing signal processing algorithms that are based on finite impulse response (FIR) filtering.

2.2(c) Optical matrix processors:

Matrix algebra provides a very broad theoretical foundation for problems that are linear and involve discrete data structures. The interconnection topologies involved in the fundamental operations of matrix algebra (scalar-vector, vector-vector, vector-matrix and matrix-matrix products) map very naturally onto parallel optical systems. The optical matrix processing architectures consist of direct and fully parallel implementation of the fundamental operations of matrix algebra. These systems can be used to implement various signal and image processing operations that are formulated in terms of matrix algebra. As noted earlier, the limited computational accuracy associated with analog data representation that is used with optical matrix processors limits the types of algorithms that can be used. The optical matrix architectures, however, can also be employed in conjunction with digital data representations and nonlinear operations to attack problems in symbolic computation as well as interconnection networks.

2.2(d): Multistage networks:

This architectural paradigm is borrowed from the electronic parallel processing architectures community and is particularly relevant to digital optical computing. It should be noted that the previous architectural categories typically involved only one or two stage of computation before the final output is generated. In the multistage network architectures, on the order of log N stages (N=number of data channels) of nonlinear switching planes are cascaded with global interconnects between them. Such a strategy is used to implement arbitrary permutation network connecting processing elements with each other or processing elements with memory in electronic parallel processing systems. In digital optical computing, such an architecture has been proposed for implementing parallel, pipelined systems for implementing the desired switching functions.

2.3 Hardware Aspect of Optical Information Processing:

The hardware aspect of optical information processing systems deals with optical and optoelectronic elements that, when assembled according to the archiitectural guidlines will implement the computational operations described in the section on algorithms. These hardware components embody the special features of optics and optoelectronics. Hence their performance is a critical determinant in the performance of the overall computing system. The components fall in two major categories: (i) active components, and (ii) passive components. The active components perform generation, modulation or detection of optical wavefronts, while the passive components transform the optical wavefronts spatially. Examples of active components include light sources (LEDs, lasers), photodetectors (bulk, vacuum, junction), and light modulators (electrooptic, magnetooptic, mechanical, nonlinear optical). The passive components include refractive (lenses, prisms), reflective (mirrors, beamsplitters) and diffractive (gratings, thin and thick holograms) elements for imaging or otherwise spatially trnasforming the optical wavefronts.

The propagation geometries of the optical components is an important aspect that will determine not only the hardware performance but also the types of allowable architectures. The free space propagation geometries lead to 3-D optical systems in which optical wavefronts are unconstrained by the components. Integrated optical geometries confine the optical wavefront in a planar geometry (1-D confinement) of slab waveguides or in channel waveguides (2-D confinement). A fiber optic waveguide is similar to a channel waveguide in that it provides 2-D confinement of optical wavefronts, but since it is not restricted to a planar substrate, it has a greater degree of freedom. On the other hand, integrated optical systems, being limited to a planar substrate, are compatible with the microfabrication techniques employed in integrated electronic circuits and therefore scale more readily to large number of parallel optical channels than do fiber optic systems.

2.4 Summary:

In this section I presented a brief outline of the algorithmic, architectural and hardware aspects of optical computing. There is no claim for the description being exhaustive and as the research in this field progresses, it is envisioned that this description will be modified and/or augmented. In order to make the previous discussionmore concrete, I will now consider some specific cases of optical computing systems and map them on the framework.

2.4(a) Optical 1-D and 2-D spectrum analyzers:

These systems are based on the electromagnetic wave propagation and the special properties of a spherical lens belong to the class of analog optical computers in the algorithmic domain. Architecturally, it belongs to the first class of Fourier optical processors. The systems performing spectrum analysis on images utilize free-space (3-D) propagation geometries, laser light sources, 2-D electrooptic or magnetooptic devices for data inputs and a 2-D detector arrays for data outputs. Optical RF spectrum analyzers deal with 1-D time signals and hence can utilize a guided wave propagation geometry (1-D confinement with slab waveguides), laser light source, surface acoustic wave devices for data inputs and a 1-D detector array of data outputs. It should be noted that the surface acoustic wave devices are used as serial-to-parallel converters to be used in a framed mode and not to be considered as tapped delay lines used in a scrolling mode.

2.4(b) Acoustooptic correlators:

These processors utilize analog data representation and handle numerical data. Hence algorithmically they belong to the second category discussed earlier. Architecturally, they belong to the class of tapped delay line architectures. These systems can use free-space or guided wave propagation gemoetries since they handle 1-D time signals. These processors can use both coherent or incoherent light sources and use acoustooptic devices in a scrolled mode as tapped delay line. The data output is via a single detector as a function of time or via a 1-D time-integrating detector array as a function of space.

2.4(c) Optical multistage switching networks:

In these systems, digital data representation is used and a multistage network architecture is adopted. A free-space propagation geometry is used with nonlinear optical switch arrays performing the routing operation. The optical switching arrays are interconnectd by fixed permutation networks, such as perfect shuffle or cross-over. The nonlinear optical switches themselves may utilize purely optical effects or use a hybrid structure where the switching itself is performed electronically and the input/output of each elment is optical.

2.4(d) Generalized optical processor description:

An analysis of the previously discussed specific optical processing sytems points to a generalized description of an optical processing system. Figure 1 shows the schematic diagram of such a system. 2-D arrays of linear or nonlinear optical switching elements are interconnected by fixed or dynamic optical interconnects. Special cases of these systems occur when the propagation geometries are restricted to 2-D. The optical switching element arrays could be electrooptic light modulators, LEDs, acoustooptic devices or even detector arrays. Although much detail about the procesor functionality i lost when considered at this level of generalization, such a view has the advantage of clearly pointing out optical systems for modifying spatial charcteristics of wavefronts to implement interconnects and the optical sytem for modifying the amplitude (intensity) distribution of the optical wavefronts as the two key technologies required by all types of optical computing systems. The optical interconnection systems components conist of lenses, mirrors, prisms, beamsplitters, and thin and thick hologram (fixed in time or dynamic). The optical switching system components that combine electronic switching circuits and optical input/output has been an active area of research during the past five years. This type of devices have been termed "smart pixels" and is the subject of next section.

3. Spatial Light Modulators to Smart Pixels

3.1 Background

It was recognized from the early days of optical processing that the problems of getting data in and out of the optical processor primarily limited its performance. In 1960's, photographic film served the multiple role of input transducer (electronic to optical, incoherent to coherent optical), data storage medium (image, holographic) and the output transducer. Although superior in resolution and sensitivity, the non-real time chemical development associated with film made the resulting systems impractical for most applications. Hence began the perrenial search for a real-time film. In the early 1970's these devices were called optical-to-optical (OTTO) devices and had the main purpose of converting an incoherent natural illumination image into a coherent image by modulating a laser wavefront. Combining this device with a CRT also provided the capability of electronic-serial-to-coherent-optical-parallel data transduction. A new class of optical devices (most notably acoustooptic bragg cells) performed this conversion more directly and hence this class of devices was given a more generic name of Spatial Light Modulators (SLMs).

In the early days, the SLMs were mostly designed for data transduction and were used in coherent optical processors. Hence, in addition to resolution and sensitivity (speed), linearity and optical quality were important performance parameters of SLMs. Soon it was realized that the nonlinearities intrinsic in the electrooptic response of materials (KDP, liquid crystal) could be utilized to perform, in parallel, nonlinear point operations on images. Furthermore, some device possessed storage characteristics either due to inherent material properties or due to capacitive storage at dielectric interfaces. Thus the range of operations performed by these devices broadened considerably. One specific nonlinearity that received much attention was the hardclipping nonlinearity (binary threshold). An SLM with such a transfer function could perform segmentation of objects from background in an image and could also implement the basic Boolean algebraic operations of AND and OR depending on the threhold level.

In late 1970's and 1980's the area of all-optical logic devices became active. These devices aimed to achieve hardclipping nonlinearity with or without hysterisis in the optical input-output response and did not involve electronic interaction at the macroscopic level. The development of nonlinear semiconductor etalons and surface emitting microcavity lasers was all towards the goal of developing 2-D arrays of optical logic devices with higher speeds and lower switching energies. The invention of self-electrooptic effect devices (SEEDs) departed from this trend in that it was a hybrid optical-electronic device that none-the-less achieved good performance through close integration of all the components. The SEED family of devices soon grew to include symmetric SEED (S-SEED) for processing dual rail data and FET-SEED which included simple electronic logic circuits (built with GaAs field efect transistor) per pixel with a dual rail optical input and output.

In the mid 1980's a new direction emerged in optical computing that focused on the use of optical interconnects within electronic processors. The optical interconnects were considered at various levels of hierarchy (board-to-board, chip-to-chip and within a chip). The competetive advantages of optical interconnects over electronics are found to be strongly dependent upon speed, distance and interconnect densities. Once again, it was the mechanism for transduction of information between electronic and optical form that represented the most severe performance bottleneck. It was realized that combining electronic switching and processing circuits with optical detectors and transmitters in a small package reduced the effects of parasitics and improved the overall efficiency of optical interconnect. This move towards optoelectronic integration was also being made simultaneously in the much larger field of fiber optic communications.

A new look at the earlier analog optical processing systems led to the view that even these processors were based on the unique ability of optical systems to implement weighted, global interconnections between 2-D data arrays using the third spatial dimension. The key difference between different branches of optical computing (analog optical image processing, optical neural nets, digital optical computing and optical interconnects) is thus seen to be in the complexity of operations performed within each pixel of a 2-D array and the nature of optical interconnects implemented between these active devices.

Initially, the idea of combining electronic functionality with optical detection and modulation in a 2-D array was termed "smart spatial light modulator" to indicate its original inspiration. In the past few years a new term "smart pixels" was coined possibly to counter the intimate association between spatial light modulators and coherent optical processors based on 2-D Fourier transforms. The "smart pixels" are therefore seen to be relevant to digital optical computing and optical interconnects as well, since neither area is necessarily based on Fourier optics.

3.2 Smart Pixels:

Smart pixels are formally defined to possess the following properties:

1. optical input,
2. optical output (light emmitting or light modulating),
3. electronic processing.

Some other properties are often implied, though never explicitly stated:

4. free-space optical I/O via surface coupling,
5. identical electronic processing in each pixel.

The origin of these last two properties is in the "smart SLMs" ancestry of "smart pixels". Neither of the implicit or explicit properties include the number of pixels as a qualifying property. Hence it is theoretically possible to design and fabricate a single "smart pixel". As a matter of fact such single element smart pixels have been in existence for quite a few years in the fiber optic communications as optical repeaters. In that case property 4 is clearly not satisfied.

Since the primary technological underpinning of smart pixels is in the intimate combination of electronic switching and optical I/O, one can argue that properties 4 and 5 are not essential for defining a smart pixel. The implicit assumption of these two properties, however, have a very strong impact on the type of optical computing architectures that are based on smart pixel active devices. It is therefore necessary to explicitly remove the properties 4 and 5 as the defining characteristics of the smart pixel devices.

3.3 Unified View:

The smart pixel devices are divided into two main categories depending on whether the optical output is via light modulation or light generation. A further implication of this category is that the light modulating variety of smart pixels could be used in coherent optical architectures but the light generating variety will be (in general) mutually incoherent. The second distinction can be made depending on whether the optical I/O is surface or edge coupled. The latter choice will automatically limit the smart pixel devices to 1-D arrays. These four categories are based on qualitative considerations. Quantitatively, the smart pixel devices are characterized by the number pixels in the array, average processing complexity per pixel (number of transistors per pixel) and the optical I/O bandwidth. It should be noted that the optical I/O bandwidth is not equivalent to the clock rate of the processing circuitry within each pixel. By overlapping processing and I/O one can obtain a large overall processing throughput even for low optical I/O bandwidth provided the processing complexity per pixel is high.

These three quantitative parameters can be represented on three orthogonal axes to create a representational space for smart pixel devices as shown in Figure 2. The axes are logarithmic allowing one to represent a large range of values for each parameters. Number of pixels in an array vary from one to 10^6 (for a 1000 X 1000 array), number of transistors per pixel vary from one to 10^6 (corresponding to a full-blown microprocessor) and the optical I/O bandwidth is shown to vary from one Hz to GHz. A point at the origin corresponds to a single transistor per pixel, one pixel "array" operating at optical I/O of 1 Hz - obviously not a very "smart" or a very fast pixel! At the other extreme is the point diagonally opposite. This represents 10^6 pixels with 10^6 transistors each with optical I/O of 1 GHz - obviously a very powerful system, albeit rather impractical! Assuming that one is considering a monolithically integrated smart pixel device, it is clear that the product of number of pixels and the avergare number of transistors per pixel will be limited by the total chip (possibly wafer) area available to us. On the other hand, the product of number of pixels and optical I/O bandwidth per pixel is limited by power dissipation considerations. Since the area of the optical I/O pads (including amplifiers and drivers) is likely to be much larger than the area of a transistor, the trade-off between pixel complexity and number of pixels for a fixed chip area will not be linear.

This 3-D space serves as a framework within which different smart pixel technical approaches as well as different applications can be mapped. For example, the region of medium pixel complexity, low pixel count and high optical I/O bandwidth will correspond to optical repeaters and the technology of monolithic optoelectronic integration with edge coupled light emmitters will have the matching performance characteristics. On the other hand, low pixel complexity, high pixel count and low optical I/O bandwidth region is seen to correspond to optical neural net or image processing applications. The appropriate technology appears to be the CMOS-Liquid Crystal device technology. The region of high pixel complexity, medium pixel count and high optical I/O bandwidth will be the realm of optical interconnects in parallel electronic systems.

It should be emphasized that the diagram shown in Figure 2 only deals with quantitative discussion of smart pixel performance parameters. When these devices are to be used for different applications, a number of other considerations enter - such as, optical quality of the device, wavelength stability requirements, mechanical packaging considerations, cost. Hence the diagram can be used as a first order guide to determine the choice of technologies that can meet the application needs. Further selection will be based on the other considerations, some which were alluded to.

3.4 Summary:

A historical view of the evolution of the optical I/O devices for optical processing was presented. The evolution was shown to proceed from SLMs to smart SLMs to smart pixels. A loose definition of smart pixels was given that will incorporate a number of existing devices, such as optical repeaters. Of particular relevance is relaxing the property of identical processing for all pixels in an array. Lifting this requirement brings the smart pixel devices in the fold of optical interconnect research. A quantitative way of describing different operating regions of smart pixel technology was described. A 3-D space with pixel complexity, pixel count and pixel optical I/O bandwidth represented along orthogonal axes was described. This space can be used to represent smart pixel technologies as well as applications and therefore can be the common meeting ground for these two communities. A user feedback describing their requirements in terms of these parameters and other special considerations will be useful in determining the technology-application mapping. It will be equally important to request the user to identify the requirements that are "soft" and the ones that are "hard" and therefore provide guidance for establishing a cost-performance trade-off for different smart pixel technologies.

4. Summary

Optical information processing is a highly diverse research area that includes research in device physics, parallel computing architectures and algorithms. Inspite of such a diversity, it is possible to define a common foundation that interrelates research in these different disciplines. Such a generalized view led to the identification of "smart pixels" as one of the key technologies or optical information processing systems. Processing complexity of each pixel, total number of pixels and optical I/O bandwidth per pixel were identified as the key performance prameters that are determined by the technology adopted. These same performance prameters were also identified as characterizing different optical information processing system applications. This view, therefore helps to map technology onto aplications.

Bibliography

Following is a list of text books, edited books, journal special issues in chronological order. These sources cover most of the aspects of optical information processing discussed in this paper.

J.W.Goodman, *Introduction to Fourier Optics*, New York: McGraw-Hill, 1968

Proceedings of IEEE, Special Issue on Optical Computing, January 1977.

Proceedings of IEEE, Special Issue on Acoustooptic Signal Processing, January 1981.

Proceedings of IEEE, Special Issue on Optical Computing, July 1984.

R. A. Athale, "Optical Matrix Processors", in "Optical and Hybrid Computing", H. H. Szu ed., Proc. of SPIE., Vol 634, pp. 96-111, (1986).

J. L. Horner, *Optical Signal Processing*, Academic Press, 1987.

D. G. Feitelson, *Optical Computing*, MIT Press, Cambridge, MA, 1988.

Proceedings of IEEE, Special Section on Optical Computing, October 1989.

"Digital Optical Computing", R. A. Athale ed., SPIE Optical Engineering Press, 1990.

J. A. Neff, R. A. Athale, and S. H. Lee, "Two Dimensional Spatial Light Modulators: A Tutorial", Proceedings of IEEE, Vol. 78, No. 5, pp. 826-855, May 1990.

A. B. Vander Lugt, *Optical Signal Processing*, Wiley Interscience Pub., 1992.

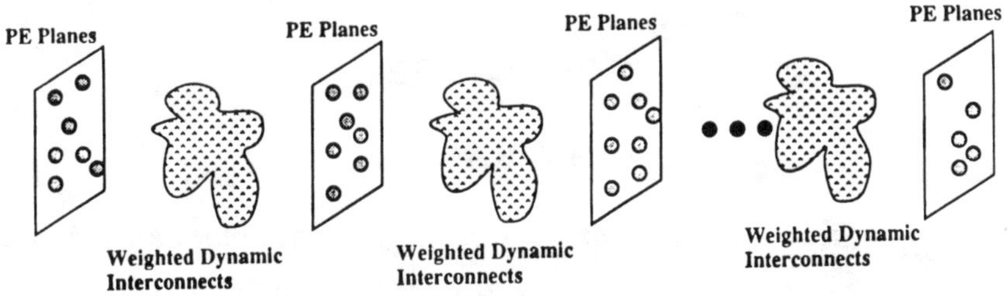

Figure 1: Generalized schematic diagram of an optical processing system.

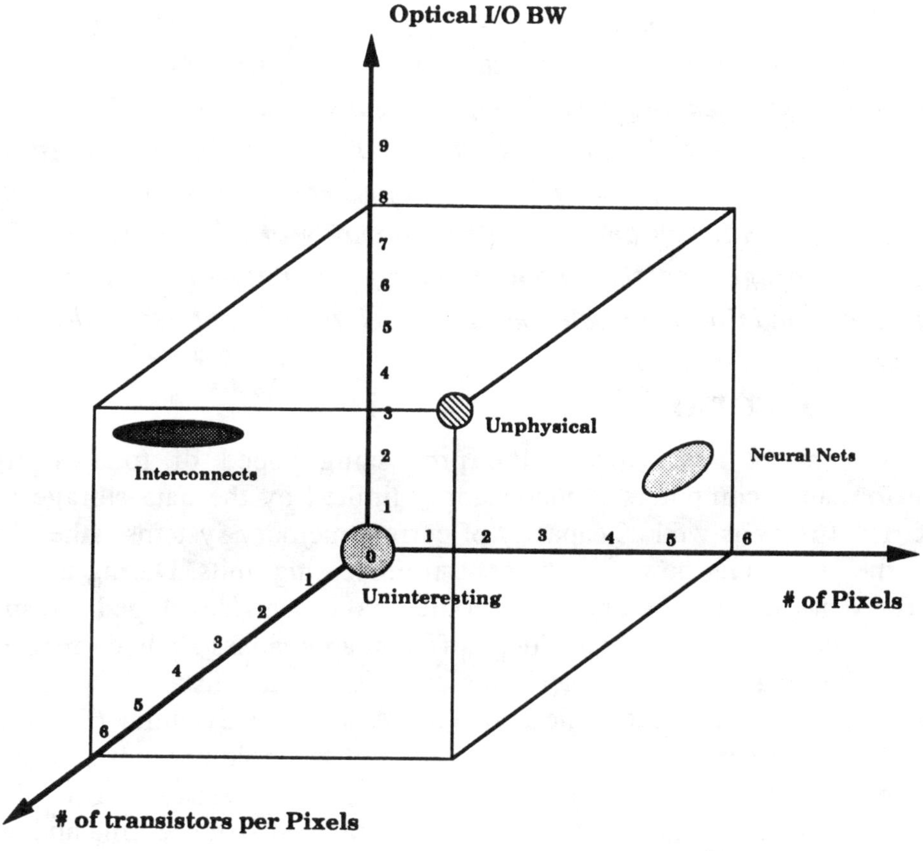

Figure 2: The 3-D parameter space characterizing "smart pixels"

Optical Data Storage and Retrieval: Research Directions for the 90's

Sadik Esener, Joseph Ford and Susan Hunter

Electrical and Computer Engineering Department, 0407
University of California San Diego, La Jolla, CA 92093
(619) 534-2732

ABSTRACT

The computational power of current high-performance computers is increasingly limited by data storage and retrieval rates rather than the processing power of the central processing units. No single existing memory technology can combine the required fast access and large data capacity. Instead, a hierarchy of serial access memory devices has provided a performance continuum which allows a balanced system design. Conventional memory technology can only marginally support the needs of high performance computers in terms of required capacity, data rates, access times and cost. Significant gaps in secondary and tertiary storage have emerged which make storage hierarchy design increasingly difficult. This paper reviews a radically different approach to data storage using the parallelism and three dimensionality of optical storage. 3-D optical storage has the potential to significantly alter the present hierarchy and fill the pressing need for high performance secondary and tertiary storage systems.

1. INTRODUCTION

For many applications, the processing speed of today's high performance computers is increasingly limited by the data storage and retrieval rates as well as capacity of current memory systems rather than by the processing power of the central processing units. During the past fifty years many memory technologies have been developed. Despite intense competition, several widely different approaches are in current use, including magnetic and optical tape and disks (hard disks, floppies, and disk stacks)[1], and electronic static (SRAM)[2] and dynamic (DRAM)[3] random-access memory. This proliferation of technologies exists because each technology has different strengths and weaknesses in terms of its capacity, access time, data transfer rate, storage persistence time and cost per megabyte. No single technology can achieve maximum performance in all these characteristics at once. Instead, modern computing systems use a *hierarchy* of memories rather than a single type.[4] The memory hierarchy approach utilizes the strong points of each technology to create an effective memory system that maximizes overall computer performance/cost.

Volume holographic storage concepts were first put forward in the sixties. Early attempts in optical archival storage were unable to compete with the rapidly improving electronic storage technologies.[5] Until recently, computer technology was incapable of making full use of a major advantage of optical volume storage: the high data rate possible with massively parallel access.

In this paper, we briefly describe the present challenges in designing memory systems with currently available data storage systems and the performance required by future memory systems. We then explore means by which optical storage systems may meet these requirements. We first discuss the present capabilities of optical disk systems and evaluate the potential benefits and means of developing higher density and parallel accessed optical storage systems. We then focus on 3-D optical storage, beginning with the underlying fundamentals and then moving to the various storage materials. We illustrate aspects of system design using one particular approach- two photon storage. Finally, we extract the performance potentials of this system and show how it can alleviate some of the problems currently encountered in hierarchical memory system design.

2. A REVIEW OF PRESENT STORAGE HIERARCHY

In standard sequential computer architecture there are four major levels of the storage hierarchy: cache, main, secondary and tertiary (archival) (see Figure 1). In parallel computers, the difference between main and cache storage becomes less distinct and they are jointly called primary memory. Primary memories are currently implemented in silicon: cache memory as local storage within the processing chip and main memory as RAM and DRAM chips located on the same board. The access

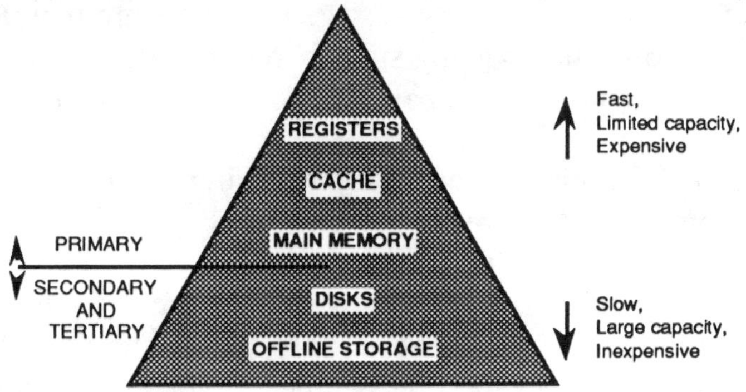

Figure 1: Memory hierarchy in existing computing architecture.

times of primary memories are comparable to a 10 ns clock cycle, but their data capacity is limited (10-100MByte for main), although it has been doubling every year. Secondary memories, such as magnetic or optical disk drives, have significantly increased capacity (10GByte) with significantly lower cost per MByte. However, the access times are on the order of 10 ms, and this access time is for many computing applications presently the major performance limiting factor. Archival (or tertiary) storage holds huge amounts of data (Terabytes), however the time to access the data is on the order of minutes to hours. Presently, archival data storage systems require large installations based on disk farms and tapes, often operated off-line. Archival storage does not necessarily require many write operations and write-once read many (WORM) systems are acceptable. Despite having the lowest cost per Mbyte, archival storage is typically the most expensive single component of modern supercomputer installations. Special computing applications such as image processing may also use a buffer memory for fast parallel data acquisition before downloading into more permanent storage. Buffer memories tend to be quite application specific, and will not be discussed in this paper.

In Figures 2, 3 and 4 we show the current memory technologies in terms of their critical characteristics. As one can observe from Fig. 2 there is a four order of magnitude performance gap in access time between electronic RAMs and secondary storage devices such as disks. The width of this gap is doubling each year, forcing the development of new secondary storage systems. In addition, because the processing power doubles every year and the memory density only every two and half years, an increasingly large gap also exists in terms of memory capacity. A similar situation exists between secondary storage systems and archival systems (such as tapes) that suffer from slow access times. In addition, Fig. 3 shows that the data rates of current secondary storage technologies are significantly lower than required for full use of parallel processors. Techniques to enhance secondary storage data rates are therefore crucial. Finally, Fig. 4 shows that existing storage systems are too costly for wide use in applications such as databases that require high capacity and fast access times.

Existing serial memory technologies using planar storage media, including electronic RAM, magnetic disk, and optical disk storage, are inadequate to bridge the gaps described in Fig. 2. The data capacity is fundamentally limited by their two-dimensional nature to the storage area divided by the minimum bit size. Similarly, as the storage area increases for higher capacity so does the access time. Another drawback is that the data transfer rates are limited by their sequential nature to the I/O channel bandwidth.

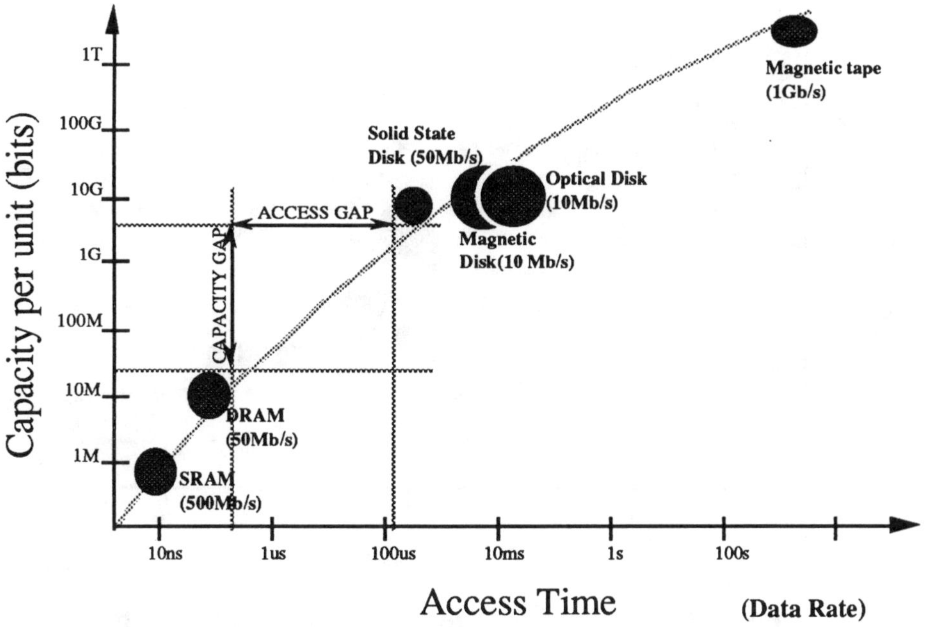

Figure 2: Access speed and capacity of existing memory technologies.

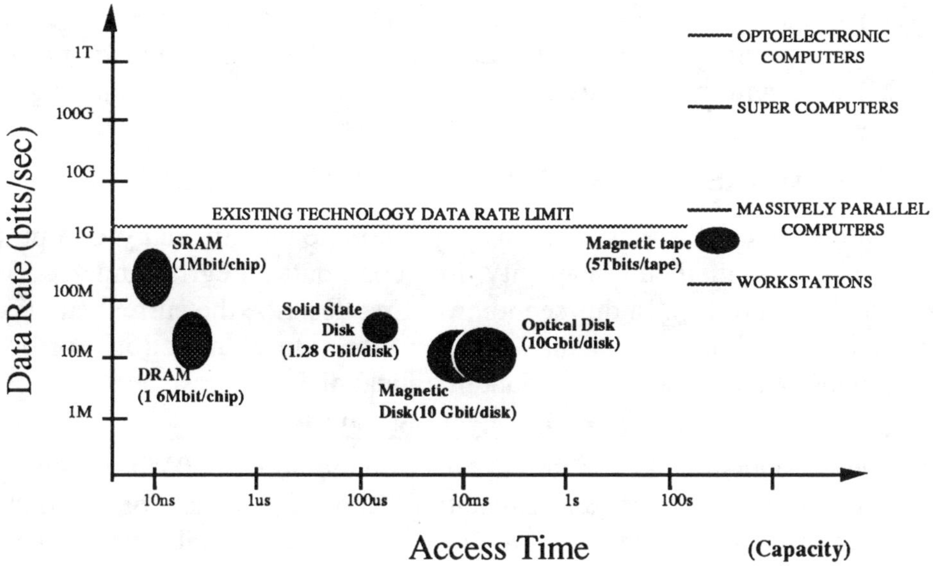

Figure 3: Data rates of existing primary storage technologies.

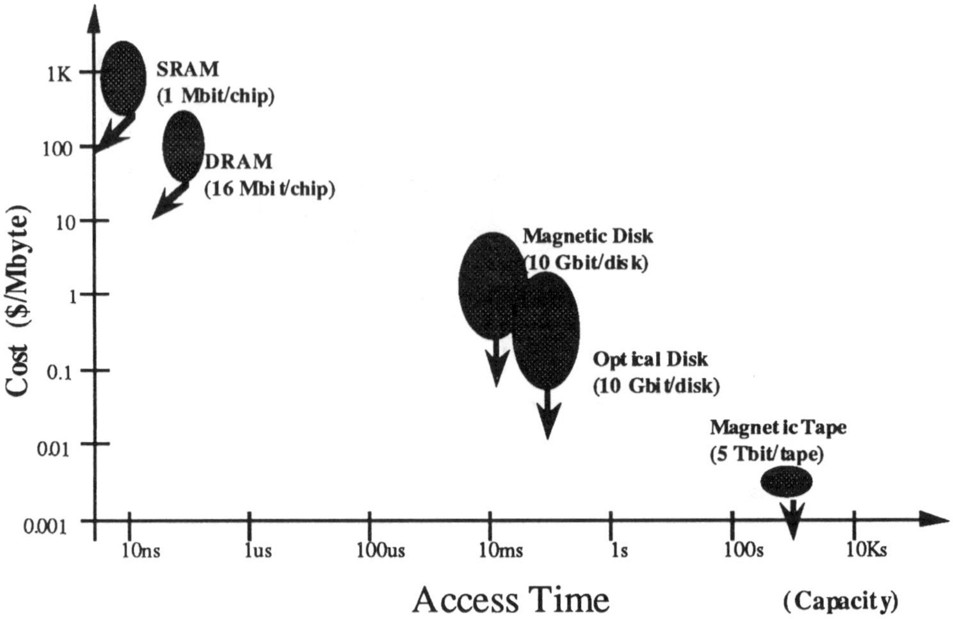

Figure 4: Cost per Mbyte of existing primary storage technologies.

In the following section, we review the current limitations of optical disk storage and explore how they can be modified to increase data density and how higher data rates can be achieved by exploiting parallelism. In the subsequent sections, we will discuss how three-dimensional memories (3DM) surmount the capacity and access time limitations by extending the storage into the third dimension.

3. OPTICAL DISK STORAGE: PRESENT AND FUTURE

Optical disks have become a major component in many computer systems due to their high capacity, low cost, data integrity and security (due to removability). In this section, we first describe the current status of optical disk technology. Then, we present the future projections for both serial and parallel access of the data in optical disk systems.

3.1 Performance of current serial optical disk systems

Present optical disk storage systems provide about 500Mb/in^2 storage density with a total maximum capacity of about 10GBytes for 14" disks. The data density is limited through diffraction by the wavelength (780-830 nm) at which data is currently recorded on optical disks. Research on short wavelength integrated lasers and on the use of superresolution in the optical read/write channel is presently being conducted to reduce the physical bit size and reach data storage densities approaching 10Gb/in^2. Simultaneously, research is being carried out to develop the necessary higher resolution optical disk media capable of read-write-erase operation.

The main disadvantages of commercially available optical disk systems are their low data rate and poor access time performance. Low data rates are due to the relatively slow disk rotation speeds (<3600 rpm) which are mainly limited by the tracking and focusing servo loops. The slow access times are mainly due to the weight of the heads and the slow rotation speeds that are respectively responsible for large seek times and large latency times. The detection speed is governed by the limited source power, the detector sensitivity, and the write and damage thresholds of the disk material. Present development efforts are aiming towards low mass optical heads which reduce the response time and increase the allowed disk rotation speed. These new heads incorporate holographic or integrated optical elements with substantially less total weight than their bulk optic counterparts. In addition, efficient error correction codes and higher power lasers (e.g. array lasers with combined beams) are used to increase the detection rates and avoid limiting the maximum data rate. Solutions derived from magnetic disk technology have also been envisioned (e.g. flying heads or air bearing drives) but have not been pursued since their adoption would eliminate one of the main advantages of optical disks: removability.

It is expected that these developments will result in a continued improvement of the optical disk performance. However, the resulting performance increase is not expected to satisfy the ever rising demands of future computers which will require terabits of data capacity accessed in sub-milliseconds with data rates approaching Tb/s. Therefore, optical disk storage research must seek alternative approaches such as the use of parallel data access to substantially improve upon the current data rates.

3.2. Parallel access optical disk systems

A major improvement in the data transfer rates of optical disks can be achieved by accessing the data in parallel. The concept was first proposed in 1980[6] and later by several others.[7,8,9,10] The research efforts are divided in two main lines of work: one approach uses either a single multiple beam head or multiple independent integrated laser heads to read several tracks in parallel, while the other consists of reading coded data blocks with a single laser beam. In this latter case, both imaging[3] and Fourier transform[2] systems have been proposed. In the following, we describe a hybrid approach currently under development at UCSD.[4]

The system is designed to read 1-D data blocks distributed radially on the disk's active surface and generate 2-D output binary images. It has the unique advantage that no mechanical motion of the head is required for data addressing, focusing or tracking. This allows a data rate of up to 1 GByte/sec for standard disk rotation speeds of 2400 rpm.

The data blocks are stored on the disk as 1-D Fourier-transform

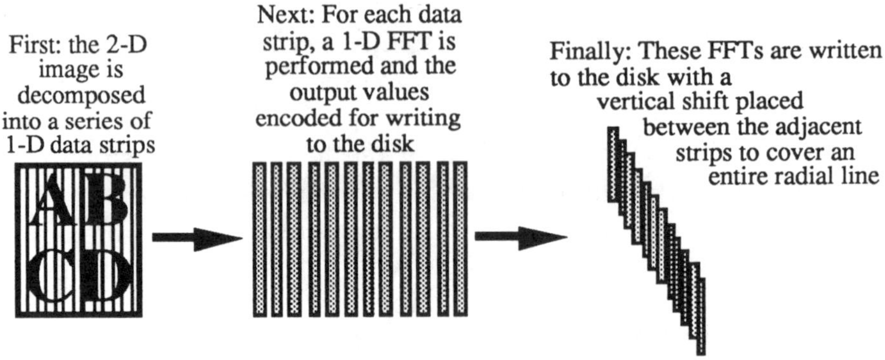

Figure 5: Image encoding in UCSD's parallel readout optical disk system.

computer generated holograms (CGH). Each one of them is calculated to reconstruct one column of a 2-D output image. During the sequential recording process, all the CGH for a given 2-D image are laid-out side by side and radially shifted from one another until they span the entire radius of the disk active surface as shown in Figure 5. During the parallel readout process, all of the data blocks of a single image are illuminated at once. Data addressing is achieved solely through the disk rotation so that the entire memory can be read in one rotation. The Fourier direction of the holograms is along the radial direction of the disk. Thus, due to the shift invariance properties of Fourier transform holograms, the tracking servo requirement can be eliminated. Under-illuminated Fourier Transform holograms will still reconstruct, although with a loss in output signal-to-noise ratio. Therefore, by slightly under illuminating the hologram block the focus servo requirement can be virtually eliminated.

The optical readout system is shown in Figure 6. It consists solely of three stationary cylindrical lenses. The first illuminates an area on the disk that corresponds to the data blocks of a single output 2-D image. The other two lenses image the data blocks of the disk along the tangential direction and perform a Fourier transform along the radial direction. In the actual system, these two lenses have been replaced by a single hybrid diffractive-refractive optical element for easier system alignment, reduced aberrations, and improved resolution. A similar optical parallel read out system can also be applied to increase the data rates of optical tapes.

A scaled-down prototype using a commercially available plastic 5.25" WORM disk[11] has been implemented that reads out 16x16 pixel images at a maximum rotation speed of 30 rpm. The complete characteristics of the prototype system have been reported.[12] Experiments performed on the prototype system showed that a full scale system could operate with a data transfer rate of 10GBytes/sec and a disk capacity of 300MBytes.

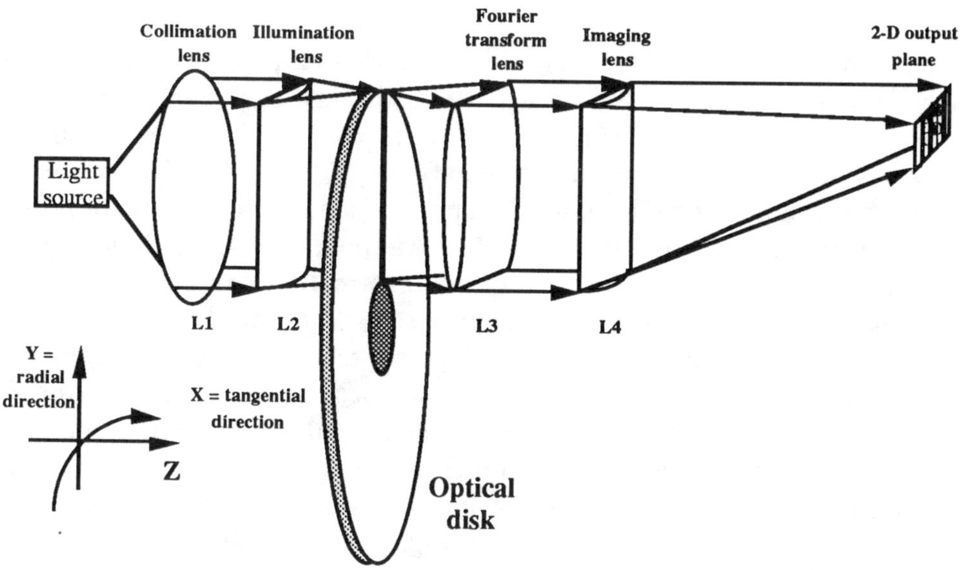

Figure 6: Parallel readout optical disk system. Cylindrical lenses combine imaging and Fourier transforming to reconstruct the recorded image.

These results indicate the optical storage systems with parallel read-out capabilities may shortly enter the realm of practice. Unfortunately, by exploiting parallelism for higher data rates the storage capacity of the system is reduced. Therefore, parallel accessed two dimensional storage systems should be useful primarily when the operations to be performed on the data can be parallelized. For example, an associative memory system based on the parallel accessed optical disk system described above is currently being constructed at UCSD.[12]

To become useful on a larger span of applications, however, optical storage systems must provide higher capacity for faster data rates and access times. This can be achieved by employing 3-D optical storage systems.

4. PHYSICAL FUNDAMENTALS OF 3-D OPTICAL STORAGE

We will briefly consider the fundamental principles underlying optical storage and retrieval of information from three dimensional media. These concerns are largely independent of the particular materials used, which will be discussed in the Section 5.

4.1 Definition and classification of 3-D optical memories

Present optical memories store one-dimensional information sequences in a two-dimensional space using two spatial coordinates (either rectangular {x-location, y-location} or polar {radius, angle}) to assign a precise physical location to each bit. The maximum storage capacity is given by A/λ^2, where A is the storage area and λ is the optical wavelength. A 3-D memory can be defined as a single memory unit where three independent coordinates are used to specify the address of the information. These coordinates may be entirely spatial {x-location, y-location, z-location} but may also use other physical parameters (e.g., {radius, angle, wavelength}). In 3-D optical memories, information can be partitioned into binary bit-planes (or images) that are stacked in the third dimension. One memory operation (write, read, or erase) is performed on the entire plane of bits, thus achieving 2-D parallel access of the data in a single cycle of operation. As in the parallel access optical disk system described in Section 3, this results in a significant increase in the maximum achievable data transfer rate over conventional systems. Since, in addition, optical 3D memories can achieve very high density (10^{11}-10^{12} bits/cm^3) they offer the potential for larger storage capacity systems with lower access times and higher data transfer rates than conventional 2-D memories.

3-D memories can be classified as using either bit-oriented (localized) or holographic (distributed) storage, and can be further classified by the type of physical coordinates used to address the data, as is shown in Figure 7. The physical mechanisms which can be used for recording and addressing in 3-D optical memories are discussed in Section 4.3. First, however, we discuss the characteristics of bit-oriented and holographic data storage formats.

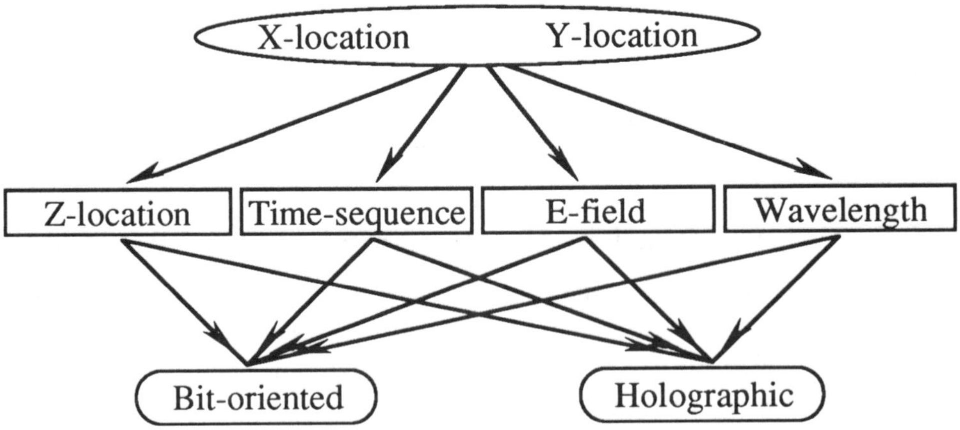

Figure 7: Classification of 3-D optical memories. Two spatial dimensions are used for the 2-D images. A third dimension is defined by an additional physical addressing parameter. The information can then be stored in either bit-oriented or holographic format.

4.2 Physical principles of 3-D information access

4.2.1 Bit-oriented(local) storage

Probably the most direct approach to volume data storage is to partition a storage volume into sub-volumes, each containing one datum. The traditional 3-D storage technology --books-- uses this approach. The location of each letter in a book can be described by its x-y coordinates on the page and its depth in the media (its page number). If you imagine a book printed on transparent pages, permanently stuck together, then the concerns associated with optical storage and readout become obvious.

Two dimensional optical wavefronts can access data in three spatial dimensions only by involving an additional mechanism. Otherwise, the image simply propagates through the storage medium, superimposing the effects of all the stored pages. The selection of a particular position in the third dimension comes from an additional parameter. In two photon materials, the additional requirement is that photons of a second optical wavelength must be present in order for the first beam to interact with the material. In spectral hole burning, the additional parameter is the wavelength. And in photon echo or free-space storage, the additional addressing parameter is time.

Diffraction effects enforce a theoretical limit to the maximum achievable spatial resolution of any optical memory. The minimum spot size is conventionally given as $\delta = 1.22\lambda F$, where λ is the optical wavelength and F is the f-number of the system (focal length divided by clear aperture).[13] The resolution in the third dimension depends on the type of addressing parameter used, as discussed in Section 4.3.

4.2.2 Holographic (distributed) storage

In bit-oriented memories the data resides in a restricted region. This means that if any portion of the storage media is damaged or blocked, the data which was stored in that region is lost, but the other data is completely unaffected. In more concrete terms, if half of your material is destroyed, half of your data is lost. This is not the case for holographic storage, where the information about each stored bit is distributed throughout a large region. If a portion of the storage media is damaged or blocked, instead of causing catastrophic loss of some of the data, all of the data is partially degraded. If half of the material is destroyed, all of the data remains to some degree. Whether *any* of the data is still legible depends on the extent of the damage. For common types of damage, such as surface dust or smudges, holograms are remarkably robust. This has generated interest in holographic data storage, and despite the more complex optical systems and the coherent sources required there has been continued research in the field since the early 1960s[5,14]. Excellent texts on holography are available.[15,16,17] We will only provide a brief description of hologram characteristics as they relate to data storage.

Holograms are created by recording the interference pattern of two optical wavefronts. The storage media can record the fringes as index and/or amplitude modulation. When the recording is illuminated by one of the wavefronts (the reference beam), the other wavefront (the object beam) is reproduced. Planar holograms, which record only a 2-D interference pattern, have quite different behavior from volume holograms, which also record the variation through a depth significantly greater than the fringe spacing. Planar holograms are theoretically limited to 6.25% efficiency if recorded in amplitude, and 33.9% efficiency if recorded in phase.[15] The maximum diffraction efficiency of a volume absorption hologram is theoretically limited to less than 7.2%. Pure phase volume holograms, however, can approach 100% efficiency.

The most important difference between planar and volume holograms is their selectivity with respect to the readout (reference) wavefront. When the reference wavefront is tilted, planar holograms still diffract with nearly full efficiency, producing a reconstruction tilted by a similar angle from the original. For a volume hologram to diffract efficiently, however, the original recording wavefront must be closely reproduced. This effect, called Bragg selectivity, allows a volume hologram to independently store and recall multiple superimposed holograms provided that each has a sufficiently distinct reference beam. The detuning angle to drop to the first zero of diffracted intensity is approximately Λ/T radians, where Λ is the grating wavelength and T is the hologram thickness.[15] For a 1 cm thick hologram, this angle is on the order of hundredths of a degree. Similarly, a volume hologram can have a 1 Å wavelength selectivity.[18]

If an additional physical parameter is used for addressing, such as time sequence, a 3-D optical memory can be constructed using planar holograms. The following discussion applies only to volume hologram multiplexing, as for example in photorefractive crystals.

The theoretical upper limit of the storage capacity of a volume hologram[19]is given by V/λ^3. Constraints arising from optical system[20]and storage media characteristics[21]significantly reduce the achievable volume holographic storage density. However, storage densities in excess of 10^9 b/cm^3 can realistically be achieved.[22] The information which can be extracted from the volume at a particular instant is still limited by the hologram aperture to A/λ^2 for each wavelength used for readout. The large theoretical capacity must be accessed in pages, multiplexed in the volume by wavelength or phase.

The goal in multiplexing holograms is to maximize storage capacity (the number of images, their resolution, and their diffraction efficiency) while minimizing crosstalk. Volume holograms can be multiplexed by wavelength and phase (and polarization in some materials), and of course by the volume illuminated. Phase multiplexing includes reference beam tilt or curvature as well as more complex (or even random) phase patterns. Polarization multiplexing is not used for data storage since there are only two orthogonal polarizations. However, volume and phase multiplexing have been extensively studied for storage applications.[23,24,25] More recently, with the availability of effective color tunable lasers wavelength multiplexing has been a topic of investigation, especially in reflection volume holograms, which are highly color selective.[18]

Crosstalk in volume holograms arises from two major sources, Bragg degeneracy and higher order diffraction. Bragg degeneracy comes from the fact that in addition to the exact reference beam angle, there is a cone of vectors which match the Bragg condition. Similarly, if the readout wavelength (or angle) is changed, a simple plane–wave hologram can still be read at full efficiency by using an appropriate angle (or wavelength) which matches the Bragg condition. This behavior effectively vanishes as the hologram becomes more complex, because the Bragg condition can only be satisfied simultaneously for the entire angular spectra by the original reference wavefront. Scattering from locally matched portions of the grating can only produce background noise, whose intensity becomes negligible as the hologram thickness increases. In currently available materials, the storage capacity seems to be limited more by the material dynamic range rather than by crosstalk resulting from multiplexing.

4.3 Physical mechanisms of 3-D optical storage

Optical recording can occur only when photons are absorbed by matter. If this absorption changes any of the material's optical properties, then data can be recorded and read out optically. The basic characteristics of an optical wavefront are its wavelength, amplitude (intensity), polarization, and phase (the direction of a beam is determined by the wavefront's phase profile). Any optical storage media affects one or more of these characteristics. For example, developed photographic film becomes opaque where it was illuminated. Other materials can change their index of refraction (e.g. photorefractive crystals and dichromated gelatin film), their surface reflectance (write-once optical disks), their absorption spectra (spectral hole burning materials), their fluorescence spectra (photochromic materials), etc., in response to optical illumination.

The 'location' of stored data is given by the set of physical parameters required to retrieve that data. This may the three spatial coordinates in a volume, but might also include the readout wavelength and polarization, an external applied voltage, an external magnetic field, and even the timing or sequence with which these parameters are applied. A set of address coordinates or dimensions is said to be orthogonal if they are totally independent of each other, as are, for example, the three spatial coordinates. The effective extent of each dimension can be defined as the range over which storage is possible divided by the resolution (the smallest resolvable division) maintained over that range. These considerations may be limited by material or experimental constraints, or by more basic principles. Polarization is a dimension whose extent is only the two orthogonal polarizations. The extent of the horizontal and vertical spatial dimensions of an image can be huge, however, with a range of centimeters and a resolution of microns.

In 3-D memories, the coordinates that specify the location of the information can be spatial, spectral, or temporal, giving rise to a variety of 3-D memory concepts using different materials with various properties. For example, 2-photon absorption materials can form a true volume memory while spectral hole burning materials provide a 3-D storage medium with two spatial and one spectral dimension. Optical memories of even higher dimension (4-D or 5-D memories) are possible provided that appropriate materials are identified and the necessary systems are developed. The image bearing character of light and the planar nature of detector arrays make three the magic number for optical memories, since this allows a time sequence of 2-D images to be retrieved, detected and processed in parallel.

Multiplexing with parameters such as wavelength or time-sequence is in effect a way of volume multiplexing at sub-wavelength scales. A cubic micron is the smallest optically resolvable volume. However, there are trillions of distinct atoms within that volume and each may have a different optical response. For example, spectral hole burning materials use the fact that similar molecules within a bulk material have different absorption bands due to variations in local stresses.[46] There may be billions of molecules with each characteristic absorption spectrum within a single cubic micron, so that each different wavelength seems to illuminate untouched material. Temperatures near absolute zero help prevent transfer of information between adjacent molecules. However, the material sensitivity and readout efficiency are probably not as high as they might have been if the material was composed of a single pure species with uniform response. The difference between simple volume multiplexing and these more exotic alternative addressing techniques is that only in volume multiplexing at optically resolvable scales can *all* of the local material be entirely devoted to recording a single bit of information.

5. 3-D OPTICAL STORAGE MATERIALS

5.1 Photorefractives

Photorefractive crystals (PRC) are a reversible phase storage media which have been extensively studied for holographic storage applications since their discovery in 1965. The sensitivity approaches that of photographic emulsions, and the index modulation is large enough to record nearly 100% efficient holograms. Crystal sizes range up to many cm^3, depending on the particular material. PRC have three characteristics which combine to make them different from any other holographic recording media: 1) in–situ recording and development, 2) energy coupling between recording beams, and 3) write–erase capability with infinite cycling. The basic mechanisms of the effect have been uncovered and analyzed, and a number of excellent overviews of photorefractive processes and materials are available.[26,27,28] The process can be qualitatively described as follows.

Consider two plane waves crossing in a photoconductive and electro-optic material. They produce a sinusoidal fringe pattern I(z) (see Figure 8). In regions of high intensity, photoabsorption excites charge carriers to the conduction band, allowing them to drift or diffuse towards darker areas, where they are trapped. The resultant space charge distribution generates an electric field E(z). This field modulates the refractive index by the electro-optic effect, creating a sinusoidal phase grating $\Delta n(z)$. If the writing beams are cut, the phase grating remains stored in the material until thermal or optical excitation redistributes the trapped charge carriers. If the crystal is uniformly illuminated, the index modulation is erased by the same process. The cycle can be infinitely repeated.

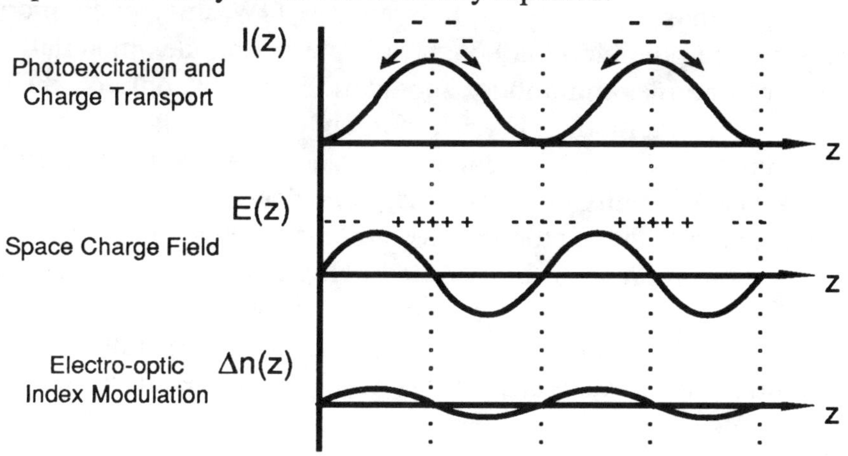

Figure 8: The photorefractive effect.

A large number of photorefractive materials have been identified. They can be broken roughly into three groups; ferroelectric oxides, cubic oxides (sillenides), and semi–insulating compound semiconductors. For storage applications materials with long relaxation times are preferable, making ferroelectric oxides the material of choice. They include Lithium Niobate[29], Potassium[30] and Potassium-Tantalum[31] Niobate, Barium Titanate[32], and Strontium Barium Niobate.[33] Their large electro–optic coefficients produce a large saturation index modulation. Their low photoconductivities ensure long dark storage lifetimes, but also result in lower sensitivity (the index change per unit absorbed optical energy per unit area). For example, the sensitivity of $BaTiO_3$ is about 10^{-3} cm^2/J, producing a response time of 0.1–1 second for 1 W/cm^2 input power. Another photorefractive with potential storage applications is ceramic Lead–Lanthanum-Zirconium–Titanate[34]

Photorefractive holograms in ferroelectric oxides are of high optical quality, and have been shown capable of reconstructing images of 10^6 bits or more.[28] As many as 5000 angle multiplexed images have been recorded in a single $LiNbO_3$ crystal.[35] The fundamental limitations on PRC sensitivity[36] come from the quantum efficiency of the photoabsorption, the carrier lifetime and mobility, and the electro–optic coefficient. The physical resolution is high enough to record even the highest frequency reflection grating accurately, although the strength of the response does depend on grating frequency and orientation. In the absence of an external applied field,[32] the self-developing index modulation response follows exponential write and erase curves, with a response time τ which is inversely proportional to the writing intensity. While the photorefractive effect is generally thought of as slow (on the order of milliseconds or longer), with sufficiently high intensity (GW/cm^2) even picosecond responses are possible,[37] although full saturation index modulation is not achieved. The maximum index modulation depends only on the contrast between the interacting beams, not on the absolute intensity of the interacting light (provided the intensity is above a threshold determined by the spontaneous redistribution of charge carriers).

The fact that the photorefractive process is reversible means that there is erasure during readout of recorded holograms. If the ratio of recording to readout intensities is large enough, the holograms can persist for a long time. The fundamental limit to storage lifetime is the time for the charge distribution to spontaneously relax to a uniform equilibrium, which is determined by the material's dark conductivity. This can range from hours in $BaTiO_3$ up to years for $LiNbO_3$. The photorefractive index grating can be semi–permanently stored, or "fixed", so that the hologram can be read out at high intensity for extended periods without significant erasure.

Fixation has been accomplished by two-photon absorption in $LiNbO_3$, KTN, and PLZT, or by ionic charge compensation, where the electron distribution is converted into a similar distribution of non–mobile ions. Permanent fixation by charge compensation has been demonstrated in $LiNbO_3$.[38] However, the materials and mechanisms of photorefractive fixation are not fully understood, and some of the early impressive results have proven difficult to reproduce.

When holograms are multiplexed in a photorefractive crystal, each exposure tends to partially erase the existing information in the crystal. To superimpose holograms and end with a specified relative diffraction efficiency - usually equal efficiency for all holograms - requires some kind of recording timetable. In scheduled recording,[22] the first hologram is recorded to the maximum achievable efficiency. The exposures for each subsequent hologram are shorter and shorter, so that at the end a set of equal efficiency holograms is obtained. The recording timetable is calculated from the material's index response curve. An alternative approach, called incremental recording,[24] records each hologram with a series of exposures which are very short compared to the crystal's response time. During recording, each image and reference pair is sequentially displayed, repetitively cycling through all the images until the process reaches saturation. The final diffraction efficiency of the multiplexed holograms decreases as the number of superimposed holograms grows. The diffraction efficiency η of N superimposed volume phase holograms is approximately $\sin^2(\pi \Delta n T / N \lambda \cos\theta)$, where Δn is the index modulation, T is the thickness of the crystal, λ is the optical wavelength, and θ is the internal angle between the recording beams. The writing and erasing processes have been assumed to be exponential with the same time constant. For large N, the efficiency drops as the square of N. Increasing T increases both hologram efficiency and selectivity, making thick crystals desirable for data storage applications.

Although significant progress has been made at the system level, the characteristics of existing PRC remain far from satisfactory for reliable low cost systems. Major improvements in crystal dimensions, cost, and dynamic range (modulation of the index of refraction) need to be realized before such systems can become competitive. Recently, photorefractive effect has been observed in low cost materials such as plastics and this direction may bring some hope of reducing the cost of PRC materials.

5.2 Irreversible volume holographic materials

Many other materials have been used to record multiplexed volume holograms. Dichromated gelatin[39] can record a permanent index hologram with high efficiency (approaching 100%), but the thickness is limited to around 100 μm. Photochromic materials, which produce a reversible color change record in response to light, can record an absorption hologram. However, the maximum diffraction efficiency of an absorption hologram is theoretically limited to 7.2%.[15] Photopolymers[40,41] show promise as a high–efficiency moderately thick (25-50 μm) recording media, particularly since they can be developed in–situ (under certain conditions) without wet processing. However, their diffraction efficiency is a strong function of the fringe spacing, and the fringes are sometimes a combination of phase and surface relief modulation.

5.3 Bacteriorodhopsin

A biological material, the bacteriorhodopsin which is a photochromic protein may present an inexpensive alternative to PRC and other more conventional volume holographic materials. The photochemistry associated with bacteriorhodopsin is well documented and the reader is referred to references [42,43, for example] for an in depth treatment. The bacteriorhodopsin has a light absorbing chromophore which is bound to the protein through a protonated Schiff base linkage. Any change in the electronic environment of the binding site of the chromophore results in a change in the spectral characteristics of the overall protein. Such changes can be induced by light absorption at proper wavelengths and temperatures resulting in a photocycle. The photocycle of bacteriorhodopsin is comprised of at least five thermal intermediate states. It has been suggested that at least three of these intermediate states (**bR**, **K** and **M**) have potentials for optoelectronic device applications. Each key intermediate exhibits a unique absorption spectrum. The initial state, **bR**, is characterized by a large absorption maximum in the yellow region of the visible spectrum. At low temperatures (77°K), through light absorption by the chromophore, a nearly instantaneous shift of electron density affects the chromophore-protein link, resulting in photo-isomerization and leading to the formation of the **K** intermediate state. The **K** intermediate exhibits a large absorption in the red. The isomerization speed is very high (psec). For storage applications, the low temperatures required for this process may complicate system packaging and the **M** intermediate gains importance. The formation of the **M** intermediate occurs at room temperature as a result of a series of protein conformational changes. During this process, the proton of the chromophore is transferred to an amino acid of the protein resulting in a highly blue-shifted absorption spectrum. However, the speed of formation of the **M** intermediate is much slower (0.1 msec) then the **K** intermediate and is a limiting factor in the memory writing process. In addition, under normal biological conditions, the **M** intermediate is not stable and reverts to the **bR** intermediate within

10 msec, seriously limiting the memory persistence time. However, by suspending bacteriorhodopsin in a polymer matrix in the presence of certain chemicals the lifetime of the **M** intermediate can be increased to about 30 minutes. This property, coupled with good quantum yields (0.64) in both directions, high photocyclicity (>10^6 cycles), low cost and room temperature operation may make this protein competitive to PRCs and attractive for some primary storage applications that require only short memory persistence times.

The use of a bacteriorhodopsin doped polymer as a holographic medium was originally proposed by Bunkin and coworkers.[44] Due to the different absorption spectra of the intermediates according to the Kramers-Kronig relationship, both amplitude and phase information associated with an optical 3-D interference pattern can be recorded. In places of constructive interference **bR** is driven to the **M** state and in regions of destructive interference no photochemistry is initiated. More recently, the volume holographic storage properties of bacteriorhodopsin in such polymers have been investigated experimentally by Birge and his collaborators who have recorded volume holograms with reasonable diffraction efficiencies (6%) in polymers hosting the protein.[45] The diffraction efficiency was mainly limited by the absorptive nature of the recorded volume gratings preventing the multiplexing of large numbers of holograms. Since phase only gratings will provided very high diffraction efficiency some of the present research effort seeks to eliminate absorption and to record phase only holograms at suitable wavelengths. If and when such holograms can be permanently recorded in bacteriorhodopsin, this highly light sensitive material may become the holographic recording material of choice.

5.4 Spectral hole burning

In this section, a second type of wavelength multiplexing is described which concentrates on photo induced transformations between the various ground states of a molecule and is generally called persistent spectral hole-burning (PSHB).[46] This technique can be called a true wavelength multiplexing scheme because there is complete independence between the information stored at each wavelength. In contrast, wavelength multiplexing in photorefractive crystals (see section 5.1), can suffer from crosstalk problems between the individual wavelength-multiplexed holograms. This crosstalk is due to the fact that the Bragg gratings created by a hologram recorded at one wavelength is visible by all other optical wavelength although they may not Bragg match the gratings.

The types of molecules that are being investigated for PSHB experiments have a large inhomogeneously broadened absorption band

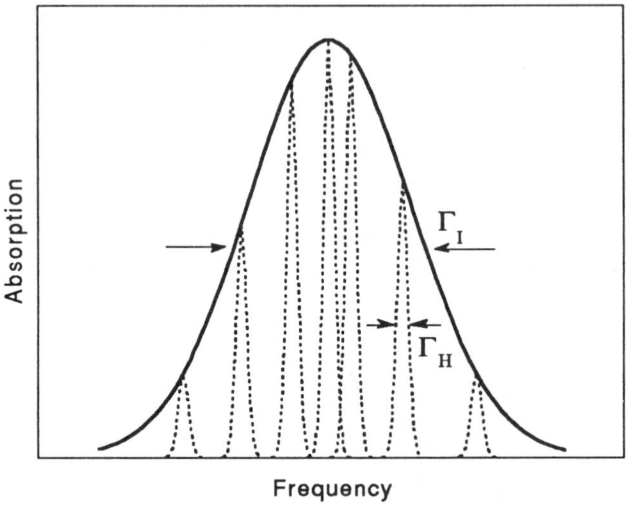

Figure 9. The absorption profile for an idealized PSHB material. The overall linewidth is defined by the inhomogeneously broadened curve of width Γ_I which is composed of a distribution of homogeneously broadened lines of width Γ_H.

comprised of a large number of narrow absorption lines as shown in Figure 9. The solid line shows the overall absorption spectrum of the material as a composite of all of the narrow homogeneously broadened absorption curves (shown as dashed lines).

To record information into the PSHB materials, a tunable wavelength, narrow bandwidth, optical source is used to illuminate the material (where source bandwidth $\Delta\omega << \Gamma_H$). The illuminating light induces many photophysical transformations which dramatically modify the absorption profile near the source energy. Figure 10 shows the absorption profile of a molecule before and after illumination with light frequencies of ω_1 and ω_2. The originally smooth profile has been altered into one with sharp dips near the illuminating frequencies, making it clear why this process has become known as *spectral hole-burning*.

The extent to which the wavelength can be used to store multiple bits of information in a PSHB material is given by the number of distinct holes that can be burned into the absorption profile. This is given by the ratio Γ_I/Γ_H, where Γ_I and Γ_H are the inhomogeneous and homogeneous linewidths, respectively (see Fig. 9). This ratio can be as high as 10^6, but only at extremely low temperatures (< 1K). This is because, in general, the homogeneous linewidth Γ_H increases with temperature[47] and can reach the point where $\Gamma_I \approx \Gamma_H$ for room temperature operation. Recent research efforts have focused on developing new materials that have a large number of spectral holes at room temperature.[48,49]

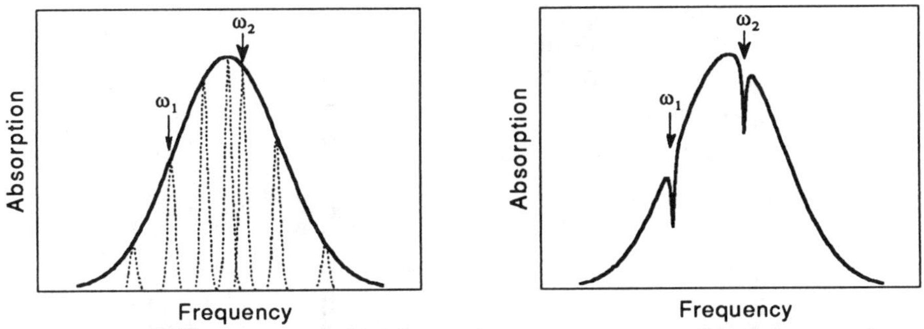

(a) Before illumination (b) After illumination

Figure 10: The effect of spectral hole burning on the absorption profile. The frequencies of the illuminating sources at ω_1 and ω_2 are indicated with arrows. (a) The smooth absorption profile for an unwritten molecule. (b) The modified absorption profile after illumination.

The original memory system proposed for this technology was a bit-oriented memory that required a thin sheet of PSHB material, a wavelength tunable laser and x-y beam deflectors. Using the two spatial dimensions and one wavelength dimension, a true 3-D memory device was achieved. The spectral holes were burned using an intense beam (write operation) and detected using a much lower laser intensity (read operation). Typically, the presence of a hole is assigned the logical value "1" and the absence of a hole the value "0". One problem with this system design is that it is quite difficult to determine if a spectral hole is present without using a variable threshold detector or other complicated scheme. Recently, absorption holography has been implemented to reduce the background intensity and greatly improve the contrast ratios.[50] Other recent advances have been made to reduce crosstalk by applying an external electric field and sweeping the optical frequency and phase while recording the hologram.[51]

PSHB is an active area of research despite the requirement of low temperature. It is hoped that future advances will raise the operating temperature and provide a storage capacity improvement of three two four orders of magnitude over that of present optical disk systems.

5.5 Photon Echo

This section describes the effect called stimulated photon echo (SPE) and how it can create 3-D memories using time as the third degree of freedom. The coherent transient effect of the SPE was first described by Mossberg as a means of storing/retrieving rapidly varying data in parallel.[52] Figure 11 shows a generalized timing sequence for storing and retrieving a 1-D time-modulated data sequence.

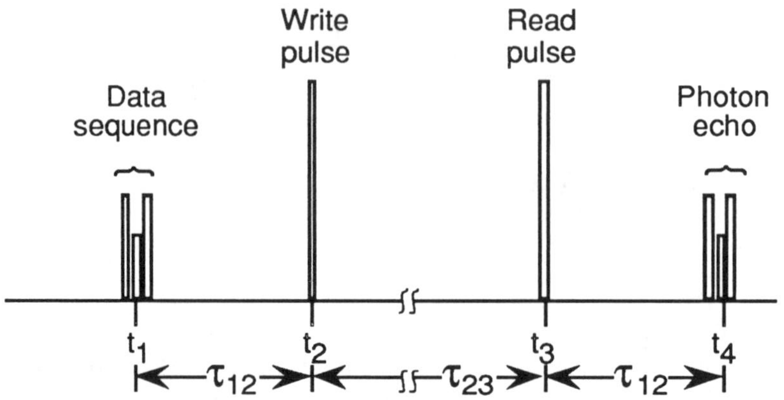

Figure 11: The timing sequences required for the storage and retrieval of time-modulated optical data using stimulated photon echo. The writing process is shown on the left as the data sequence followed by a short write pulse. The data is retrieved by applying a read pulse to stimulate the photon echo that will reconstruct the original data.

The time-modulated data sequence reaches the material at a time t_1 and excites the material according to the optical and temporal frequencies that are present in the signal. A write pulse must be applied to the material, within the homogeneous dephasing time (T_2) of the material, to create a temporal hologram within the ground- and excited- state population densities. This hologram now contains information on the temporal structure as well as the spectral structure of the data sequence. By applying the read pulse at a much later time ($\tau_{23} > T_2$), a photon echo will be emitted by the material with a delay of precisely τ_{12}. The echo can be stimulated multiple times, but each time the readout signal becomes weaker because of the disruption from the energy deposited by the readout process. Eventually, the input signal must be refreshed or the sample will return to the original ground state population density.

There are several conditions that must be satisfied to create a photon echo that will accurately represent the input data. The first requirement is that the material must be very cold (less than a few degrees K) so that all of the excitations and transitions are caused solely by the optical pulses. The second requirement is that the material have an inhomogeneously broadened absorption line (see section 5.4 on persistent spectral hole burning) that is capable of recording both spectral and spatial information.

The types of memories that have been proposed using the SPE are called spatial-temporal memories and usually store information in two spatial dimensions (x, y) and the dimension of time to locate any single memory bit. The extent to which time domain can be used is given by the amount of temporal data that can be stored/retrieved in a single echo. This value depends on the characteristics of the material in the following way. First, the minimum pulsewidth that can be recorded must be longer that the phase relaxation time T_2^*. Second, the entire data sequence and write pulse must occur in an interval of less than the overall coherence time of the excited state T_2. The ratio of T_2/T_2^* is typically in the range 100-1000.[46]

There are similarities between stimulated photon echo and persistent spectral hole burning in that they use the same types of materials and require very low temperatures. One important difference is that SPE does not require a tunable laser source. It does however require a modulated laser beam and detector that can attain speeds of up to 100 MHz.

5.6 .Multi-wavelength storage materials

Another approach to volume data access is to use the difference generated in the absorption characteristics of written molecules which absorb at different wavelengths. This mechanism is used in the multi-frequency optical volume memory[53] that is being developed in Japan. This memory consists of many layers of J-aggregate photochromic Langmuir-Blodget thin films having sharp and distinct absorption bands as shown in Figure 12. By pre-exposing the films with appropriate UV radiation the required sharp difference in the absorption spectrum of each layer can be synthesized. To write a bit, molecules are excited with a UV beam while the reading of stored bits is performed by selecting the appropriate storage layer using a wavelength tunable laser.

The capacity and density of such a memory are ultimately determined by how sharply the absorption bands are synthesized and how well the relative positions in the spectra are controlled. The success of this concept depends also on the availability of tunable laser sources with linewidths compatible with the material requirements.

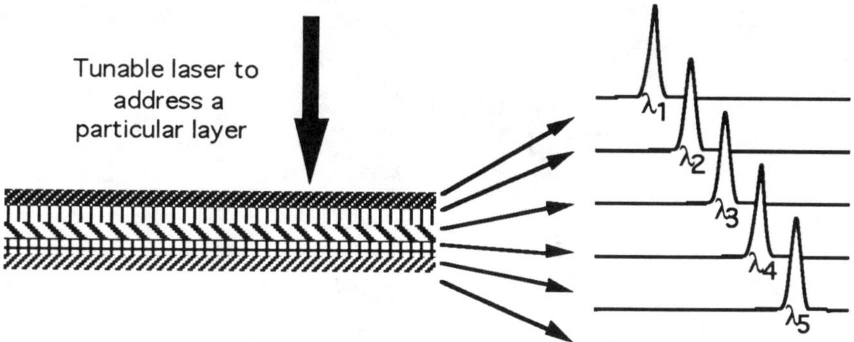

Figure 12: The hypothetical spectra of the various films used a multi-wavelength memory with many layers.

5.7 Two-photon 3-D memory concept

Two photon three dimensional memory[54,55,56] is a data storage technique that requires the simultaneous absorption of two photons in order to store information in the material. A simple diagram showing how this method can create a high capacity memory is given in Figure 13.

The physical process at the heart of the three dimensional memory (3DM) is a molecular change caused by a two photon optical absorption, as shown in Figure 14 for a spiropyran molecule. A molecule in the ground (unwritten) state is excited to a higher energy state by the simultaneous absorption of two distinct photons, one red (1064 nm) and one green (532 nm). The energy required to reach the excited state is greater than either photon alone can provide, but when two photons interact simultaneously they are absorbed, resulting in a bond dissociation. The molecular geometry (structure) is changed into a new, written, molecule with an entirely different absorption spectrum. The intensity of the infrared beam can be high, because a two photon absorption of the infrared beams has insufficient energy to write the material. However, a relatively low intensity of green light, in the presence of the intense infrared beam, can write the material.

The written bit can be read by illumination with two photons of a different energy than those absorbed by the unwritten molecule (i.e., at 1064 nm, as shown in Figure 14). The written molecule absorbs the two 1064 nm photons and fluoresces in the red, at around 700 nm. Using two photon absorption to write and read the memory makes it possible to identify a single bit anywhere within a three-dimensional volume by simply intersecting two optical beams at that point. The capacity of such a bit-oriented volume memory is fundamentally limited only by the memory volume divided by the optical spatial resolution, leading to an upper bound in data storage density as high as 10^{12} bits/cm^3, compared to 10^8 bits/cm^2 for planar storage media.

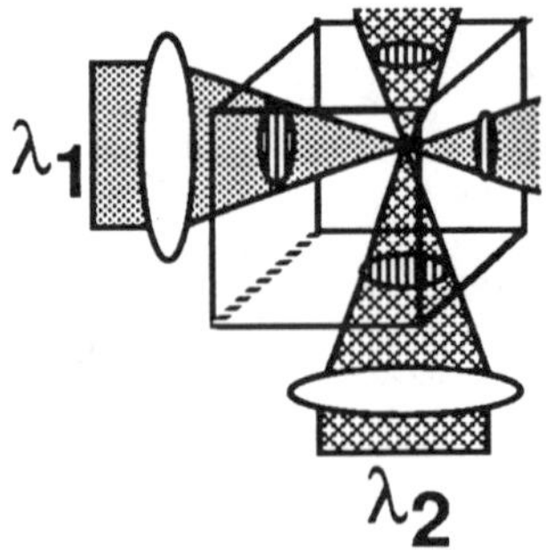

Figure 13: The schematic diagram for using the two photon effect to write throughout a volume of material.

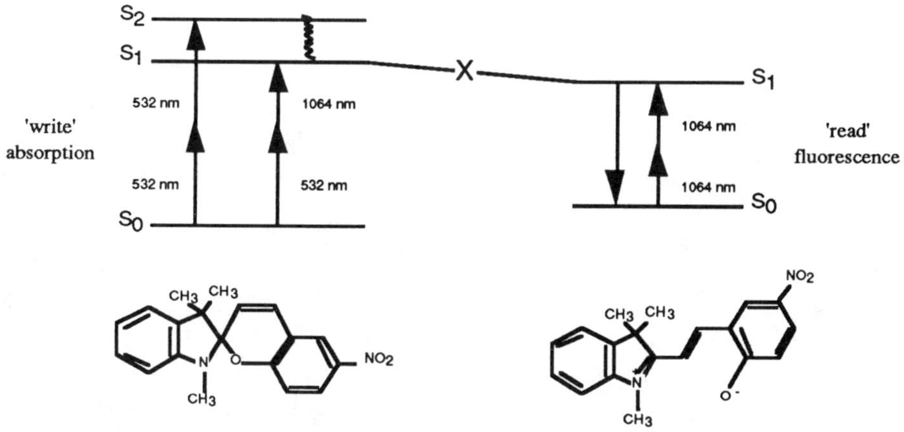

Figure 14: Two photon absorption (left) and readout (right) processes for a spiropyran molecule.

Most of the past research for the secondary storage materials was based on bond dissociation of spirobenzopyran (SP) molecules held in a polymer host. A more detailed discussion of these results can be found in a recent publication[57]. The initial challenges in materials development were to find materials that (a) could be written and read using the two photon effect and (b) exhibited the long term stability of the written form critical for secondary storage. Several promising two photon materials have been identified and incorporated into a polymer host material (PMMA) which can be shaped into any form appropriate for the optical system. The absorption and fluorescence spectra of SP1 in a polymer host are shown in Figure 15. The recording energy required for such materials is about 10 pJ/μm^3, and the fluorescence efficiency is on the order of 1%. The storage lifetime depends on temperature and the readout intensity. The dark storage lifetime is in the months at 3°C, and years at 77°K.

The written form of SP1 is a polar molecule, with a positive and negative charge on the ends of the open ring. The storage lifetime can be extended indefinitely by chemically binding the charged ends to another polar molecule or by anchoring them to the polymer matrix. We have demonstrated nearly infinite stability of the written form by using HCl to bridge the written (open) forms. These materials were tested and bits were recorded in the memory cube. The absorption and fluorescence spectra of the bridged SP1 molecule are shown in Figure 16. Note that the fluorescence peak is very well separated from the absorption peaks. This is important for practical application of the material, since if the fluorescent output were absorbed by the written material, crosstalk would occur between a recalled output and other planes of stored data. The bridged materials will be suitable for a write-once, read many times memory and can be used for secondary and archival storage systems.

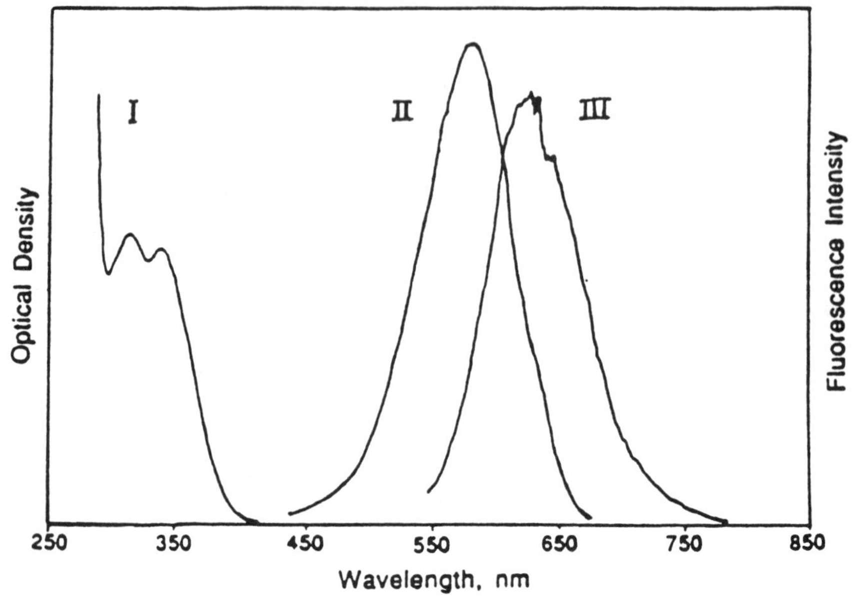

Figure 15: Spectral response for the absorption of the unwritten (I) and written (II) forms of SP1, as well as the fluorescence spectrum of the written form (III).

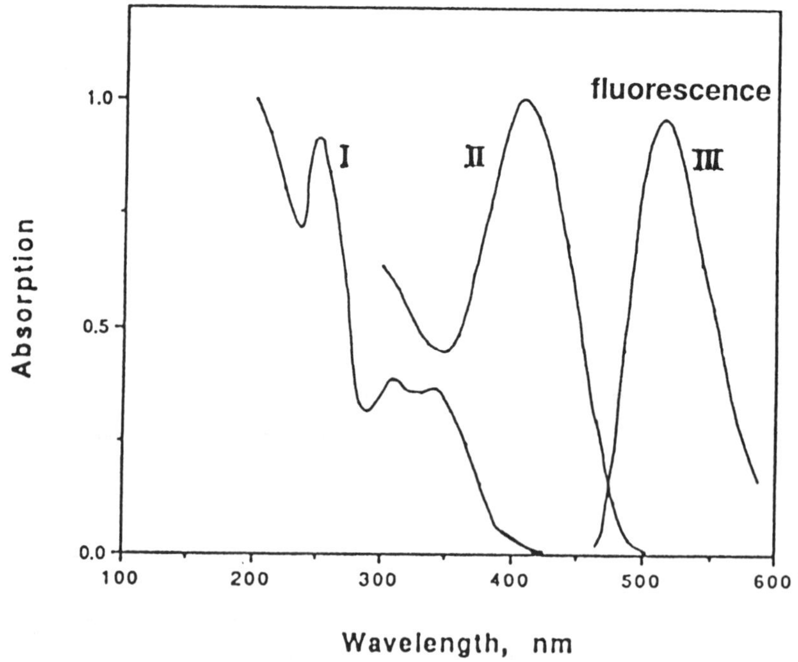

Figure 16: Spectral response for the bridged SP1 material, showing the absorption of the unwritten (I) and written (II) forms of SP1, as well as the fluorescence spectrum of the written form (III).

A second approach to using two photon absorption for 3-D memories that modifies the refractive index change in the material has also been demonstrated.[58] This method writes the information into the material using one sharply focused beam to cause two photon absorption near the focal point. Differential interference contrast microscopy is used to convert the phase modulation into an intensity modulation and read the data. Both reading and writing of the material are limited in the amount of parallel data that can be written using this second approach.

6. SYSTEM DESIGN: TWO PHOTON 3-D MEMORY

In the previous section, we have introduced several approaches to 3-D memories. In this section we focus on one approach, that of two-photon 3-D memory, and investigate potential system architecture and the required devices that will be suitable for its implementation. A two-photon 3-D memory can be addressed by simply intersecting two focused beams, but this only allows bit sequential readout. For parallel address, the data is organized into images or binary bit-planes. These bit planes can be stored within the memory using two different data addressing methods: orthogonal and counterpropagating pulse collision addressing.

6.1. Orthogonal Addressing system

Orthogonal addressing (beam intersection) is based on the interaction of a 2-D optical image containing the data to be stored entering the memory cube from one face with a second enable 1-D field that defines the bit plane at orthogonal incidence, as shown in Figure 17. This scheme can in principle use cw lasers as light power sources, although two photon processes are much more efficient at the higher intensities characteristic of pulsed laser sources. Figure 18 shows in more detail the components necessary to build an actual memory device. The principle components

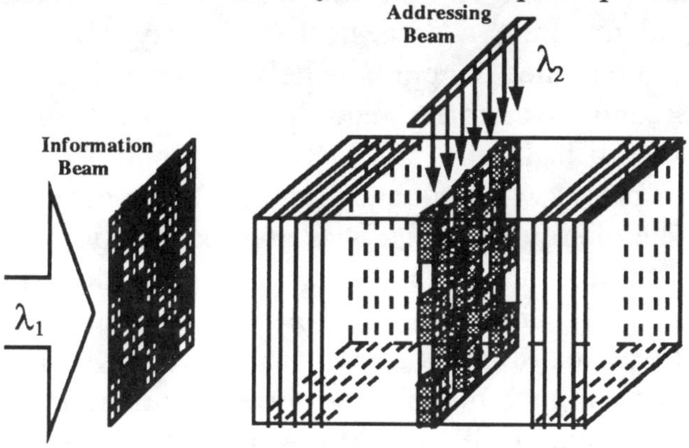

Figure 17: Image storage in two photon 3-D memory materials using orthogonal beam addressing.

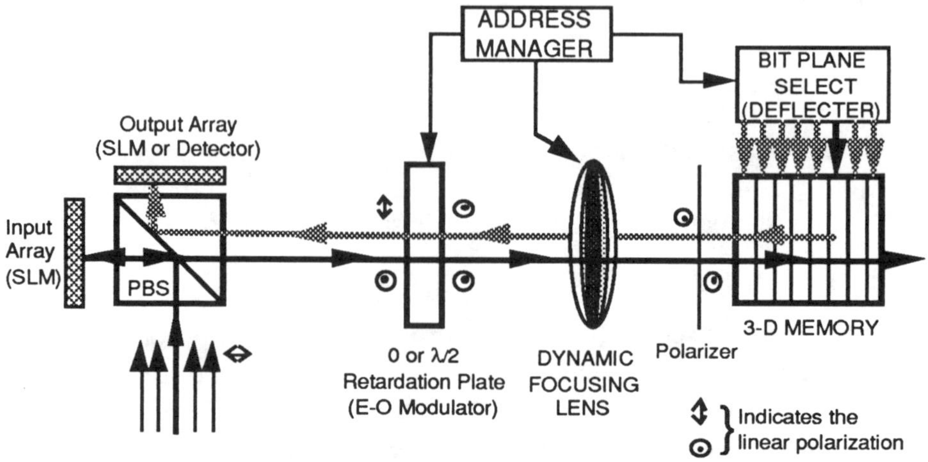

Figure 18: System diagram of orthogonal 3D memory system using orthogonal beam addressing.

required are a spatial light modulator to compose the data and a dynamic focusing lens to image the data onto the desired memory plane, a beam deflector to direct the enable beam onto the storage planes, and an output detector array.

It should be possible to store a one dimensional line of data with diffraction limited resolution and storage density. For two dimensional arrays, however, orthogonal addressing can not simultaneously provide high capacity and high data rates because the minimum width of the enable beam is restricted by Gaussian beam propagation of the cylindrical 'enable' beam. A low F/number lens will illuminate a small area with a narrow line, or a larger F/number lens can illuminate a large area with a deeper beam. The deeper the 'enable' beam, the higher the parallelism possible, but the larger the volume which must be dedicated to each bit. For example, to maintain a data storage density of 1 Gbit/cm^3, the parallelism must be limited to approximately 128x128.

The experimental system for recording an image using two photon absorption is shown in Figure 19. A flashlamp pumped Nd:YAG laser producing 30 ps pulses was frequency doubled to produce output at 1064 and 532 nm. The energy per pulse was approximately 1 mJ in both the green and the infrared pulses. However, the green output was used to illuminate a relatively large area input plane (about 100 mm^2) while the infrared output was focused to a line approximately 8 mm by 100 μm so that its intensity was much higher. The secondary memory material used was an unpolished cube of SP1 in PMMA polymer. The input image was transmitted through the cube and imaged onto the CCD camera for recording the output.

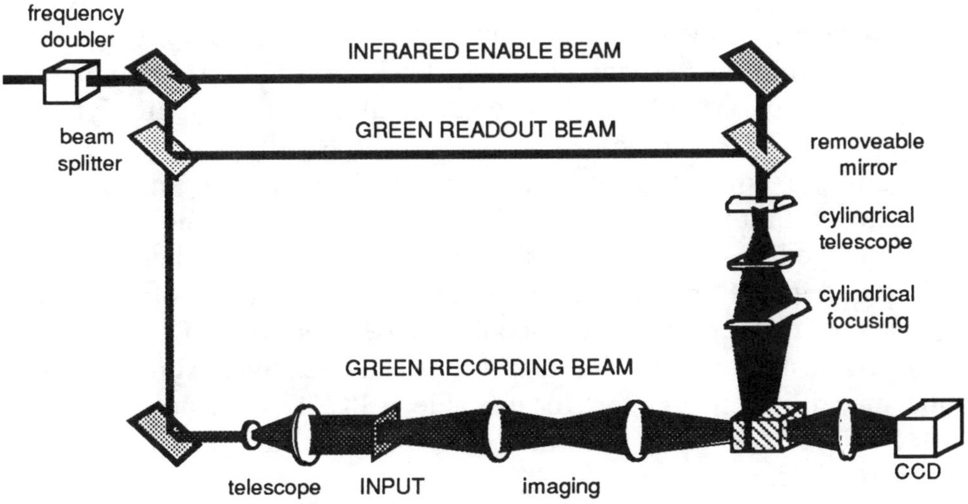

Figure 19: Experimental image recording set up.

Figure 20: Transmitted (left) and recalled (right) image stored by two photon absorption.

Figure 20a shows the transmitted image. The recording plane was selected by intersecting the infrared line with the green image beam within the cube. The images could in principle be recalled using two photon absorption of the infrared beam. However, to increase output intensity a small fraction of the green beam was diverted along the infrared path to allow recall by single photon absorption. The recalled image is shown in Figure 20b. The resolution seemed to be comparable to that of the transmitted image, although there was a significant amount of background intensity which reduced the contrast ratio.

We have chosen a dual approach in constructing the dynamic focusing lens, combining a continuously variable liquid crystal lens[59] with a discrete stepping holographic lens.[60] The holographic lens can provide quick changes between a small number of widely separate focal planes, while the liquid crystal lens provides slower access to a continuum of more closely spaced planes. The concept of the holographic dynamic focusing lens (HDFL) is shown in Figure 21. Basically, a number of lens functions are computed as computer generated holograms. Each is multiplied with a unique random phase pattern. Then the holograms are summed together and the result is fabricated as a surface relief pattern to yield a multiplexed phase-only hologram. The individual holograms are accessed by imaging one of the encoding phase patterns onto the hologram plane. The justification for this approach is that a relatively simple electro-optic spatial light modulator (for example, a PLZT wafer patterned with surface electrodes) can access the high resolution holographic lens in a very short time.

Figure 22 shows the experimental results for 4 holograms multiplexed into a single HDFL. The photo on the left is a spot array imaged through the lens illuminated with the code corresponding to the position of the CCD camera. The photo at center is the result of changing to a different code recalling a holographic lens defocused by 5.5 mm. The photo at the right is the background output generated when the lens is illuminated by a

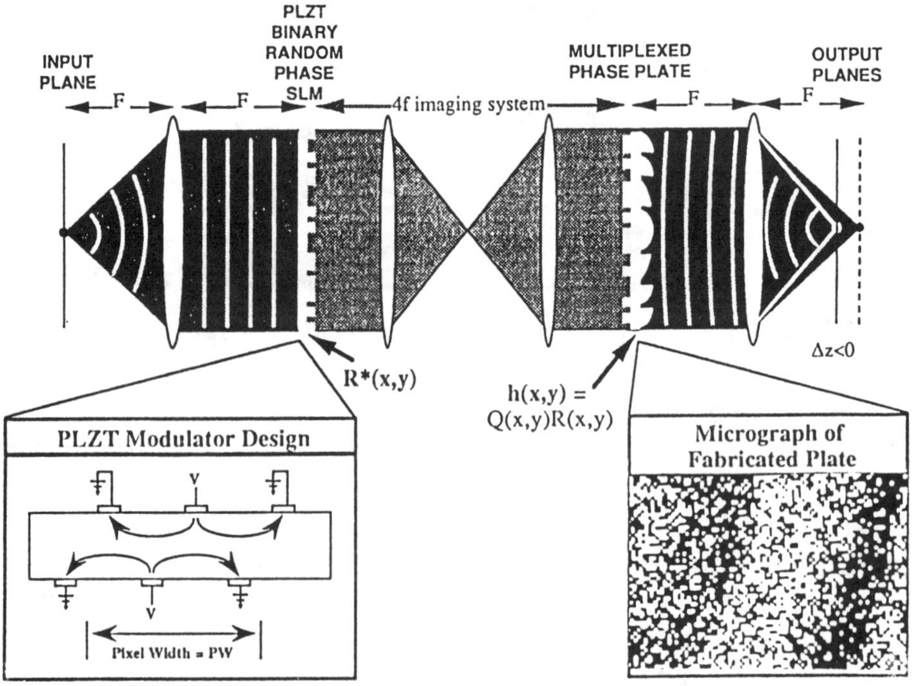

Figure 21: The concept of the holographic dynamic focusing lens.

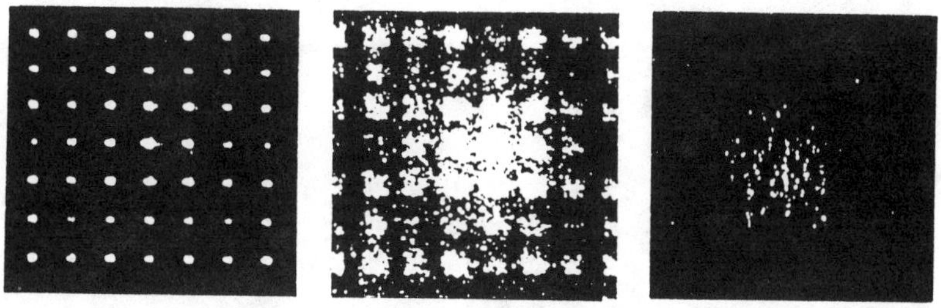

Figure 22: Experimental output of the holographic dynamic focusing lens.

phase code which doesn't match any of the stored patterns. The phase code addressing for these tests was provided using etched glass phase plates, rather than a electro-optic SLM. Using a single point input an output spot size of 8.5 μm microns was measured with a signal to background intensity ratio of greater than 100:1. The resolution and uniformity seen in Figure 22 were limited by the holographic spot array generator used rather than by the HDFL.

6.2 Pulse collision addressing system

Pulse collision addressing was devised to provide higher data transfer rates using more parallel data channels. An ultra-short pulse tunable infrared Ti:Sapphire laser provides the illumination. Infrared pulses 100 fs (1 fs = 10^{-15} s) long are routinely generated by such lasers, and pulses as short as 17 fs have been demonstrated.[61] A fraction of the laser output is frequency doubled into the green and used to illuminate a spatial light modulator (SLM) displaying the data plane for storage. This bit-plane is sent into one face of the two photon material. The rest of the laser output is sent into the memory from the opposite face. The two pulses intersect within the material, and the information is stored. The thickness of the stored plane is determined by the pulse widths. Since the speed of light in a material of index n is 0.3/n μm/fs, the intersection of the two 100 fs pulses in a material with an index of 1.5 is approximately 20 μm thick. The position of the intersection is determined by the path length of the two pulses. Each plane is addressed by applying a corresponding physical delay to one of the beams. Memory readout is also accomplished by pulse collision, using wavelengths corresponding to the absorption spectra of the written bits. An additional advantage to short pulse lasers is the increase in

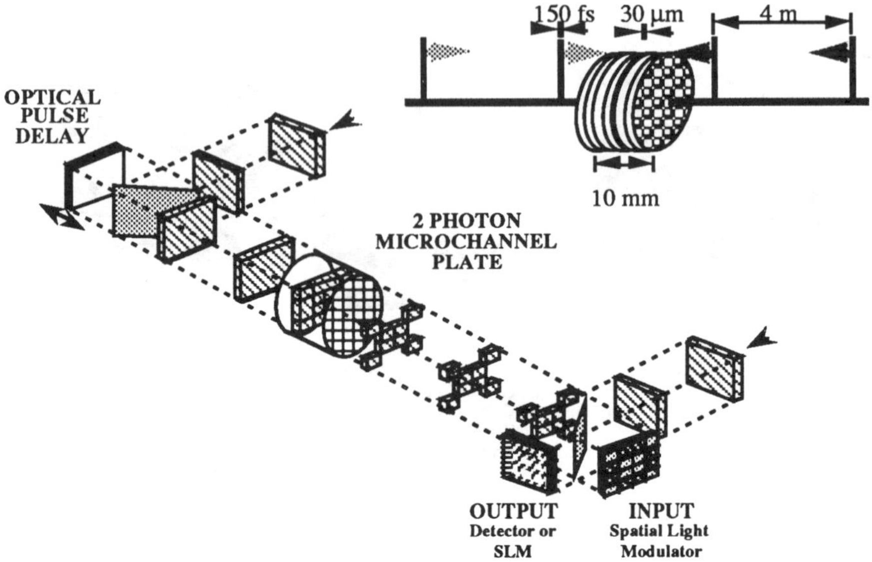

Figure 23: Two photon storage using ultrashort laser pulse collision addressing.

two photon absorption probability which occurs at the high peak intensities characteristic of such lasers. The geometry for pulse collision recording is shown in Figure 23.

In addition to the increase in data density, a major advantage of pulse collision is that the dynamic focusing lens can be eliminated without loss of resolution by incorporating the storage material into a waveguide array. Microchannel arrays used for image amplifiers are made by etching the central cores from a glass optical fiber bundle, leaving an array of hollow channels. Theliquid storage material can be drawn into the channels then allowed to harden. The resulting waveguide array will transmit an image from the front surface to the back without blurring the image.

Pulse collision addressing within the microchannel or bulk material is accomplished by applying a physical path delay to one of the recording or readout beams. One possible method of doing this uses an acousto-optic modulator to reflect the beam from stepped mirror. The short pulse beam has a significant color spectral width ($\Delta\omega$ is about 1% for a 100 fs beam) which increases as the pulse width decreases. The acousto-optic modulator changes this color spectrum into an angular spectrum which could smear the beam. However, using a two-pass geometry corrects for this dispersion exactly. The result can be a cascadeable random access pulse delay with up to 1 MHz access speeds.

For the two photon memory using pulse collision addressing, the data retrieval (access) time is limited by the speed at which the position of the colliding pulses can be controlled. We propose to develop an optical pulse

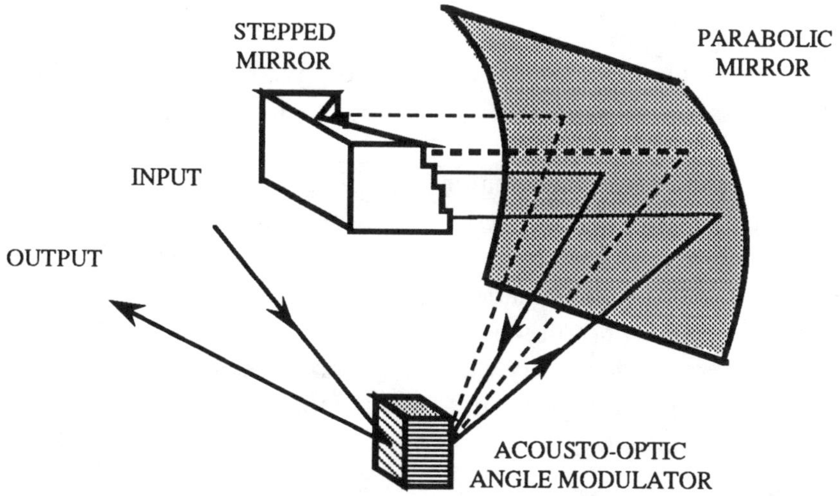

Figure 24: Achromatic short-pulse optical pulse delay using acousto-optic modulation.

delay (OPD) that can provide an access time down to 0.1μs. The characteristics of an ultra-short pulse laser output (high peak intensity and a 5-10 nm wavelength bandwidth) render most approaches impractical. Figure 24 shows a proposed OPD device which can provide the necessary performance. The input pulses diffract from a small aperture acousto-optic deflector. The diffracted light reflects from a parabolic mirror onto a V-grooved stepped mirror constructed with the required delay settings. The reflected light is re-diffracted by the A-O cell, automatically correcting for diffractive pulse dispersion. Currently available commercial A-O cells are available with 5 MHz access time and 112 resolvable spots. Using two such units, cascaded together, an OPD could yield over 1024 distinct intersection planes within the memory material. For high speed operation near the 100 MHz range, standard A-O cells have fewer resolvable spots (<10). Therefore several units must be cascaded to increase the total number of addressable memory planes to reach the desired number of planes. Ultimately, the entire device could be constructed with integrated optics using surface acoustic wave (SAW) modulators to increase operation speed, reduce the size, and enhance reliability.

7. CONCLUSIONS

It is expected that over the next few years for some specific applications areas such as associative memories a fast transition from sequentially addressed optical disk systems to parallel accessed disk systems will occur to satisfy their high data rate and medium capacity requirements. However, 3-D optical memories may have a profound impact on the present storage hierarchy. A considerable amount of research and development over the next decade will be needed however, to take 3-D memories from concept to technology. We strongly believe that such an effort is well justified by the applications in fast, high capacity storage.

At this early stage, when major improvements in 3-D optical material characteristics are expected, it is premature to assess the far reaching technological implications of many of the 3-D memory concepts. However, the fast progress made over the past few years in PSHB, multi-wavelength and two-photon storage raises our hope that by the turn of the millennium Terabyte capacity optical memories could become available with access times of less than a millisecond and data transfer rates exceeding Terabits/sec.

8. ACKNOWLEDMENTS

The authors would like to acknowledge the following people for their contributions to this paper: W. Randall Babbitt, Robert Birge, Alan Craig, Sasha Dvornikov, Philip Henshaw, Al Jambordino, Ravindra Kachru, Philippe Marchand, Ram Piyaket, Demetri Psaltis, Peter Rentzepis, James Strickler, and Walt W. Webb.

9. REFERENCES

1 A. E. Bell, "Critical issues in high-density magnetic and optical data storage," Laser Focus **19**, 61-66, (1983).

2 K. Sato, K. Fujita, M. Miyazawa, M. Shirai, K. Kobayashi, M. Ishihara, T. Nakao, "A system-integrated ULSI chip containing 11 4Mb RAMs, 6 64kb SRAMs and an 18k gate array," 1992 IEEE International Solid-State Circuits Conference, 52-53, (1991).

3 H. Suguira, E. Morita, S. Nagasawa, "F6631 solid state disk: high speed virtual disk unit," Fujitsu Scientific and Technical Journal 26, 296-305 (1991).

4 S. A. Przybyski, *Cache and Memory Hierarchy Design*, Morgan Kaufmann Publishers, Inc., San Mato, CA, 1990.

5 E. G. Ramberg, "Holographic Information Storage," RCA Review **33**, 5-53 (1972).

6 K. Kubota, Y. Ono, M. Kondo, S. Sugama, N. Nishida, and M. Sakaguchi, "Holographic disk with a high data transfer rate," Applied Optics **19**, 944-051 (1980).

7 D. Psaltis, M. Neifeld, A. Yamamura, and S. Kobayashi, "Optical memory disk in information processing," Applied Optics **29**, 2038-2057 (1990).

8 J. Rilum and A. Tanguay, "Utilization of optical memory disk for optical information processing" in *OSA annual Meeting*, Vol. **11** of 1988 OSA Technical Digest Series (Optical Society of America,

Washington, D.C., 1988), paper M15.

9 P. Marchand, A. Krishnamoorthy, K. Urquhart, P. Ambs, S. Esener, and S. H. Lee, "Motionless-head parallel readout optical disk system," Applied Optics **32**, 190-203 (1993).

10 W. P. Altman, G. M. Claffie, and M. L. Levene, "Optical storage for high performance applications in the late 1980s and beyond," RCA Engineer **31**, 46-55 (1986).

11 5 1/4" Write-Once, Read-Many Optical Disk Specification Sheet, Daicel Chemical Industries, Torrance, CA, 1991.

12 A. Krishnamoorthy, P. Marchand, G. Yayla, and S. Esener, "Opto-electronic associative memory using parallel readout optical storage," UCSD Internal Rep. (University of California, San Diego, La Jolla, CA, 1991) and submitted for publication in IEEE Transactions on Neural Networks.

13 W. J. Smith, *Modern Optical Engineering*, Chapter 6, (McGraw-Hill, Inc. 1990).

14 R. G. Zech, "Volume hologram optical memories," Optics and Photonics News, 16-24, (Aug. 1992).

15 R. Collier, C. B. Burckhardt, and L. Lin, *Optical Holography*, (Academic Press, 1971).

16 L. Solymar and D. J. Cooke, *Volume Holography and Volume Gratings*, (Academic Press, 1981).

17 P. M. Hariharan, *Optical Holography*, (Cambridge University Press, 1984).

18 F. T. S. Yu, S. Wu, A. Mayers, S. Rajan, and D. A. Gregory, "Color holographic storage in LiNbO3," Opt. Comm. **81**, 348-352 (1991).

19 P. J. van Heerden, "Theory of optical information storage in solids," Appl. Opt. **2**, 393-400 (1963).

20 I. J. Cox and C. R. J. Sheppard, "Information capacity and resolution in an optical system," JOSA A, **3**, 1152-1158 (1986).

21 K. Bløtekjaar, "Limitations on holographic storage capacity of photochromic and photorefractive media," Applied Optics **18**, 57-647 (1979).

22 F. Mok, D. Psaltis, and G. Burr, "Spatially- and angle- multiplexed holographic random access memory," SPIE Proc. **1173**, 334-345 (1992).

23 D. Psaltis, D. Brady, and K. Wagner, "Adaptive optical networks using photorefractive crystals," Applied Optics **27**, 1752-1759 (1988).

24 A. Marrakchi, W. M. Hubbard, S. F. Habiby, and J. S. Patel, "Dynamic holographic interconnects with analog weights in photorefractive crystals," Opt. Eng. **29**, 215-224 (1990).

25 Y. Taketomi, J. Ford, H. Sasaki, J. Ma, Y. Fainman, and S. H. Lee, "Incremental recording for photorefractive hologram multiplexing,"

Opt. Lett. **16**, 1774-1776, 1991.

26 P. Gunter and J. P. Huignard, Eds., *Photorefractive materials and their applications I and II*, (Springer–Verlag, 1988 and 1989).

27 G. C. Valley and M. B. Klein, "Optimal properties of photorefractive materials for optical data processing," Opt. Eng. **22**, 704-711 (1983).

28 G. C. Valley, M. B. Klein, R. A. Mullen, D. Rytz and B. Wechsler, "Photorefractive materials," Ann. Rev. Mater. Sci. **18**, 165-188 (1988).

29 H. Kurz, "Photorefractive recording dynamics and multiple storage of volume holograms in photorefractive $LiNbO_3$," Optica Acta **24**, 463-473 (1977).

30 P. Gunter and A. Krumins, "High-sensitivity read-write volume holographic storage in reduced $KNbO_3$ crystals," Appl. Phys. **23**, 199-207 (1980).

31 D. Von der Linde, A. M. Glass and K. F. Rodgers, "High–sensitivity recording in KTN by two–photon absorption," Appl. Phys. Lett. **26**, 22-24 (1975).

32 J. E. Ford, Y. Fainman, and S. H. Lee: "Enhanced photorefractive performance from 45°-cut $BaTiO_3$," Applied Optics **28**, 4808-4815 (1989).

33 J. E. Ford, J. Ma, Y. Fainman, S. H. Lee, et al, "Multiplex holography in SBN:60 with applied field," JOSA A **9**, 1183-1192 (1992).

34 J. W. Burgess, R. J. Hurditch, C. J. Kirby, and G. E. Scrivener, "Holographic storage and photoconductivity in PLZT ceramic materials," Applied Optics **15**, 1550-1557 (1976).

35 F. Mok and H. M. Stoll, "Holographic inner product processor for pattern recognition," Proc. SPIE **1701**, 312 (1992).

36 R. V. Johnson and A. R. Tanguay, Jr., "Fundamental physical limitations of the photorefractive grating recording sensitivity," Chapter 3, *Optical Processing and Computing*, (Academic Press, 1989).

37 A. J. Smirl, K. Bohnert, G. C. Valley, R, A. Mullen, and T. F. Boggess, "Formation, decay, and erasure of photorefractive gratings written in $BaTiO_3$ by picosecond pulses," JOSA B **6**, 606-615 (1989).

38 D. L. Staebler, W. J. Burke, W. Phillips, and J. J. Amodei, "Multiple storage and erasure of fixed holograms in Fe-doped $LiNbO_3$," Appl. Phys. Lett. 26, 182-184 (1975).

39 B. J. Chang, "Dichromated gelatin holograms and their applications," Opt. Eng. **19**, 642-648 (1980).

40 W. K. Smothers, B. M. Monroe, A. M. Weber, and D. E. Keys, "Photopolymers for holography," *Practical Holography IV, SPIE OE/Lase Conference Proceedings* **1212-03**, (1990).

41 A. M. Weber, W. K. Smothers, T. J. Trout, and D. J. Mickish, "Hologram recording in Du Pont's new photopolymer materials," *Practical Holography IV, SPIE OE/Lase Conference Proceedings* **1212-04** (1990).

42 R. R Birge, "Photophysics and molecular electronic applications of the rhodopsins," Annu. Rev. Phys. Chem. **41,** 683-733 (1990)

43 R.R. Birge, Biochim. Biophys. Acta **1016**, 293-327 (1990)

44 F. V Bunkin, N.N. Vsevlodov, A.B. Druzhko, B.I. Mitsner, A.M. Prokhorov, V.V. Savranskii, N.W. Tkachenko and T.B. Shechenko, Sov. Tech. Phys. Lett. **7**:630-631 (1981).

45 R.B. Gross, K. C. Izgi and R.R. Birge, SPIE **1662**, Image Storage and Retrieval Systems, pp 186 (1992).

46 W.E. Moerner, ed., *Persistent Spectral Hole-Burning : Science and Applications* (Springer-Verlag, Berlin 1988).

47 C. De Caro, A. Renn, U. P. Wild, "Spectral hole-burning - applications to optical image storage," Berichte Der Bunsen Gesellschaft Fur Physikalische Chemie **93**, 1395-1398 (1989).

48 C. Brauchle, "Spectral hole burning at room temperature and with a single molecule - 2 new perspectives," Angewandte Chemie-International Edition in English **31**, 426-429 (1992).

49 S. Arnold, C. T. Liu, W. B. Whitten, J. M. Ramsey, "Room-temperature microparticle-based persistent hole-burning spectroscopy," JOSA B-Optical Physics **9**, 819-824 (1992).

50 C. De Caro, A. Renn, U. P. Wild, "hole burning, stark effect, and data storage 2. Holographic recording and detection of spectral holes," Applied Optics **30**, 2890-2989 (1991).

51 B. Kohler, S. Bernet, A. Renn, U. P. Wild, "Holographic optical data storage of 2000 images by photochemical hole burning," Persistent Spectral Hole-Burning: Science and Applications, OSA Conference Monterey, California **16**, 46-49 (1991).

52 T. W. Mossberg, "Time-domain frequency-selective optical data storage," Optics Letters **7**, 77-79 (1982).

53 International Symposium on Future Electron Devices - Bioelectronic and Molecular Electronic Devices - [FED BED/MED SYMPOSIUM], November 20-21, 1985, Tokyo.

54 D. M. Parthenopoulos and P. M. Rentzepis, "Three-dimensional optical storage memory," Science **245,** 843-845 (1989).

55 S. Hunter, F. Kiamelev, S. Esener, D. A. Parthenopoulos, and P. M. Rentzepis, "Potentials of 2-photon based 3-D optical memories for high performance computing," Applied Optics **29**, 2058-2066 (1990).

56 A. S. Dvornikov, S. Esener, and P. Rentzepis, "3-Dimensional Optical Storage Memory by means of two-photon interaction," Optical Computing, S. H. Lee and J. Jahns, Ed., Wiley Academic Press

(1993).

57 P. M Rentzepis, A. S. Dvornikov, "Three dimensional optical memory by means of two-photon processes," presented at OE-LASE 1993, SPIE Proceedings **1153**, Los Angeles, CA (1993).

58 J. H. Strickler, W. W. Webb, "3-dimensional optical data storage in refractive media by 2-photon point excitation," Optics Letters **16**, 1780-1782 (1991).

59 P. Brinkley, S. Kowel, and C. Chu, "Liquid crystal adaptive lens: beam translation and field meshing," Applied Optics **27**, 4578 (1988).

60 S. Hunter and S. Esener, "Dynamic focusing lens for volume memory applications," SPIE Proceedings **1773**-15 (1992).

61 C. P. Huang, M. T. Asaki, S. Backus, M. Murnane, H. Kapteyn, and H. Nathel, "17-fs pulses from a self-mode-locked Ti:sapphire laser," Optics Letters **17**, 1289-1291 (1992).

"OPTICAL PATTERN RECOGNITION"

DAVID CASASENT
Carnegie Mellon University
Department of Electrical and Computer Engineering
Center for Excellence in Optical Data Processing
Pittsburgh, PA 15213

ABSTRACT

We discuss how one optical processor (a correlator) can be used for all levels of scene analysis (low, medium, and high-level computer vision). This is achieved by the use of different filter functions for the different levels of a hierarchical inference system. New optical processor filter research that allows such flexibility is advanced and examples of each of these filters are provided. For large class problems, feature extracted from each region of interest in a scene are fed to a neural net processor which performs recognition. New algorithms for optical neural net classifiers are also required. We conclude that present hardware optical correlator advances can significantly benefit from such processing.

1. INTRODUCTION

We consider the general pattern recognition problem in which multiple objects are present in different orientations in a clutter background. The problem is to locate each object and to identify its class. We consider a hierarchical inference processor approach to this problem and the role for optics in each level of this system.

1.1 Architecture

The major optical architecture we consider is a correlator (with an adaptive filter (Figure 1a) or with a set of different filter functions fixed on film and with the filter used selected by the input laser diode (at P_0) activated (Figure 1b) [1]. A portion of a present DARPA program [2] considers optical correlator hardware on which the new filters to be described can be implemented. The use of different filter functions to address the different levels of scene analysis is very attractive, since the same hardware can then be used and thus the system becomes multifunctional [3].

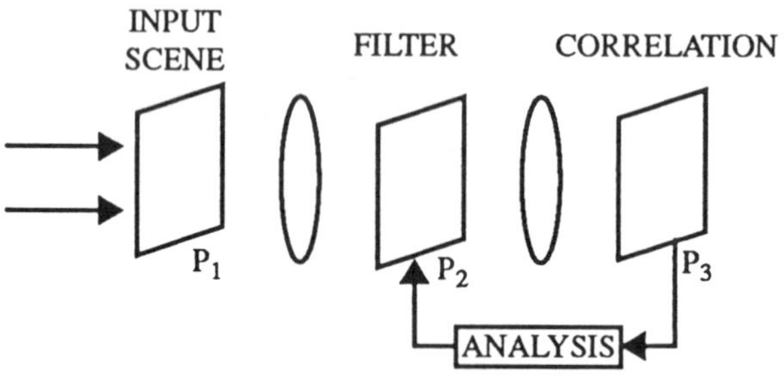

(a) Adaptive filter

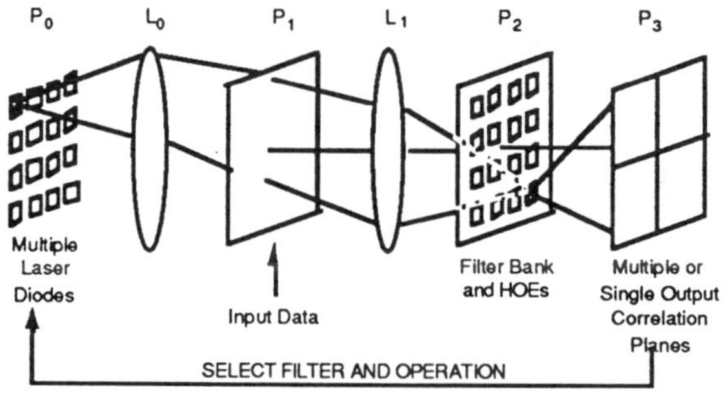

(b) Fixed filter bank

Fig. 1. Optical correlator architectures.

1.2 Hierarchical Inference Approach

Figure 2 shows the levels of processing used in this approach. They are similar to those [4] used in automatic target recognition (ATR), but their implementation and the specific algorithms are novel because of their correlation implementation. Sections 2-5 highlight each level of processing in Figure 2 and provide initial examples of the data for each. A key new aspect of this presentation is that all required operations in the different levels of the system in Figure 2 can be realized on the same architecture (a correlator) with the different new filter functions described. This is vital as it yields a multifunctional general purpose correlator processor that is programmable (with different filter functions) for all necessary image processing functions. Section 6 notes the same optical correlator system of Figure 1 can also generate different feature spaces. These with classifiers allow the identification of a larger set of objects. Section 6 also discusses several net neural net classifiers, their use in large class problems and the role for optics in neural nets.

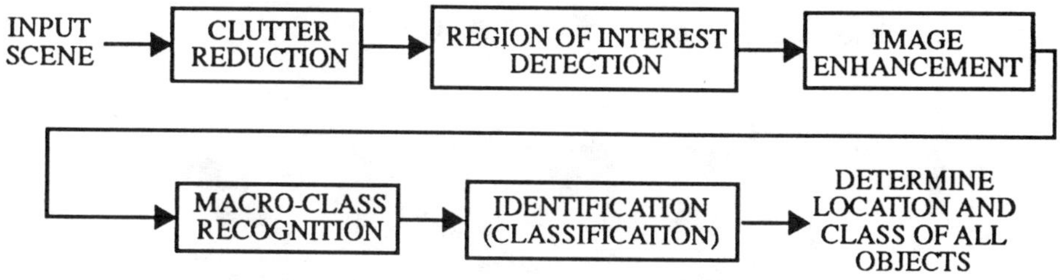

Fig. 2: Hierarchical inference processor levels for correlation filters.

2. CLUTTER REDUCTION

For the most difficult problems of interest, the first step is to reduce clutter in the input scene. This simplifies all future steps, especially detection. For this level, we have found a sequence of nonlinear morphological operations and wavelet transforms to be most attractive. These are detailed elsewhere [5] in this conference. They require gray scale operations. The morphological operations alone are suitable for low-level clutter and to decrease large fluctuation differences on the background (e.g. mountains and sky). They also have the advantage that they convert hot (white on black) and cold (black on white) objects into hot (bright) targets, thus greatly simplifying future steps (especially detection). Figure 3 shows data from such clutter reduction processing. The input (Figure 3a) contains objects of different contrast and background differences (mountains and sky). The preprocessing removes such background variations and converts all objects to bright (white on black) as shown in Figure 3b.

In high clutter cases (Figure 4a), morphological processing alone is not sufficient (Figure 4b). But, wavelet processing locates clutter (Figure 4c) and allows it to be subtracted from the scene improving results (Figure 4d) and enabling subsequent processing (Section 3) to easily locate only the 6 objects with few false alarms (Figure 4e).

These low-level computer vision functions require shift-invariance (a correlator), since the same clutter removal (morphological) or clutter detection (wavelet) filters must be applied to each local scene region and the output data is 2-D.

(a) Input scene

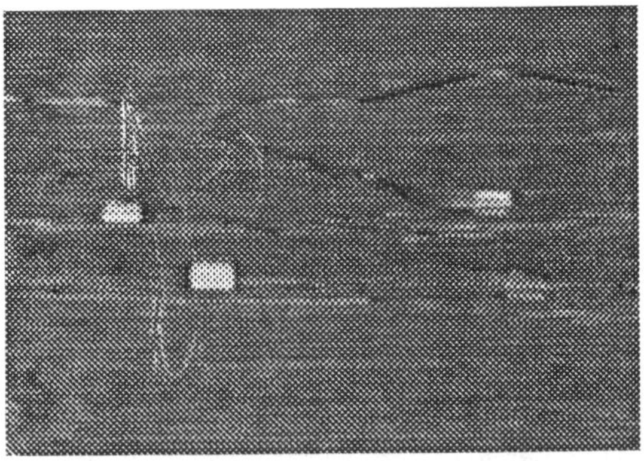

(b) Clutter reduced and consistent object polarity.

Fig. 3: Low clutter removal and consistent object polarity preprocessing.

3. DETECTION

In this stage of the processor, we detect candidate regions of interest in the scene. This is necessary to reduce data (bandwidth compression) regions and image areas that must be subsequently processed for locating and identifying objects. The goal of the detection stage of our hierarchical processor is to locate candidate regions of the scene that may contain objects of interest. This must be achieved in the presence of object distortions and must be achieved for all objects (i.e. we do not attempt to identify the object class in this level of our processor).

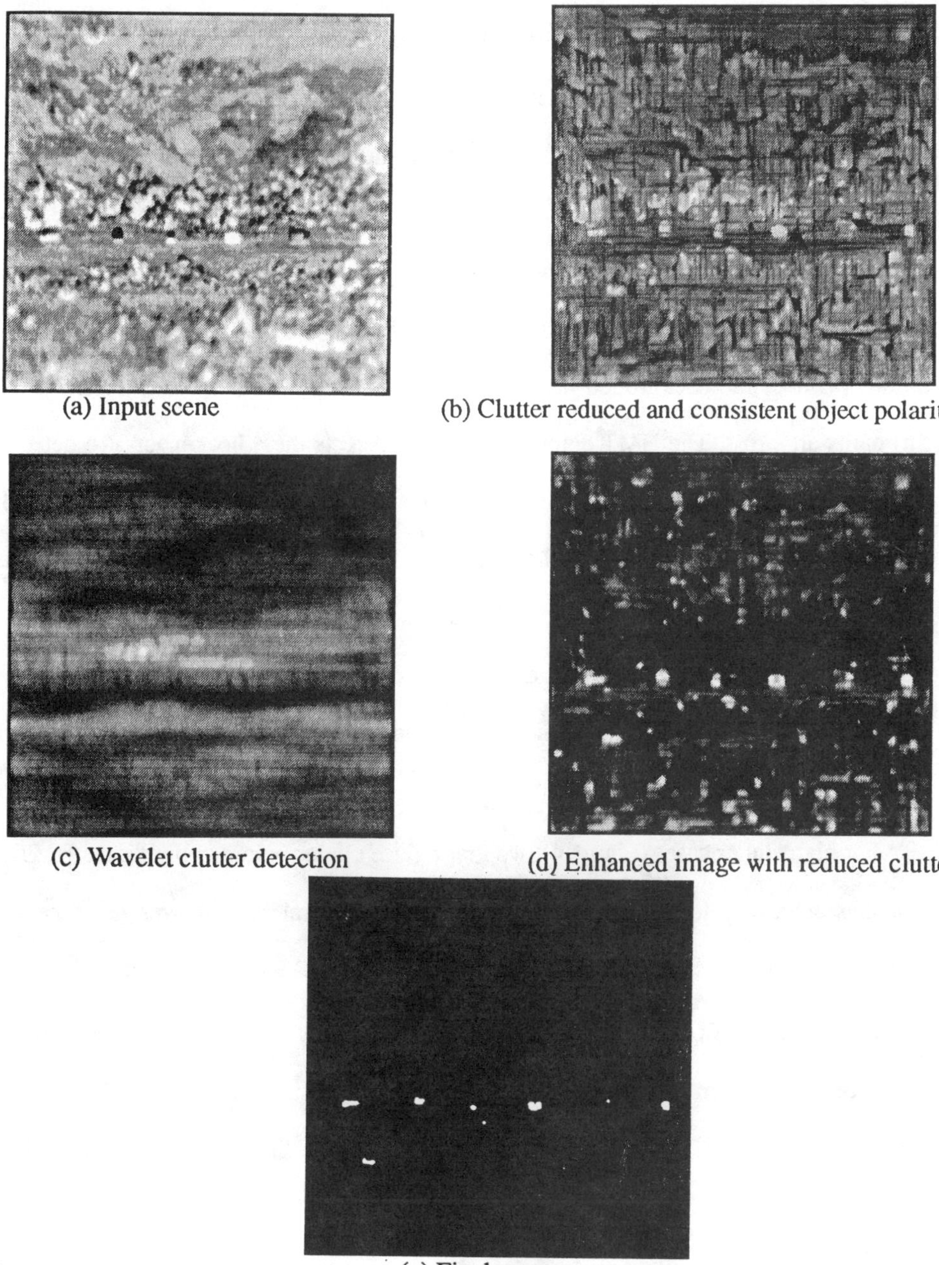

(a) Input scene

(b) Clutter reduced and consistent object polarity

(c) Wavelet clutter detection

(d) Enhanced image with reduced clutter

(e) Final output

Fig. 4: High clutter morphological and wavelet processing.

The techniques employed include the hit-miss transform (HMT) [6]. This involves the correlation of an input scene (Figure 5a) thresholded above the mean (Figure 5b) with a hit filter (Figure 5c). The result (Figure 5d) locates all scene regions of size equal to the central hit filter's rectangular area or larger. The complement of the input image (Figure 5e) is then correlated with a miss filter (Figure 5f) and its output (Figure 5g) locates all image regions equal to the larger miss central rectangular area or smaller. The intersection (AND) of the hit and miss outputs locates all bright object regions between the smallest and largest (Figure 5g). This filter works and significantly improves any correlator, since it uses contributions from both the object and the background border region around it. As seen from Figure 5d, the standard single rectangular object filter produces many false alarms (as does the background filter in Figure 5g). However, their intersection (Figure 5h) omits all errors. The HMT output (Figure 5h) detects the 4 hot objects. To detect the cold objects, we threshold the input below the mean and apply a miss-hit transform (MHT) to it. This detects all 3 cold objects. The union of this and Figure 5h contains 7 peaks at all object regions (Figure 5i). To facilitate detection of low contrast targets, we threshold the original scene at $\mu \pm \sigma$ (and $\mu \pm 0.3\sigma$, where μ and σ are the mean and standard deviation (local) of the scene). The hit and miss correlation thresholds are varied and thus this is a rank order HMT [6].

For high clutter cases (Figure 4), we first reduce clutter before applying the HMT and can thus achieve detection in severe clutter cases. Several false alarms are acceptable (Figure 4e) since subsequent levels of the system (Figure 2) will remove them.

Since this stage of our hierarchical processor requires searching all scene regions, a shift-invariant system (a correlator) is necessary. Due to the wide range of objects and object distortions and the presence of clutter, only simple filters (not recognition filters) are employed.

A second technique that we have found to be of use in extracting targets is the Gabor transform [7]. The Gabor functions are a complex linear exponential modulated by a

(a) Input scene

(b) Thresholded above mean

(c) Hit filter (h)
(not to scale)

(d) Hit correlation (regions ≥ H)

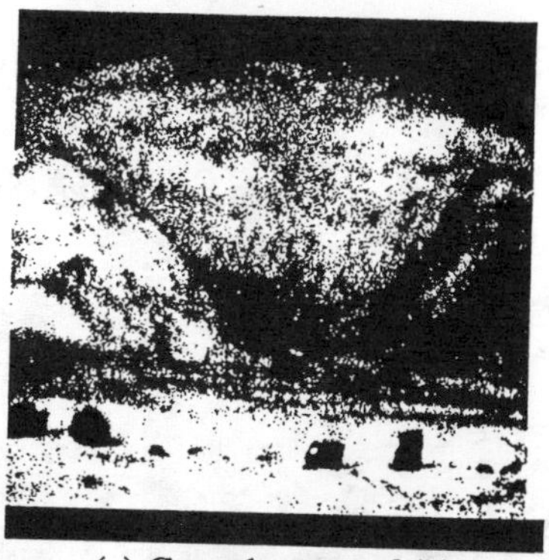

(e) Complement of (b)

(f) Miss filter (M)
(not to scale)

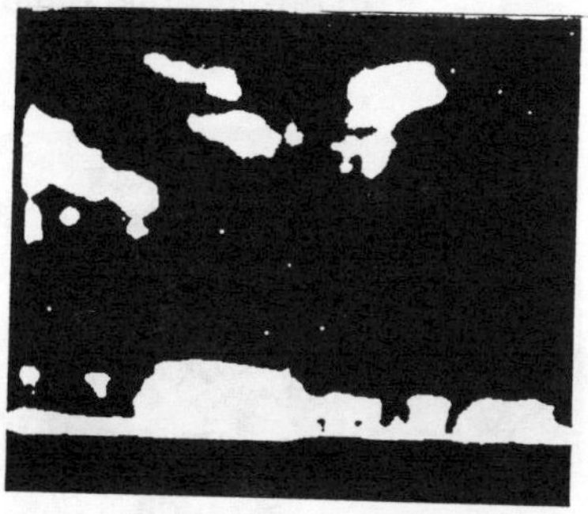

(g) Miss correlation
(regions ≤ M)

(h) Intersection of (d)
and (g)

(i) Union of HMT and MHT

Fig. 5: Hit miss detection of candidate regions of interest.

Gaussian envelope. This linear filter produces an output at each local scene region proportional to the amount of a given radial spatial frequency at a given angle in a given local region (these are parameters of the Gabor function). We use a linear combination of several such Gabor functions and have found this to be most necessary to reduce the number of 2-D output planes to be processed and thus to simplify output data analysis. We have achieved detection of all distorted versions of several different objects with the same combination filter and have demonstrated detection of very low contrast objects also and have noted the similarity of this filter to the HMT and the advantages of it over the HMT. We presently advocate use of both filters and fusion of the output data from both.

4. IMAGE ENHANCEMENT

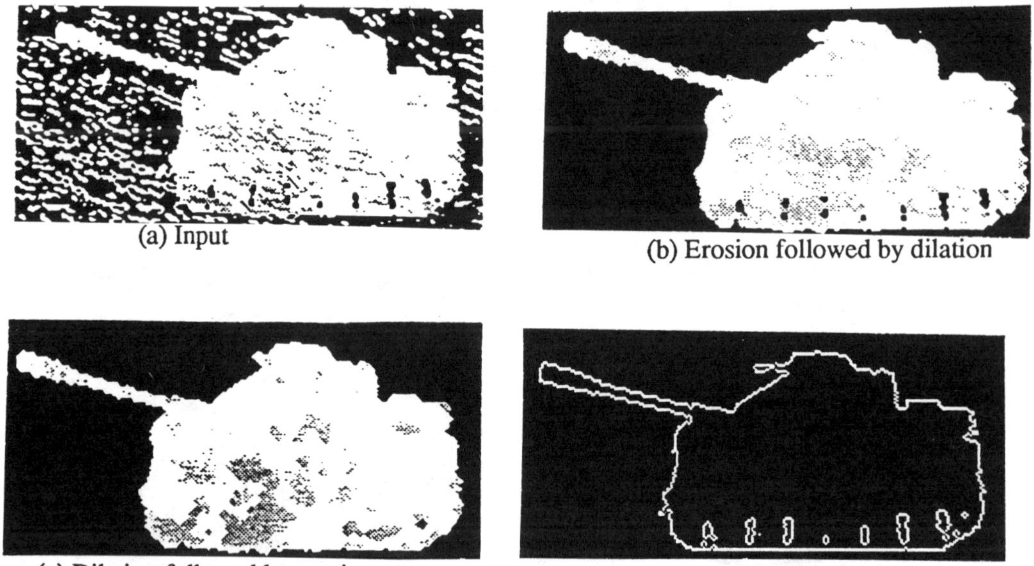

(a) Input

(b) Erosion followed by dilation

(c) Dilation followed by erosion

(d) Edge image

Fig. 6: Morphological image enhancement filters.

Once candidate regions of interest (ROIs) have been located (in detection), we then enhance each. This stage employs the morphological operations of erosion and dilation and combinations of them (opening and closing). These operations are also used to reduce clutter. They comprise all low-level computer vision functions and thus their implementation on a correlator [8,9] is quite noteworthy. Figure 6 shows examples of the different basic operations on the input image of Figure 6a. To remove noise, we perform an erosion. This involves a correlation with a structuring element (SE) filter and a high nonlinear thresholding of the output correlation plane. The size of the SE is chosen to be larger than the size of the noise clutter removed (the SE is chosen to be less than the size of the smallest feature on the object, the gun). Since an erosion distorts the object boundary, we follow it with a dilation (a correlation with the same SE with a low nonlinear threshold). The result is shown in Figure 6b. To fill in holes on the object, we dilate with an SE filter whose size determines the size of the holes to be filled in on the object. To restore the boundary, we follow the erosion with a dilation. The result (Figure 6c) has object holes filled in. Edge enhancement is also possible (Figure 6d) by forming the difference of an eroded and dilated image. These noise removal and image enhancement operations are generally necessary prior to determining the class of the object in each ROI and if each ROI contains an object or clutter. With the full scene at the input of a correlator, we can apply these operations to all ROIs in parallel (i.e. we need not isolate individual input ROIs, rather we only consider specific output correlation plane ROI regions). This is attractive, since the correlator input f remains unchanged and only the filter h changes and the local nonlinear morphological operations are applied in parallel to all local scene regions.

5. MACRO-CLASS RECOGNITION AND CLASS IDENTIFICATION

For the most general large class problems, one would isolate the individual ROIs, extract features from them, and feed these features to a classifier for identification of the object in each ROI location [3]. In this paper, we now advance how correlators with new advanced distortion-invariant filter sets can be used to aid ROI location and to further process each ROI (to determine macro-class information such as object size or orientation) and to provide class identification in small class problems. The ability to provide macro-class information is attractive as it significantly reduces the search space required of the classifier.

The filters we consider are our new minimum noise and correlation energy (MINACE) filters [10]. All advanced distortion invariant filters use a training set of images of representative true class objects to be located and impose a correlation peak constraint that the filter $\underline{H}$ (in the frequency domain) give a correlation peak value of 1.0 (the elements of the vector $\underline{u}$ are all 1) for all training set images (the columns of the matrix $\underline{\underline{X}}$)

$$\underline{H}^T \underline{\underline{X}} = \underline{u} \,. \tag{1}$$

The MINACE filter includes a constraint that $\underline{H}$ be chosen to minimize signal energy E_s and noise energy E_n in the output correlation plane

$$E = E_s + cE_n = \underline{H}^+ \underline{\underline{D}} \underline{H} + c\underline{H}^+ \underline{\underline{N}} \underline{H} \,. \tag{2}$$

The control parameter c is varied to produce filters with properties (minimizing noise more or less than signal energy). The diagonal matrices $\underline{\underline{D}}$ and $\underline{\underline{N}}$ are varied for different models of the signal and noise spectrum. The resultant filter that satisfies (1) and minimizes (2) has the solution

$$\underline{H} = \underline{\underline{T}}^{-1} \underline{\underline{X}} (\underline{\underline{X}}^+ \underline{\underline{T}}^{-1} \underline{\underline{X}})^{-1} \underline{u}, \tag{3}$$

where the diagonal matrix $\underline{\underline{T}}$ has elements that are the maximum envelope of $\underline{\underline{D}}$ and $c\underline{\underline{N}}$ at each spatial frequency. This has been shown to be preferable to other filters that minimize the average signal spectrum.

A unique aspect of these filters is our use of $\underline{\underline{D}}$ and $\underline{\underline{N}}$ to model object distortions and real clutter. These filters must exhibit good intra-class recognition (of all distorted versions of the true-class training set objects), rejection of clutter (this has been shown to be modeled well by correlated noise, CN), and good inter-class discrimination (rejection of false class objects). All of these properties are not simultaneously obtainable in general and thus an inference hierarchy set of filters is required (each filter output infers more knowledge and information about the objects and regions in the scene). For objects viewed from a nearly top down depression angle of 90°, we can quantify the filter set well. We show this by example in Figure 7.

Figure 7a shows an input scene with 13 objects in 5 different classes present at a number of different rotations (Figure 7b) in considerable CN clutter. As an adjunct to the other methods to provide the different information required in the different levels in Figure 2, we use different MINACE filters with different training sets and different choices for c (and for $\underline{D}$ and $\underline{N}$). For detection, we train the filter on the smallest object (the ZSU-23) in the set (since if the filter recognizes the smallest object, it will also recognize the largest objects), and we use a white Gaussian noise model for N (since it models these rotation distortions well) and we use a large value of c (this reduces high frequencies and thus improves intra-class recognition). The results for this detection filter (Figure 7c) are excellent. All 13 objects are detected with only one false alarm (lower left) and with most clutter rejected. This output provides ROI locations that can be fused with other results (Sections 2-3). To demonstrate macro-class recognition, we apply a second filter intended to locate only the large objects. This filter is trained on the large objects (Scud and Frog) and uses a medium value of c (to allow increased emphasis on high spatial frequencies to reject other objects). The outputs from this filter are fused with the ROI regions in Figure 7c. The results (Figure 7d) show peaks at only the 5 locations of the 5 large (Frog and Scud) objects. The lower c value rejects the other target locations and use of the ROI information in Figure 7c rejects peaks due to clutter. In cases where two objects are very similar (e.g. the Scud and Frog), identification of one object class may require use of all object pixels (rather than a feature space). In this case, a correlator is optimal. To demonstrate identification of all 3 Scuds and rejection of the 2 Frogs in Figure 7a, we formed a third filter trained only on the Scuds and used a low value of c (to emphasize high frequencies more and to thus improve discrimination of Scuds versus Frogs). This

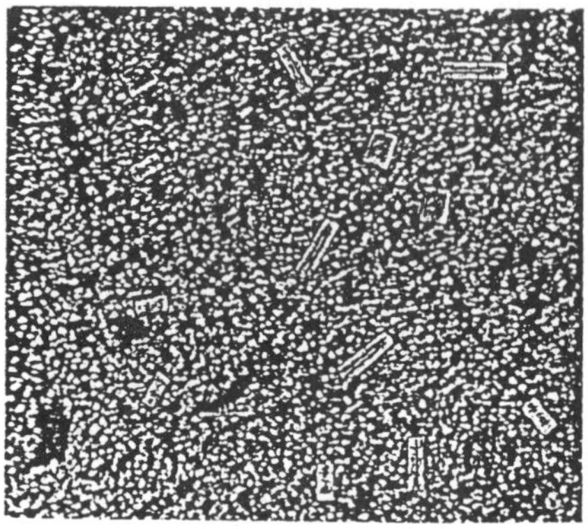

(a) Input scene

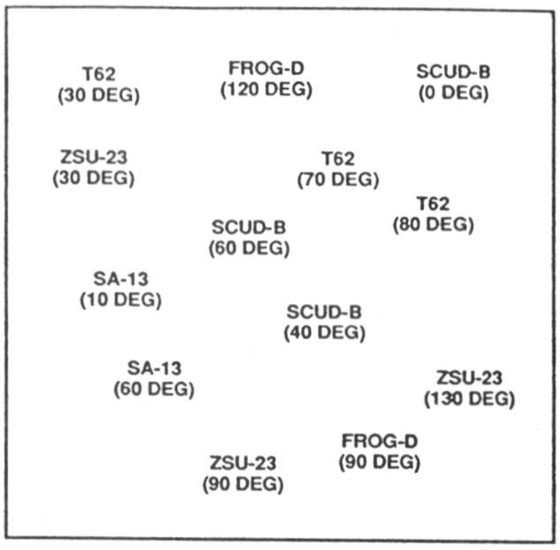

(b) Vehicle code

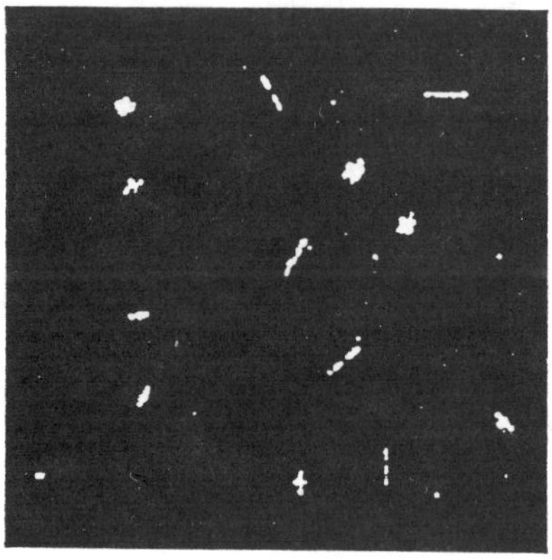

(c) Detection (c = 0.1)

(d) Macro-class recognition of large objects (c = 0.005)

(e) Identification and Rejection of similar objects (c = 0.001)

Fig. 7: Hierarchical inference filter sets as adjuncts to detection to provide macro-class recognition, and for difficult identification/discrimination problems.

correlation output is fused with the large object ROI data in Figure 7d to produce the result in Figure 7e with peaks at only the locations of the 3 Scuds.

The use of a sequence of distortion-invariant correlation filters is an attractive adjunct to other detection and clutter reduction filters. It also provides valuable macro-class information and it can provide identification in difficult discrimination cases. These filters are attractive from an implementation standpoint since they can employ the same full input scene (i.e. there is no need to window each ROI and thus the input data to the correlator need not be changed).

Many other variations and sequences of filters are possible. If clutter is not severe, we can use the distortion spectrum B of the training set (this is the maximum departure of the spectrum of the training set from its average at each spatial frequency) for D and minimize distortion energy [11]. If clutter is severe, clutter reduction techniques (Section 2) are necessary prior to this (since the B and CN spectra are nearly opposite and both cannot be simultaneously minimized). In another advanced filter [12], we have included distortion-invariance as a specific constraint in filter synthesis. In this case, we can optimize clutter rejection with a CN model for noise [13]. Figure 8 shows representative data to demonstrate these issues. The input contained 2 distorted true class objects to be recognized and a similar false class object (on the right) to be rejected with different background noise present. In Figure 8a, the objects are present in a white Gaussian noise (WGN) background. The filter was synthesized using a WGN model for N (thus, it is optimal for performance in white noise). The correlation plane shows 2 large peaks of height 1.0 for the true class objects and a low peak of 0.36 for the false class object. However, when the targets were placed in clutter (Figure 8b), performance was poor with a large false peak and sidelobes greater than 0.5. To obtain better performance in real clutter, we used a CN model in filter synthesis and obtained the much better results in Figure 8c (the false peak is now only 0.2). These results show the ability to optimize the filter for clutter reduction. It also shows that real clutter is not WGN (WGN models sensor noise whose level is always well below that of clutter) and that <u>a filter which is best in WGN is not best in real clutter.</u>

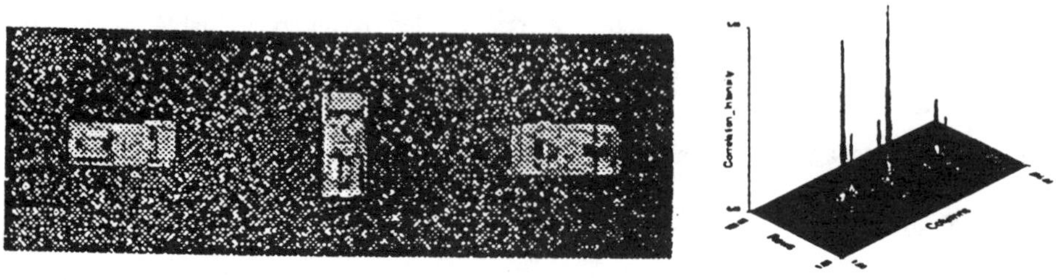

(a) Filter synthesized with WGN correlated with a WGN background

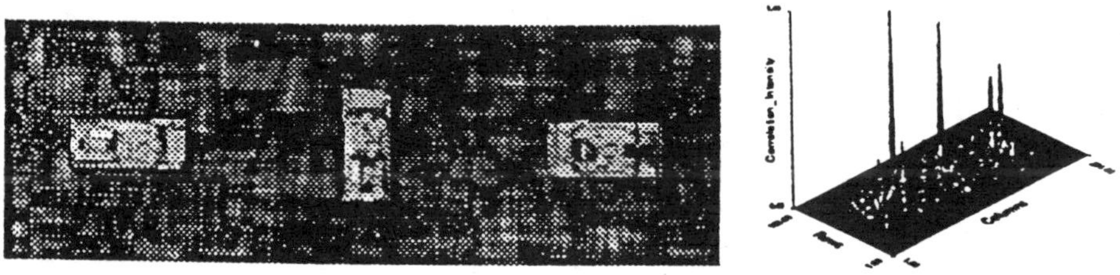

(b) Filter synthesized with WGN correlated with a real background clutter

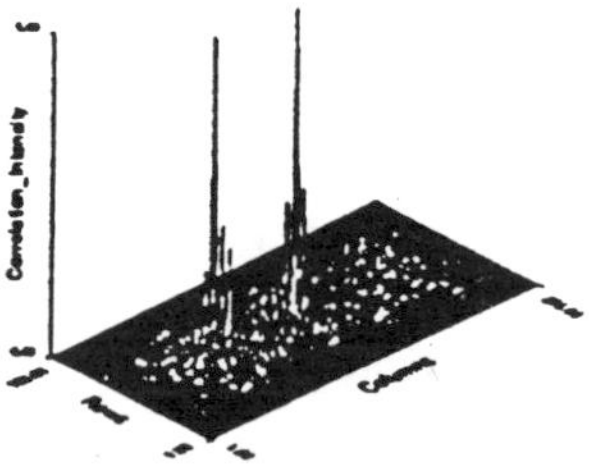

(c) Filter synthesized with CN correlated with real background clutter
Fig. 8: Filter optimized for clutter rejection

6. LARGE CLASS PROBLEMS

When the number of object classes to be identified is large, it is preferable to first extract features from each ROI and to feed these to a classifier to achieve identification.

6.1 Feature Extraction

Many different features can be calculated on a wide variety of hardware and software. Several feature spaces that can be calculated with a correlator with different filter functions used include [14]: the Fourier transform, wedge ring Fourier transform samples (these are very attractive since the Fourier transform magnitude is shift-invariant, the wedge samples are scale invariant, and the ring samples are rotation-invariant), moments (these can be combined to produce shift, scale and rotation invariant features), Hough transforms, etc. Feature spaces allow a smaller number of inputs to the classifier. If the features have some distortion-invariant, they simplify the number of training images required and the training time.

6.2 Classifiers

We consider neural net (NN) classifiers, since they allow simultaneous calculation of discrimination functions and combinations of them into discriminant surfaces and they can perform better than other classifiers. The standard 3-layer neural net (Figure 9) is considered with N_2 hidden layer neurons. The P_1 input neurons are the N_1 features for each ROI. Each hidden layer neuron can be viewed as producing a decision boundary in feature space allowing class separation and identification of the object in each ROI (by the P_3 output neuron activated). An additional input P_1 neuron with a constant input of 1 is used [15] to allow the threshold for all P_2 neurons to be adapted. This allows each decision boundary to be shifted (otherwise it is required to pass through the origin). Figure 10 shows the piecewise linear decision surfaces produced by our adaptive clustering neural net [15] for the case of 2 input feature space neurons (the two axes of the feature space shown) and 6 hidden layer neurons (the 6 linear decision boundaries shown correspond to these 6 neurons). As seen, these boundaries separate the 3 classes of data (shown by different symbols) even though one class is split and has part of another class in the center of it.

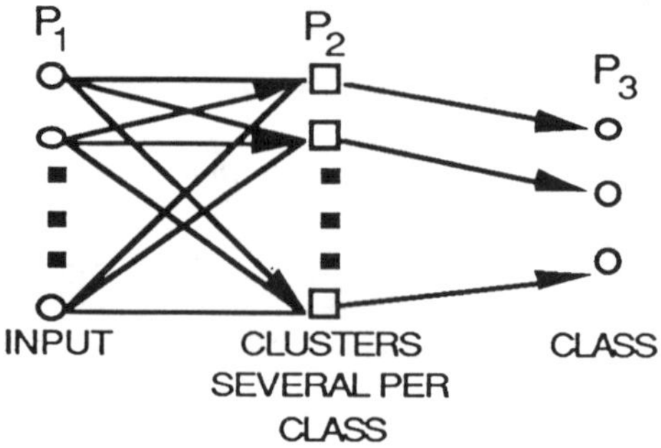

Fig. 9. Three layer neural net.

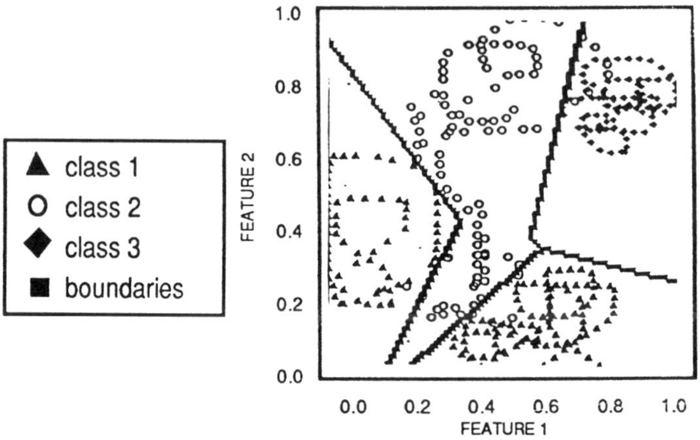

Fig. 10. Piecewise linear decision surfaces.

More complex (higher-order) decision surfaces are desirable to provide better neural net performance and storage. By increasing the number of hidden layer neurons N_2, one can produce more complex decision boundaries (however, increasing N_2 can lead to memorization of the training set and hence poor generalization performance on test data). One preferable solution is to use additional input neurons. To show this, we consider the case of 3 input neurons $\underline{x} = [1 \;\; x_1 \;\; x_2]^T$, whose images are two features (x_1 and x_2) and a constant input of 1. We add one extra input neuron whose input is $x_1^2+x_2^2$. If we denote the weights from these 4 input neurons to one hidden layer neuron by $\underline{w} = [w_1 \;\; w_2 \;\; w_3 \;\; w_4]^T$, then the full input to that hidden layer neuron is $\underline{w}^T\underline{x}$. Setting this equal to zero describes the decision surface for that P_2 neuron as

$$w_1 + w_2(x_1^2 + x_2^2) + w_3x_1 + w_4x_2 = 0 \tag{4}$$

which is the equation for a circle. Thus, this simple addition allows one to insure inherently hyperspherical decision surfaces (with $N_1 > 3$). With $N_2 > 1$, we obtain piecewise nonlinear hyperspherical surfaces between classes. Figure 11 shows an example of such surfaces for the case of two input feature space neurons and $N_2 = 6$ hidden layer neurons using a binary threshold at P_2. As seen, 3 different piecewise circular decision regions are defined that separate the 3 classes of data shown. A threshold nonlinearity at P_2 yields several "no decision" regions (the white areas in Figure 11) and yields a neural net with the best (lowest) probability of error (P_e). If a winner take all nonlinearity is used at P_3, all regions of feature space have outputs and there are no "no decision" regions. The decision boundaries occur between the different class boundaries shown in Figure 11. Figure 12 shows these boundaries for this problem. This neural net provides the best probability of correct recognition (P_c). Thus, two different types of neural nets with different advantages (P_c and P_e) with piecewise hyperspherical decision surfaces are possible one the same neural net with the same inputs and weights with different P_2 nonlinearities. We refer to this as a hyperspherical NN.

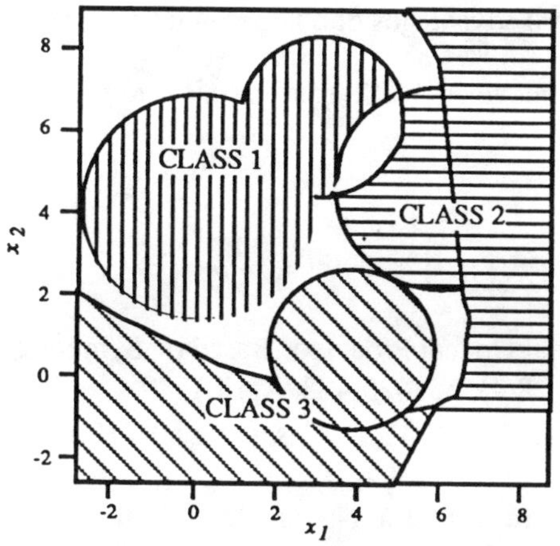

Fig. 11. Piecewise hyperspherical decision surfaces (best P_e).

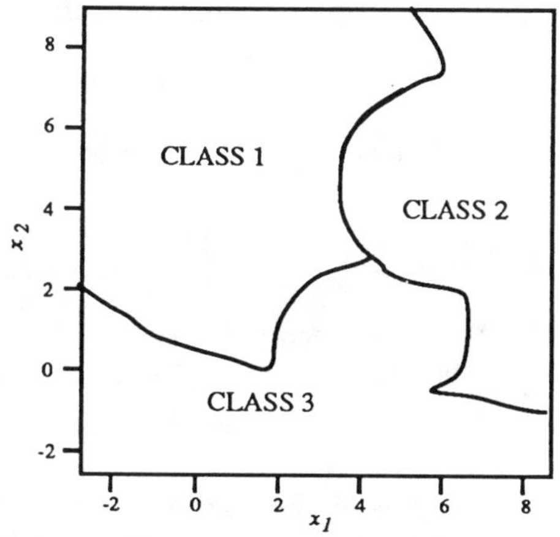

Fig. 12. Piecewise hyperspherical decision surfaces (best P_c).

Another new approach to producing higher-order decision surfaces uses a different P_2 nonlinearity (a square law nonlinearity) and different types of weights (complex-valued). These are both easily achieved in optics (detectors automatically provide square-law detectors). To derive a general expression for the decision surfaces produced in this case we consider 3 input neurons $\underline{x} = [1 \;\; x_1 \;\; x_2]^T$ and complex weights $\tilde{w} = w_{ij} + jv_{ij}$. Forming the outputs $\underline{x}^T \underline{w}_i$ for two (i = 1,2) hidden layer neurons and equating these outputs $\underline{x}^T \underline{w}_1 = \underline{x}^T \underline{w}_2$, we obtain the general equation for a decision surface for one hidden layer neuron in a 2-D space (x_1,x_2) to be

$$ax_1^2 + bx_2^3 + cx_1x_2 + dx_1 + ex_2 + f = 0. \tag{5}$$

This is the general equation for any quadratic decision surface. Thus, with $N_2 > 2$ hidden layer neurons, each hidden layer neuron is this new neural net produces any general piecewise hyperquadratic decision surface. We refer to it as a *Piecewise Quadratic Neural Net*. It and the hyperspherical neural net allow better classification performance and larger storage for any neural net and they achieve this with only simple additions (one additional input neuron, as a square law nonlinearity, complex weights) rather than by adding many more hidden layer neurons or input neurons or by use of higher-order weights, etc.

Optical implementations offer the standard advantages of a larger number of neurons, and more interconnections than one can achieve in VLSI (due to pin limitations and crosstalk in electronic interconnections).

REFERENCES

1. D. Casasent and E. Botha, "Multifunctional Optical Processor Based on Symbolic Substitution", *Optical Engineering*, **Vol. 28**(4), pp. 425-433, April 1989.
2. Optical Pattern Recognition IV, *Proc. SPIE*, **Vol. 1701**, D. Casasent and T.H. Chao, eds., April 1992.
3. D. Casasent, "An Optical Correlator, Feature Extractor, Neural Net System", *Optical Engineering*, **Vol. 31**(5), pp. 972-978, May 1992.
4. B. Bhanu, "Automatic Target Recognition: State of the Art Survey", *IEEE Trans. Aerosp. and Elec. Syst.*, **AES-22**, pp. 364-379, July 1986.
5. R. Schaefer, D. Casasent and A. Ye, "Optical Morphological Processors: Gray Scale with Binary Structuring Elements, Detection, and Clutter Reduction", *Proc. SPIE*, **Vol. 1825**, November 1992.
6. D. Casasent, R. Schaefer and R. Sturgill, "Optical Hit-Miss Morphological Transform", *Applied Optics*, 1992.

7. D. Casasent, J-S. Smokelin and A. Ye, "Wavelet and Gabor Transforms for Detection", *Optical Engineering*, 1992.

8. D. Casasent and E. Botha, "Optical Symbolic Substitution for Morphological Transformations", *Applied Optics*, **Vol. 27**, pp. 3806-3810, 15 September 1988.

9. D. Casasent, "Optical Morphological Processors", *Proc. SPIE*, **Vol. 1350**, pp. 380-394, July 1990.

10. G. Ravichandran and D. Casasent, "Minimum Noise and Correlation Energy (MINACE) Optical Correlation Filter", *Applied Optics*, **Vol. 31**, pp. 1823-1833, 10 April 1992.

11. D. Casasent and G. Ravichandran, "Nonlinear Distortion-Invariant Filter Performance with False Targets", *Proc. SPIE*, **Vol. 1772**, July 1992.

12. G. Ravichandran and D. Casasent, "Advanced In-Plane Rotation-Invariant Filter Results", *Proc. SPIE*, **Vol. 1567**, pp. 466-479, July 1991.

13. D. Casasent, "Hierarchical Scene Analysis for Object Recognition", *Proc. SPIE*, **Vol. 1699**, April 1992.

14. D. Casasent, "Computer Generated Holograms in Pattern Recognition: A Review", *Optical Engineering*, **Vol. 24**, pp. 724-730, September/October 1985.

15. D. Casasent and E. Barnard, "Adaptive Clustering Optical Neural Net", *Applied Optics*, **Vol. 29**, pp. 2603-2615, 10 June 1990.

Aerospace Sensors

Chair
Edward R. Washwell
Lockheed Missiles & Space Company, Inc.

Star sensors for autonomous attitude control and navigation

Roelof W. H. van Bezooijen

Lockheed Palo Alto Research Laboratory
3251 Hanover Street, Palo Alto, California 94304

ABSTRACT

The attitude control and navigation systems of future advanced spacecraft will be characterized by a high degree of autonomy, very high accuracy, efficient commandability, and fast fault recovery. These characteristics are incompatible with the constraints of conventional star sensors which mandate a-priori definition of all onboard attitude fixes and work only if attitude uncertainties remain small. With the availability of accurate, anti-blooming capable CCDs, fast microprocessors, high density memory chips, and star pattern recognition algorithms, it is now feasible to fabricate miniature Autonomous Star Trackers (ASTs) capable of (1) determining their attitude rapidly and reliably while having no a-priori attitude knowledge, (2) autonomous attitude updating, and (3) providing their attitude at rates up to typically 40 Hz. In addition to providing the functionality needed for future missions, ASTs can also be exploited to improve the reliability, mass, power, and cost of spacecraft and reduce the cost of operating them.

This paper describes star identification schemes used in the past, it discusses a number of star pattern recognition algorithms, and provides the main characteristics of current CCD star trackers. A number of specific functions enabled or enhanced by an AST are described including fast attitude acquisition, rapid fault recovery, attitude safing, gyroless/cheap-gyro attitude control, autonomous target acquisition by astronomy telescopes, autonomous optical navigation, and precision pointing to terrestrial targets. The AST being developed at Lockheed uses a fast, memory-efficient, and highly robust star pattern recognition algorithm based on matching groups of stars. The algorithm, which is also applicable to star scanners, is described along with a realistic simulation program for testing its performance. It is shown that an AST with an 11.3 degree FOV diameter, a database of 4100 guide stars, a 25 MHz MC68030 class microprocessor, and 800 Kbytes of memory will be capable of determining its attitude in 0.45 seconds with a success rate greater than 99.98% when using an optimal guide star selection method. Compute time and memory are found to be inversely proportional to the FOV area. The paper also reports on AST development by other organizations in a number of countries.

1. INTRODUCTION

The attitude control, pointing control, and navigation systems of future advanced spacecraft will be characterized by a high degree of autonomy, very high accuracy, efficient commandability, and fast fault recovery. These characteristics are incompatible with the constraints of conventional star trackers which mandate a-priori definition and careful preparation of each on-board attitude fix, work only when attitude uncertainties are small, and, for some applications, require manual star identification. In addition to being incapable of fulfilling the needs of future spacecraft systems, conventional star trackers are costly to use because their lack of intelligence necessitates the expenditure of significant human, ground software, uplink, on-board storage, and telemetry resources.

With the availability of accurate, anti-blooming capable CCDs,[1] fast microprocessors, high density memory chips, and a fast, memory-efficient and highly robust star pattern recognition algorithm,[2,3] it is now feasible to build an Autonomous Star Tracker (AST) capable of determining its attitude rapidly and reliably with no a-priori attitude knowledge. In the past, analysis was used to show that an AST with an 11.3 degree field of view (FOV) diameter, a 10 arcsec (1 sigma) spatial accuracy, an all-sky database of some 4,100 guide stars, and a 12 Mhz MC68000 class microprocessor, is capable of determining its attitude in approximately 1 second with a success rate very close to 100%.[2,4]

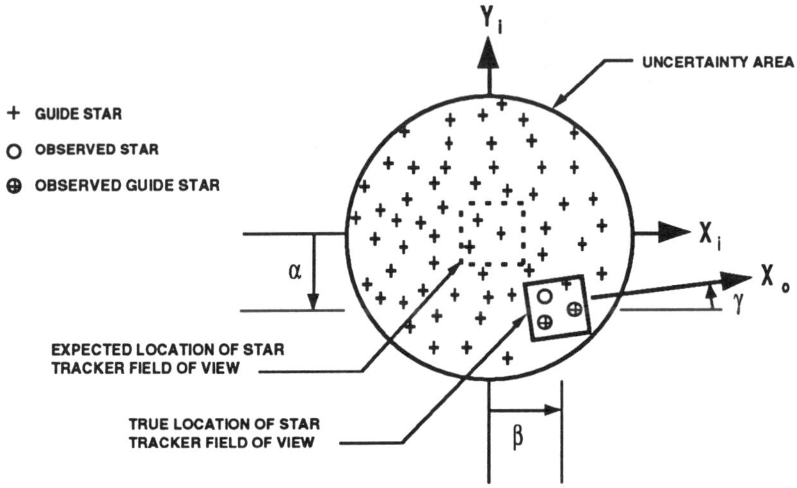

Fig. 1. Attitude determination problem.

The attitude determination problem is illustrated in Fig. 1. Due to attitude errors, the star tracker coordinate system (X_o, Y_o, Z_o) deviates from its expected inertial orientation (X_i, Y_i, Z_i). Employing geometric and brightness criteria, a star identification algorithm determines which of the stars observed by the tracker (the "observed stars") correspond to which of the a-priori selected "guide stars". If at least two of the observed stars are guide stars, and these stars are correctly identified, the attitude about all three axes (angles α, β, and γ) can be computed. The area in which the guide stars are selected, referred to as the "uncertainty area," needs to be sufficiently large to ensure that it will include the field of view of the star tracker. As is illustrated by Fig. 1, some of the guide stars within the FOV of the star tracker may not be observable, while some of the observed stars may be non-guide stars. The star identification algorithm needs to be sufficiently robust to cope with this situation.

The uncertainty area is generally not much larger than the star tracker FOV when updating the attitude of gyro-based spacecraft attitude control systems. It typically measures 10 times the FOV area when astronomy telescopes acquire celestial targets using just a (narrow FOV) Fine Guidance Sensor (FGS).[2,5] The uncertainty area covers the entire sky when performing attitude acquisition with an AST. Only in the latter case is the on-board star identification referred to as being "fully autonomous." In all other cases (uncertainty area does not cover entire sky), the term "limited autonomy" is used.

If V-slit star scanner type sensors are used, the area of the sky from which observed stars are obtained is typically a complete or partial small circle band on the sky. For this type of star sensor, the relative position of the observed stars in the scan direction is obtained from the star image passage times over the slit detector that is perpendicular to the scan direction and the known scan rate. The relative star positions in the cross-scan direction are obtained by measuring the time it takes a star image to move from the "perpendicular" to the "skew" slit detector and translating this time interval into an angle using the known scan rate, the distance between the two slits, and the angle between the two slits (typically 30 to 40 degrees).

The Lockheed Palo Alto Research Laboratory is currently developing an AST prototype designed to operate in three different modes. In the "acquisition mode," the tracker is required to establish its attitude while having no a-priori attitude knowledge. In the "update mode," it determines its attitude, given its expected attitude and attitude error. The AST prototype will output its attitude continuously at frequencies up to 20 Hz when operating in the "tracking mode." In addition to building and testing the prototype, the research program also involves the evaluation of different guide star selection algorithms and the development of a realistic computer simulation program.

The research concentrates on autonomous star trackers with a moderate FOV and most of the simulations have been carried out for a "Reference AST" with a FOV diameter of 11.3 degrees (area of 100 square degrees). While this FOV is small enough to achieve high spatial accuracy, it is sufficiently large to ensure fast and reliable star identification at the cost of a moderate amount of memory. Although they have mass and memory advantages, ASTs with a larger FOV suffer from a number of disadvantages, including: (1) higher probability of interference from the Sun, the Earth, and the Moon, (2) more complex optical design, (3) increased optical distortion, (4) more difficult to accommodate on the spacecraft, and (5) lower accuracy. Furthermore, it was found that, for a constant focal ratio, the integration time does not decrease with an increase in the FOV.

Section 2 of this paper provides background information, including flight examples of star identification, a discussion of different star identification algorithms, and a brief description of CCD star trackers that have flown in space, are currently flying, or are scheduled to fly in 1993. The significance of an AST is pointed out in section 3 where a wide range of applications is presented. The AST prototype development approach at Lockheed is outlined in section 4, an overview of the star identification algorithm employed is given in section 5, the simulation program is described in section 6, and the simulation results are presented in section 7. AST developments at other organizations are described in section 8, while the conclusions are contained in section 9.

2. BACKGROUND

2.1. Identification examples

The various star identification techniques currently in use or used in the past for attitude determination can be divided into manual, limited autonomy, and full autonomy groups. For the first group, star identification is performed by a human operator. The identification is automated for the second and third groups. In the limited autonomy group, star identification is relatively simple because attitude information is available. Star identification is most challenging in the full autonomy group where there is no benefit of a-priori attitude knowledge. Flight examples for all three categories are described below.

2.1.1. Manual identification

Acquisition of celestial targets by ground based, airborne, or orbiting astronomy telescopes is currently often performed by a human operator. The acquisition process starts with a slew maneuver from a known attitude to the celestial target. Because of errors in the initial attitude and slew imperfections (e.g., due to gyro drift, scale factor, and alignment errors) the pointing error at the new target can be large relative to the narrow FOV of the FGS. This precludes direct acquisition of the guide star(s) used for pointing control. To remove the error, the operator first identifies the stars in the FOV of a star tracker that is coboresighted with the telescope. Following identification, the pointing error is determined and subsequently removed. The operator then identifies the stars in the FGS FOV and designates the star(s) used for pointing control.

This method is commonly used for astronomy telescopes carried aloft by balloons, it is employed by the infrared telescope of the Kuiper Airborne Observatory (KAO), while the International Ultraviolet Explorer[6] represents an example of an orbiting telescope requiring manual target acquisition. The KAO operator needs a few minutes to identify the stars. This represents a significant time penalty as each observation typically lasts 20 to 30 minutes.

2.1.2. Limited autonomy

Voyager, ANS, IRAS, and Galileo are examples of spacecraft with limited star identification autonomy. The Voyagers, launched in 1977, perform attitude acquisition by rolling the spacecraft about the sun vector while keeping the sun at its desired location in the FOV of a two-axis Sun Sensor. The sun location in the FOV is selected such that Canopus is guaranteed to move through the FOV of the star tracker (called the Canopus Star Tracker). Canopus is the second brightest star in the sky with an apparent visual magnitude (m_V) of - 0.72, which makes it unique in the small circle band scanned by the tracker. Following acquisition of Canopus or another bright star, the High Gain Antenna is pointed to Earth by adjusting the setpoints for the sun sensor and the star tracker.

The Astronomical Netherlands Satellite (ANS), launched in 1974 into a sunsynchronous polar twilight orbit, was equipped with an image dissector type star tracker integrated in the detector bay of the spacecraft's ultraviolet telescope.[7] The tracker had a 1.5 by 1.5

degree FOV and could track stars down to $m_V = 8$. Identification was based on the use of guide star pairs. While rolling the spacecraft about the sun vector at orbital rate, the tracker searched a 3 arc minute long line within its FOV in the cross-scan direction at a location where the first guide star was expected to pass. Following detection of a star brighter than the specified threshold, a 3 arc minute line was scanned at a location where the second guide star was expected to pass. Correct identification was assumed if a sufficiently bright star crossed this line within one second of time. The telescope was then pointed to the celestial target using the second guide star. When rotating about the sun vector, the angle between the telescope axis and the sun vector was maintained at its desired value (in a range from 89 to 91 degrees) using a two-axis Fine Sun Sensor, while a scanning type Horizon Sensor measured the angle of the telescope axis about the sun vector.

The Infra Red Astronomical Satellite (IRAS),[8-10] launched in 1983 into a 900 km sunsynchronous polar twilight orbit, used a two-axis Fine Sun Sensor to maintain the desired angle (in a range from 60 to 120 degrees) between the sun vector and the optical axis of its infrared telescope and, like ANS, to keep one axis of the spacecraft perpendicular to the sun vector. Redundant single axis rate integrating gyros were used for controlling the angle of the telescope axis about the sun vector. A redundant star scanner with a FOV of 30 arc minutes was an integral part of the telescope focal plane instrumentation. It was sufficiently sensitive to detect stars down to $m_V = 7.4$ while rotating at orbital rate. Each redundant half of the scanner comprised four silicon detectors arranged into two V-slit pairs. Schott BG 38 optical filters were used to ensure that the instrument magnitude of a star would not differ too much from its visual magnitude.

The scanner was used to measure the pointing direction of the telescope at a rate of typically 9 times per orbit. These measurements allowed accurate reconstruction of the telescope pointing direction. They were also used for onboard updating of the spacecraft attitude about the sun vector. Robustness was ensured by implementing updates only if the results of consecutive attitude fixes (up to 5) did correlate. One guide star was used for each attitude fix, implying that 60 guide stars were typically selected by ground software for each 12 hour observation sequence.

Galileo,[11] launched in 1989, is equipped with a radiation hard V-slit star scanner employing redundant photomultiplier tubes.[12] The 10 degree FOV scanner is mounted on the part of the dual-spin spacecraft that rotates at nominally 3.15 rpm. It has sufficient sensitivity to use the brightest 500 stars in the sky. One of the slits is parallel to the spin axis, while the other slit is at a 30 degree angle. Thus, a great circle band covering roughly 8.7% of the sky is observed in one revolution. The spacecraft carries an updatable database of 18 guide stars at all times;[13] 6 guide stars are available for the Sun pointing attitude (spin axis pointed to the Sun), 6 for the Earth pointing attitude, and 6 for a maneuver attitude. For each orientation, at least 2 of the 6 guide stars need to be identified, and these stars are subsequently used for attitude updating.

2.1.3. Full autonomy

One of the authors of the paper entitled "WFOV Star Tracker Camera,"[14] stated in a private communication that full autonomy was achieved on an experimental SDI

Brilliant Pebbles (BP) sub-orbital spacecraft flight in 1991 reported on by the press.[15] The star field images on this flight were obtained with a prototype miniature CCD star tracker with a 28 by 44 degree FOV and a 14 mm aperture. The tracker provided the digitized intensity information of its 576 x 384 pixels to the spacecraft flight computer where star identification was performed. Because of the large FOV, the guide star database could be limited to 500 stars covering the entire sky.

The wide FOV was realized using concentric optics which necessitated a fiber optic coupler assembly for transmitting the light from the spherical focal plane to the flat CCD. This coupler affects both the light transmission and the optical quality of the image at the CCD.

2.2. Algorithms for star identification

Star identification algorithms attempt to match patterns of observed stars with patterns of guide stars. They can be distinguished by the number of stars in the pattern and the level of autonomy enabled by them. In 1979, Junkins and Strikwerda described a star identification method for use with CCD star trackers.[16] The method is iterative and is based on matching star pairs. Its application is limited to cases where the attitude error is small relative to the FOV of the tracker. Bone, Parkinson, and Rodden describe an iterative method where the three parameters defining the attitude of the star tracker are varied to find the best match of a group of observed stars with a group of guide stars.[17] Their method also falls in the limited autonomy class as it requires good a-priori attitude knowledge to find a solution within a reasonable amount of time.

In 1981 Wong and Lai described a fully autonomous star identification technique for use with the aforementioned Galileo spacecraft.[18] Their method was based on matching star triplets. Since the Galileo star scanner observes approximately 8.7% of the sky during a full revolution, a total of 200 guide stars was deemed sufficient. The method was computation intensive and required a processing time of approximately 30 seconds when simulated on a UNIVAC 1108 computer. It was not implemented due to flight computer memory constraints.

A fully autonomous star identification algorithm for use with star trackers was proposed by Groth in 1986.[19] His method, based on matching star triplets, was applied to the BP experiment mentioned in section 2.1.3. All guide star triplets that fit in the FOV of the star tracker are stored onboard. This requires an enormous amount of memory, making the method practical only for rather inaccurate wide FOV star trackers. For a circular FOV it is easy to show that the number of triplets (N_{tr}) is given by

$$N_{tr} = (n-1)(4n-1)n \,/\, 3 \; A_{sky} \,/\, A_{fov} \qquad (1)$$

where n is the mean number of guide stars per FOV, A_{sky} is the area of the sky, and A_{fov} is the area of the FOV of the star tracker.

The algorithm described in this paper allows fully autonomous star identification. It is based on matching groups of observed stars with groups of guide stars.[2-5] Basically, it finds the largest group of observed stars that matches a group of guide stars and exploits the full information contents of a star pattern to maximize identification reliability.

Although designed for use with star trackers, the method can easily be adapted for use with star scanners. The algorithm, described in more detail in section 5, requires onboard storage of all guide star pairs that fit in the FOV of the star tracker. For a circular FOV the number of guide star pairs (N_{pr}) is given by

$$N_{pr} = (4n-1)n / 2 \;\; A_{sky} / A_{fov} \tag{2}$$

A comparison with (1) shows that for a given FOV, there are 2(n-1) / 3 times less pairs than triplets, which gives the method a significant advantage over Groth's algorithm and allows fabrication of highly accurate autonomous star trackers. For a typical guide star density of 10 per FOV, there are 6 times more triplets than pairs. Since 2 bytes are needed to specify the angular distance between stars and 2 bytes are required for identifying each star in a pair or a triplet, it follows that Groth's method typically requires 12 times more memory for storage.

2.3. CCD star trackers

An overview of CCD star trackers that have flown onboard spacecraft, are currently flying, or are scheduled to fly in 1993 is provided by table 1. It may be seen that the accuracy varies from 0.05 (ASTROS I and ROSAT) to 0.1 pixel (HDOS ASTRA-2).

This high accuracy, which is essential for autonomous star identification, is obtained by providing an appropriate amount of defocus (e.g., such that 80% of the star image energy falls within a 2 pixel diameter circle) and use of an interpolation algorithm. The algorithm commonly used assumes a rectangular line spread function, resulting in an S shape small scale (per pixel) error characteristic. This error is a function of the star angle and the color of the star. The former effect changes slowly with the angle, allowing its removal following careful calibration at a manageable number of points in the FOV. This technique was used for ASTROS I where the small-scale error was characterized at the 45 points of a 5 by 9 grid enabling subsequent removal in the entire FOV by means of interpolation.[25]

Table 1. CCD star trackers.

Tracker	FOV (deg)	Pixels	Accuracy (", 1 σ)	Sensitivity (mag.)	Numb. of stars tracked	Remark
JPL ASTROS I[20]	2.2x3.5	330x512	1.3	8.0	3	Flown in '89 (STS ASTRO-1)
HDOS ASTRA-2[21]	3.8x4.8	256x403	3.2	7.0	5	Flown in '89 as GAS on STS
JENA ASTRO 1[22]	5.3x8.0	520x580	3.7	6.3	7	Flying (1989-) on MIR SS
SIRA/MBB STR[23,24]	4.4x5.9	288x385	2.5	6.5	4	Flying (1990-) on ROSAT
HDOS ASTRA-1[21]	7.0x9.0	256x403	11.0	5.7	1	Flying (1992-) on TOPEX
BALL CT-601[30]	8.0x8.0	512x512	5.0	6.0	5	First flight planned for '93

The CT-601 is to be capable of a 10 Hz update rate and a 5 arcsec accuracy (1 sigma) for scan rates up to 0.3 degrees per second. The device has a mass of 8 kg (without light shade) and draws less than 12 W of power. If the base plate temperature is less than zero degrees C the power consumption drops to 7.5 W as there is no more need to operate the thermoelectric CCD cooler. The other trackers in table 1 are also equipped with coolers needed for limiting the CCD dark current to an acceptable level.

A significant reduction of dark current is obtained with multi pinned-phase (MPP) CCDs.[1] This allows the maximum CCD temperature for trackers with short integration times to be raised to a level where cooling is no longer needed.

3. AST APPLICATIONS

3.1. Attitude acquisition

Conventional attitude acquisition methods are complex, time consuming processes where typically a combination of the Sun, the Earth, and the Earth's magnetic field is employed to obtain an initial coarse attitude fix. Next, a star sensor (sometimes in combination with a fine sun sensor) is used to determine the attitude accurately. By contrast, an AST, operating in its acquisition mode, can perform accurate attitude acquisition in a fraction of a second, following stabilization of the spacecraft.

The Assured Crew Return Vehicle (ACRV) is an example of a spacecraft that would benefit from an AST. Following separation from the Space Station, the ACRV must be placed in the correct attitude to perform a collision avoidance burn. This maneuver needs to be carried out within 15 minutes, which is hard to accomplish with conventional attitude acquisition methods that rely on use of the Earth's magnetic field. The latter method is time consuming because the magnetic field vector needs to rotate over an angle sufficiently large to allow adequate 3-axis attitude determination.

3.2. Attitude safing and fault recovery

Operating in its acquisition mode, the AST enables attitude safing and fast recovery from faults that result in loss of attitude knowledge. This capability is important for spacecraft where outages are very costly, or where a loss of attitude could lead to catastrophe if not remedied quickly. JPL's Space Infrared Telescope Facility (SIRTF) and Cassini mission to Saturn are projects where an AST could be used to avert catastrophe.

Following a fallback to the safe mode operated from ROM, SIRTF[26] must be kept in a safe attitude relative to Sun and Earth to avoid damage to the cryogenically cooled telescope. This can be accomplished by bringing the telescope axis perpendicular to the sun line with the aid of the Sun Sensor, next use the AST (co-boresighted with the telescope) to determine the attitude about the sun line, and then point the telescope to either the ecliptic North or South Pole following a roll about the sun line. The 100,000 km orbit selected for SIRTF makes it impractical to use an Earth sensor for performing the safing function. Being able to acquire targets autonomously is an added benefit of using an AST on SIRTF.

In the case of Cassini,[27] the quick fault recovery enabled by an AST would reduce the length of the critical sequences associated with Saturn orbit insertion, Titan Probe release, and the Titan fly-by targeting burns needed for gravity assist. This reduction in critical sequence length translates directly into a reduced probability of catastrophic failure.

3.3. Celestial target acquisition

The observing efficiency of high-precision airborne and orbiting astronomy telescopes can be improved markedly by automating celestial target acquisition using an AST that is coboresighted with the telescope. After a slew to a new target, the AST, operating in the update mode, identifies the stars in its FOV, determines its attitude and provides this information to the attitude control system. The telescope pointing error is then reduced to a level where the FGS can acquire the guide star(s) used for pointing control.

Relative to an automated acquisition scheme that uses the FGS exclusively,[5] the addition of an AST eliminates the need for ground selection and on-board storage of a significant number of guide stars around each target (in addition to the one(s) needed for pointing control).

SIRTF, SOFIA, and KAO (see section 2.1.1.) are examples of programs that would benefit from the presence of an AST. Implementation of automated target acquisition would increase the KAO observing time by some 10%. This significant improvement can be obtained either by equipping the KAO with an AST, or by adding software that would enable automatic star identification with the existing hardware.

3.4. Attitude updating

When used to update the attitude of gyro-based attitude control systems, the AST eliminates ground selection and onboard storage of guide stars for each update. By operating in the update mode, the AST provides its attitude every time a request is received from the attitude control system. Since there is no penalty for performing frequent updates, the AST can be used either to improve accuracy, or to reduce cost by allowing the use of cheaper, less accurate gyros. Autonomous attitude updating is planned for F Sat using conventional CCD star trackers, an all-sky guide star database, and the spacecraft flight computer.[16]

The AST is also ideally suited for updating the attitude of strapdown inertial navigation systems. Due to the absence of an inertial platform, these systems need more frequent attitude updates to maintain navigational accuracy. In addition to being able to perform updates at the desired higher frequency, the all-sky capability of the AST eliminates the need to reorient the vehicle for updates.

3.5. Gyroless spacecraft

ASTs can provide their attitude at high rates (e.g., 40 Hz) through proper selection of FOV and aperture. This capability allows gyros to be eliminated from spacecraft in certain cases, thus improving reliability, weight, power, and cost.

3.6. Optical navigation

Optical navigation is employed by planetary spacecraft to determine their location relative to nearby bodies such as planets, satellites, comets, and asteroids. It involves establishing the inertial direction and rotation rate of the vector from the spacecraft to the solar body of interest using multiple images of that body against the stellar background. Being performed on the ground, optical navigation is currently hampered by long delay times, making it unacceptable for future missions like Rosetta, a joint NASA/ESA effort for returning a comet nucleus sample.

3.7. Precision pointing to terrestrial targets

Combined with a Global Positioning System (GPS) receiver, the AST, operating in its tracking mode, will enable autonomous, high-precision pointing from space to terrestrial targets as required for secure laser communication and efficient, quick-response surveillance. It would be possible to point to targets with better than 20 m accuracy from an altitude of 500 km, assuming the availability of trackers with a 1 arcsec accuracy.

3.8. Position determination

An AST, integrated with a device for measuring the direction of the local vertical (e.g., a three-axis accelerometer) can be used for autonomous position determination of planetary rovers. The AST would be used to determine the inertial direction of the local vertical. Given the time of the star field observation and the direction of the local vertical, the position of a vehicle on Mars can be computed to an accuracy of a few hundred meters. Further position refinement can then be obtained using stored terrain maps.

3.9. Rate sensing

The AST capability could possibly be extended by adding a rate sensing algorithm, allowing elimination of rate gyros for certain applications. A definition study for a CCD rate sensor has been performed for ESTEC.[28]

4. PROTOTYPE DEVELOPMENT APPROACH

The AST research program at Lockheed is depicted in Fig. 2. It involves three main tasks: guide star database generation, creation of a computer simulation program for predicting the AST performance, and demonstration of the AST performance using a prototype and the true sky. The simulation program is described in section 6.

4.1. Guide star database generation

The guide star database needed for performing star identification consists of: (1) the position and predicted instrument (or AST) brightness of the selected guide stars, and (2) all pairs of guide stars having a mutual angular distance less than a certain value, typically chosen to be equal to the greatest FOV dimension. The guide stars are obtained from the SKYMAP version 3.5 star catalog[29] which has been compiled primarily for the purpose of satellite attitude determination and provides 622 bytes of information for each of its 248,563 stars.

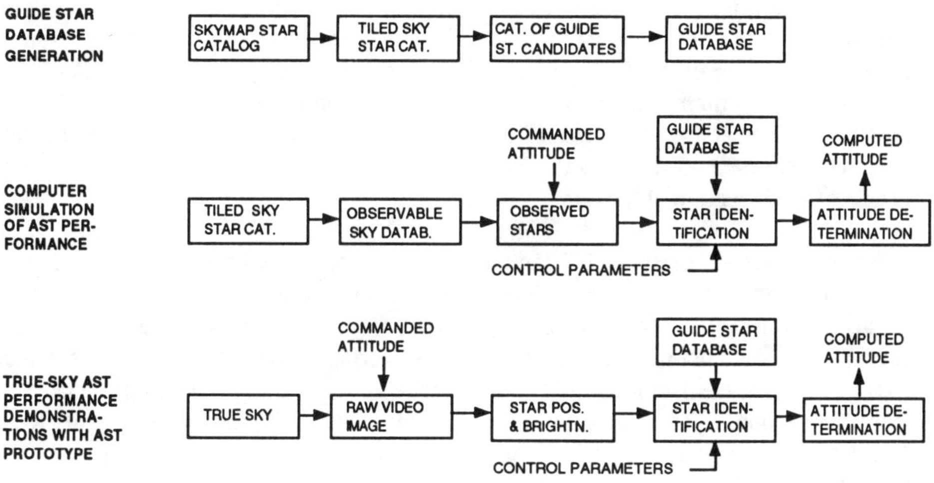

Fig. 2 AST prototype development approach.

We can either assume that the AST magnitude of the guide stars is the same as the visual magnitude given in the catalog, or we can translate the visual magnitude of the stars to the AST magnitude using the spectral response of the AST and the spectral star information provided by the catalog. In the former case we can either accept a large error in the predicted AST magnitude of the guide stars, or we have the option to minimize this error by adding an optical filter to the AST to bring its spectral response closer to that of the V filter.

Large magnitude errors make star identification more difficult and necessitates a higher guide star density. Addition of a filter makes guide star selection and identification easier at the cost of a reduced AST sensitivity. Currently, translation of the visual to instrument magnitude has not yet been implemented. However, further research will be performed to determine the accuracy of V to instrument magnitude translation, given the spectral information available in the SKYMAP star catalog.

Guide star database generation is performed using the following steps: (1) conversion of the single-file SKYMAP catalog on 9 track tape to 100 files on Macintosh readable Teac (150) tape, (2) extraction of the necessary star data, (3) organization of the stars into a tiled-sky catalog, (4) creation of a catalog of guide star candidates, (5) guide star selection, and (6) formation and sorting of guide star pairs. The data currently extracted for each star at step (2) consists of its SKYMAP record number, epoch 2000.0 right ascension and declination, position error, proper motion and visual magnitude. The selection of guide stars from the catalog of candidates (step 5) and the generation of guide star pairs (step 6) is described in section 6.

4.1.1. Tiled-sky star catalog

The sky has been partitioned into 110 slightly overlapping tiles to allow efficient generation of the catalog of guide star candidates, the guide star database, the observable sky database, and the observed stars. Within each tile, the stars of the tiled-sky catalog are sorted in sequence of increasing visual magnitude. The tiles overlap each other by at least 0.7 degrees, which permits a tile by tile generation of the catalog of guide star candidates. In this catalog, each candidate is accompanied by a list of its close neighbors out to a distance of 0.35 degrees. This structure facilitates rapid guide star selection using one of the selection algorithms.

The tiles are sized such that an AST can never extend into more than 4 tiles, provided its largest FOV dimension is less than 16 degrees. This is achieved by arranging the tiles into 5 declination bands per hemisphere and partitioning each band in an appropriate number of tiles. The core area (i.e., area minus its overlap border zone) of the first band, extends to 16.2 degrees from the pole and comprises one tile, the "polar cap." The core width of the second, third, fourth and fifth bands is 16.4, 20.7, 18.4, and 18.3 degrees, respectively. These bands are partitioned into 6, 12, 18, and 18 tiles along constant right ascension lines.

In the tiled-sky catalog, the number of stars in the core area of the tiles ranges from a low of 1024 to a high of 5033, while the number of stars in the border area of the tiles ranges from 78 to 407. It was found that a total of 797 stars (0.32%) lack visual magnitude information. Using the core stars with defined visual magnitude, an integrated star density model is generated for each tile. This density model assumes that the logarithm of the integrated star density is a linear function of the magnitude, and a least squares fit is employed to establish the two model parameters. Using the model, a "model magnitude limit" is established per tile. At this limit the number of stars according to the model equals the number of stars in the tile. It ranges from 9.35 to 9.87 magnitude.

4.1.2. Catalog of guide star candidates

Generation of a catalog of guide star candidates requires the definition of a close neighbor radius value and the desired total number of guide star candidates or an upper magnitude limit. The latter is defined as a delta relative to to the aforementioned model magnitude limit. When requesting a certain total number of guide star candidates, the number per tile is computed and extraction from each of the tiles is performed. When requesting a total of 25,000 candidates a quantity of 24,956 was extracted with magnitudes down to 8.1. For this run a "close" neighbor distance of 0.35 degrees was specified, resulting in a total of 60,567 close neighbors with a maximum of 53 close neighbors per candidate. The need for a catalog with the depth of SKYMAP is illustrated by the fact that even if the AST magnitude were equal to the V magnitude and the SKYMAP catalog were 100% complete down to the model magnitude limit, the faintest guide star candidates could have unknown close neighbors with a brightness up to 25% of that of the candidate (neighbor is only 1.5 magnitude fainter).

4.2. AST Prototype.

Since the simulations can only approximate the truth, suffer from some statistical uncertainties, and cannot possibly account for all parameters involved, it is absolutely necessary to validate the AST performance using a proof-of-concept or prototype model. Using this prototype, a large area of the sky will be mapped and the raw video data will be stored on disk or tape. In the lab, each of the frames thus obtained can be further processed and used to test the star identification capability of the AST. These lab tests will be followed by real time AST performance demonstrations at an observatory.

The AST prototype and its Support Equipment will be assembled from the following commercially available components: (1) A thermoelectrically cooled CCD camera head with a NuBus interface board for a Macintosh computer, associated software, and a camera lens adapter, (2) A 35 mm f/1.4 and a 50 mm f/1.2 camera lens, and (3) A computer controlled telescope with equatorial wedge and tripod.

The anti-blooming capable CCD camera will be equipped with a 1024x1024 pixel CCD that is used in a Frame Transfer mode, implying a sensitive area of 512x1024 pixels. Equipped with the 35 mm lens, the FOV can be selected within the range from 10x10 to 10x20 degrees. The integration time will be approximately 1.1 s using a V filter and the smallest FOV, and will be shorter by a factor of 5 without this optical filter. With the 50 mm lens, the FOV can be selected within the range from 7x7 to 7x14 deg. The larger aperture allows a reduction of the integration time to 0.4 s for the square FOV and with the V filter in place. A Macintosh IIci computer will be used to control the camera, store the video data, and perform the star identification.

The CCD camera will be mounted in a piggyback fashion to a computer controlled 8", f/6.3 Meade LX200 telescope enabling efficient mapping of the sky.

5. STAR IDENTIFICATION

Star identification is achieved using a star pattern recognition algorithm. The objective of this algorithm is to find the largest group of stars within the AST FOV that matches a group of guide stars. The brightest stars within the FOV are picked for the identification process. In the following, these stars are referred to as the "observed stars." The anti-blooming capability of the CCD camera is exploited to ensure that the desired number of stars can be obtained, even if a very bright star is present. The recognition algorithm assumes that a group of observed stars matches a group of guide stars if all of the following criteria are met:

1. The measured angular distance of each pair of observed stars matches the predicted angular distance of the corresponding pair of guide stars to within the distance tolerance (*tdis*).
2. The measured magnitude of each of the observed stars matches the predicted magnitude of the corresponding guide stars to within the magnitude tolerance (*tmag*).
3. The geometry of the group of observed stars is not the mirror image of the guide star group.

In the above, the angular distance of a star pair is defined as the angle between the two stars of the pair. In practice, the "vector distance" between two stars is used as a measure of the angular distance. The vector distance is defined as the linear distance between the stars, assuming that the stars are located on a celestial sphere with unit radius.

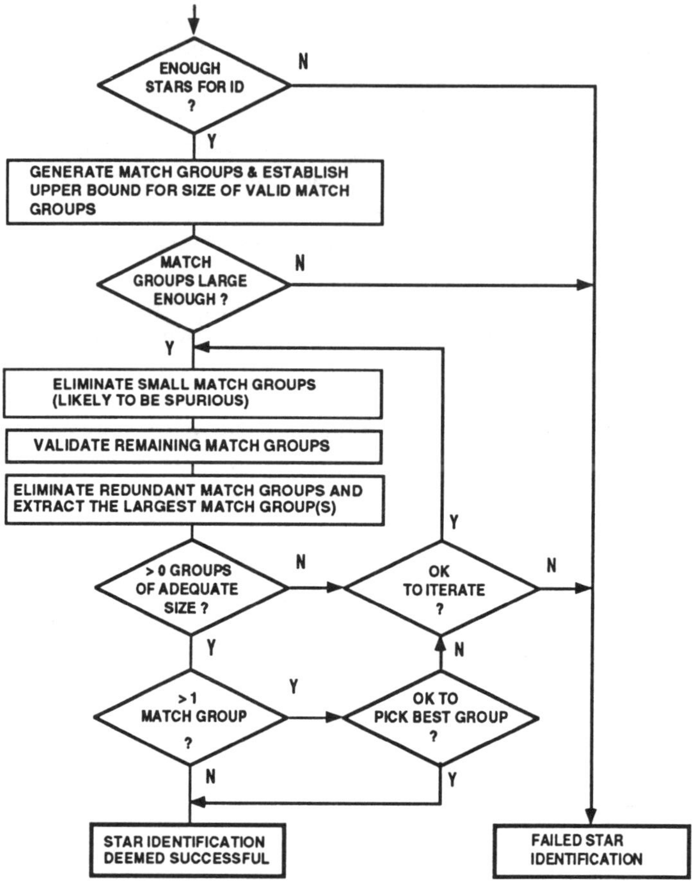

Fig. 3. Recognition algorithm flow diagram.

A high level flow chart of the star pattern recognition algorithm is depicted in Fig. 3. First it is checked if a sufficient number of stars is available to enable successful identification. For the acquisition mode this number needs to be at least 3. If a sufficient number of observed stars is available for identification, match groups are generated (see first box). This match group generation process is clarified using the example shown in Fig. 4.

In this example, 8 observed stars (O1 through O8) are available for identification. Four of the 8 stars are guide stars, namely observed stars O1, O2, O4, and O7 which are equal to guide stars G14, G26, G51, and G3 respectively. We will assume that the observed brightness of the 4 observed guide stars deviates less than *tmag* from their predicted values, while the measured 6 distances between the observed guide stars deviates less than *tdis* from their predicted values.

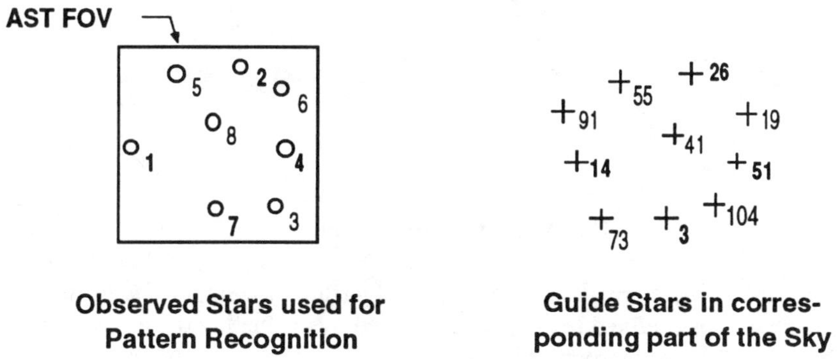

Fig. 4. Example of matching star patterns.

Following generation of all observed star pairs (28 for the example shown in Fig. 4), it is checked which pairs of guide stars match each of the observed star pairs. To make this process efficient, all guide star pairs are stored in the AST guide star database in sequence of increasing angular distance.

As a result of the star pair matching process, match groups are generated. A match group consists of a kernel and a number of members associated with the kernel. The kernel and each member is made up of an observed star and a guide star. Each match group satisfies the following conditions:

1. The measured magnitude of the observed star matches the predicted magnitude of the guide star at the kernel and each of the members.
2. The measured angular distance between the observed star of the kernel and the observed star of a member matches the predicted angular distance between the guide star of the kernel and the guide star of the member.

After having processed all observed star pairs, all match groups (typically 7,000 for the reference AST case specified in table 2) have been generated.

One of the match groups that would be generated for the example shown in Fig. 4 is depicted in Fig. 5. That match group was formed because it was found that the pair of observed stars 1 and 2 matched the pair of guide stars 14 and 26. It also turned out that the pair of observed stars 1 and 4 matched the pair of guide stars 14 and 51. During the processing of the pair of observed stars 1 and 7 it was found to match the pair of guide stars 14 and 3.

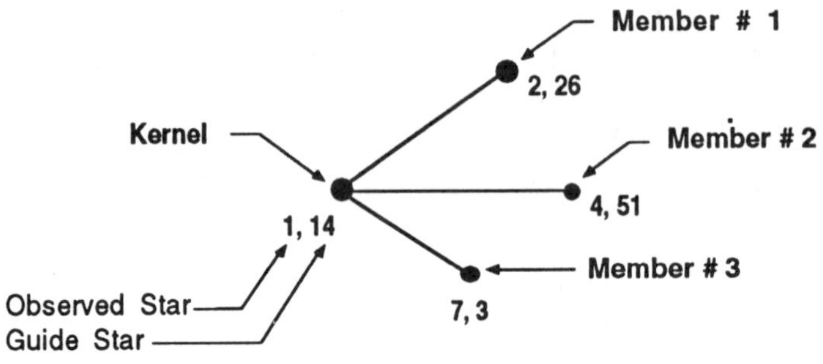

Fig. 5. Match group example

In the next stage of the star pattern recognition algorithm, the upper bound for the size of valid match groups (defined below) is determined by evaluating the match groups. In order to have a valid match group of i star matches (kernel plus i - 1 members), we need at least i match groups with i - 1 or more members as may be seen from the example of Fig. 4 where i = 4. If the upper bound of the match group size is less than the minimum value required (3 or 4 for the acquisition mode), star ID failure is reported.

Next, the threshold for the match group size is established. This threshold is one of the control parameters of the star pattern recognition algorithm. It is usually set at the upper bound of the match group size minus 2. All match groups that are smaller than the threshold are eliminated in the following operation (second box in flow diagram).

The remaining match groups are then validated. This is an operation where it is checked if each of the angular distances between the members of the match groups satisfies the distance tolerance. For the example match group shown in Fig. 5, the validation process involves the checking of three additional angular distances, which are indicated by the dashed lines in Fig. 6.

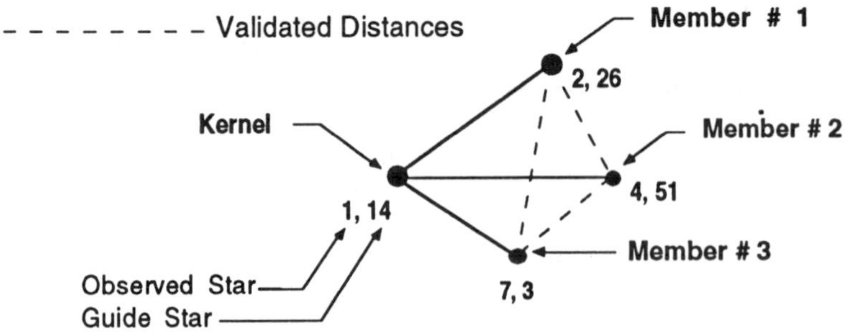

Fig. 6. Match Group Validation.

Next, redundant groups are eliminated which is followed by eliminating match groups where the pattern of guide stars is the mirror image of the pattern of observed stars. In the example shown in Fig. 4, there would be three match groups that are redundant to the one shown in Fig. 6. One has observed star 7 and guide star 3 (7,3) as its kernel with (1,14), (2,26), and (4,51) as members, while the other two have (4,51) and (2,26) as their kernels. Finally, the largest remaining match group(s) is (are) extracted.

If the size of the extracted group(s) is insufficient and iterations are permitted, the process is repeated, starting with thresholding of the initial match groups, using a threshold that is lowered by 1. Iterations are permitted only if the algorithm allows it and the threshold has not reached its minimum value yet. If iterations are not allowed, the algorithm determines that it is incapable of star identification and reports a failed star ID situation. If the size of the extracted match group(s) is sufficient and only one valid match group of sufficient size is left, the star identification is deemed to be successful.

If there are multiple, non-redundant valid match groups, and the algorithm allows ambiguity resolution by picking the match group with the highest match quality, the latter group is selected and the star identification is deemed successful. The RSS value of the distance errors is used as the criterion for finding the best match group. If picking the best match group is not allowed and iterations are permitted, the threshold is lowered by 1 and the process is repeated starting with thresholding of the initial match groups. Failure is reported if iterations are not allowed.

6. SIMULATION PROGRAM

Figure 7 gives an overview of the computer simulation program. A simulation run consists of: (1) creation of the guide star database, (2) generation of the observable sky, (3) evaluation of the star identification performance at the defined sky locations, and (4) presentation of the simulation results.

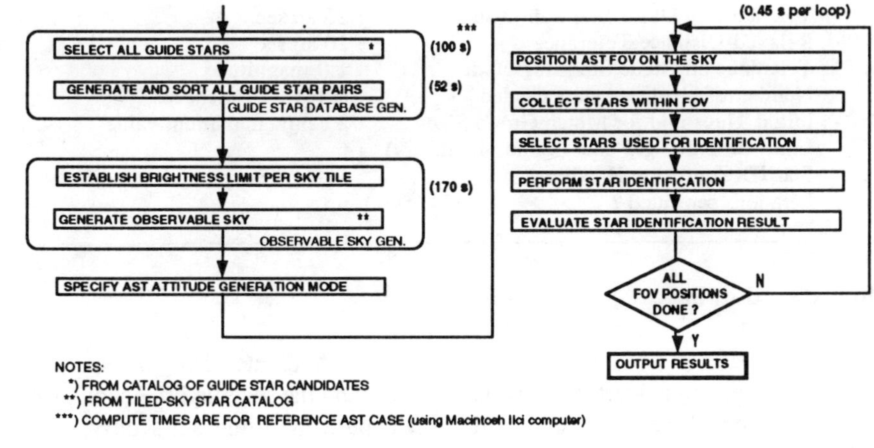

Fig. 7. Simulation overview.

The parameters used to simulate the reference AST case are shown in table 2. The numbers quoted below are for the reference AST case, while the compute times assume use of a Macintosh IIci computer without a cache card.

Table 2. Parameter values for reference AST case.

Operation	Parameter Description	Parameter Value
Guide Star Database Generation	FOV Shape	: Circular
	FOV Area	: 100 square degrees
	Guide Star Density	: 10 per AST FOV
	Close Disturbing Star Radius	: 240 arcsec
	Close Dist. Star Magnitude Margin	: 3.0 magnitudes
	Guide Star Selection Mode	: Brightest Stars in 110 tiles
	Guide Star Pair Distance Limit	: 11.3 degrees
Observable Sky Database Generation	Catalog Completeness	: 95%
	Catalog Reliability	: 95%
	Catalog Magnitude Error	: 0.0 magnitude
	AST Sensitivity Limit	: 7.5 magnitude
	AST Magnitude Error	: 0.3 magnitude (1 σ)
	AST Spatial Accuracy per axis	: 5 arcsec (1 σ)
	Star Clumping Factor	: 3x mean density
	Time from Guide Star (GS) Epoch	: 10 years
	Magnitude Threshold for Observable Stars	: 0.42 mag > faintest reach. GS
	Magnitude Margin relative to Threshold	: 0.77 magnitude
	Magnitude Limit for Close Star generation	: 4.0 mag above faintest GS in tile
Star Identification Testing	FOV Positioning Mode	: Prescribed, all sky
	FOV Positioning Pitch	: 2.0475 deg (for 10,000 steps)
	AST Dynamic Range	: 5 magnitudes
	AST Saturation Range	: 5 magnitudes
	Number of Stars used for ID	: 8 (brightest, unsaturated)
	Limit to Number of Saturated Stars	: 3
	Tolerance on Guide Star Pair Distance	: 20 arcsec
	Relaxed Distance Tolerance	: 20 arcsec
	Tolerance on Guide Star Magnitude	: 1.0 magnitude
	Number of identified stars needed	: at least 3 for success
	Initial Threshold for Match Group Size	: 2 below maximum value
	Limit to match grp size (kernel+members)	: 6
	Star ID Confusion Handling	: Select best group immediately
	Iterations permitted ?	: no

6.1. Guide star database

The guide stars are selected from the catalog of guide star candidates using the specified guide star density, the close neighbor acceptance criteria, and the specified selection mode as inputs. Stars with close neighbors will be accepted as guide stars only if the combined brightness of neighbors closer than a specified distance (the close disturbing star radius) is fainter than the guide star brightness by the specified magnitude margin (the close disturbing star magnitude margin). For the reference AST case the required number of guide stars per tile is obtained by simply picking the brightest qualifying stars. The distribution of the guide stars obtained this way is not very uniform and a better method is currently available, while an optimal method is being developed. Given the guide star pair distance limit, all guide star pairs are generated and sorted in sequence of increasing angular distance.

It takes 100 s to select the 4100 guide stars for the reference AST case. The faintest guide stars have a visual magnitude of 6.3. A highly efficient algorithm enables generation and sorting of all 80,436 guide star pairs in just 52 s. This speed allows the amount of AST ROM needed for storing the guide star database to be reduced from 552 Kbytes to 42 Kbytes for applications where an attitude acquisition delay of some 60 s can be tolerated.

6.2. Observable sky generation

The objective of this operation is to generate the AST magnitude and "true" sky position of all stars brighter than the AST threshold. The magnitude and position are computed by combining the error contributions of the AST and the catalog, making it possible to compute these parameters only once per simulation run. The brightness threshold for each tile is established by specifying it relative to the faintest reachable guide star. Guide stars within the area that covers the tile and extends up to a FOV diameter distance beyond the edge of the tile are defined as being reachable.

For the reference AST, the threshold is set at 0.42 mag, which gives a randomly selected guide star a 99% probability of exceeding this threshold, given a magnitude error of 0.3. If the AST sensitivity limit is brighter than the threshold, the threshold is replaced by the sensitivity limit. Magnitude errors are generated for all stars down to the AST threshold magnitude plus a sufficiently conservative margin. In the case of the reference AST, this margin is set at 0.77 magnitude, implying that all stars having at least a 1% probability of exceeding the AST threshold are considered.

The guide stars are processed first. A fraction of the guide stars is made non-observable at random to account for a non-perfect catalog reliability. For the reference AST case, 197 of the 4100 guide stars were eliminated because of this reason. Because of the incompleteness and limited depth of the star catalog, additional "unknown" close stars are generated randomly within the close disturbing star area of the remaining guide stars. The magnitude limit for these stars is specified relative to the faintest guide star in the tile, while their density can be biased using a "clumping factor" different from 1 to reflect the fact that most stars are members of multiple systems. Seven hundred and fifty unknown close disturbing stars were generated for the reference AST case and added to the 321 known ones.

The guide star position errors include the effect of the star position error in the catalog, the contribution of the AST, the effect of uncompensated proper motion, and the effect of any close disturbing stars. The latter effect is incorporated by computing the combined centroid location. The guide star magnitude errors are generated randomly using the 1 σ value of the sum of the error in the catalog and the AST error. The effect of close disturbing neighbors is accounted for by adding their brightness to that of the guide star.

Next, magnitude errors are randomly assigned to all non guide star catalog stars down to the aforementioned conservative limit. All stars with a "true" instrument magnitude brighter than the AST threshold are retained. For the reference AST case, a total of 6,432 of these stars were collected. Additional observable stars are generated randomly to account for the lack of catalog completeness. A total of 535 of these stars were generated for the reference AST case. The catalog completeness used for the simulation is modeled as a function of the magnitude. This model is obtained by evaluating the curve describing the logarithm of the integrated density of the stars in each tile versus magnitude and postulating that this relation should be linear.

It takes 170 seconds to generate the 11,067 observable stars of the reference AST case. Among them are 3,903 guide stars.

6.3. Star identification testing

The test sequence consists of positioning the AST FOV on the sky, collection of the observable stars within the FOV, selection of a subset of these for identification use, star identification, and evaluation of the identification result. The FOV can be positioned using the prescribed, all-sky method, the prescribed single position method, or the all-sky random method. The first method, which allows up to 10,000 evenly distributed positions covering the entire sky is very useful for detecting problematic sky areas. Having detected these difficult spots, the single position method can be used to diagnose the identification failures.

Following collection of the observable stars within the FOV, it is attempted to select the desired number for identification use (8 for the reference AST). The selection of these stars is influenced by the dynamic range constraint, the AST anti-blooming capability (defined in terms of a saturation range), and the number of stars allowed to be in saturation. In addition to the number of observed stars desired for the identification process, the tolerance on the guide star pair distance, the relaxed distance tolerance (applicable when performing the match group validation), and the magnitude tolerance need to be provided to the star pattern recognition algorithm. For the reference AST case the tolerances are selected such that the probability of meeting each of these tolerances is at least 99%. It is also necessary to specify how to resolve star ID confusion cases (more than one final match group) and whether or not to allow iterations.

The average time needed for performing one loop is 0.45 s for the reference AST case. The high speed is made possible, among other things, by storing the guide star database and the observable sky database in RAM.

7. SIMULATION RESULTS

7.1. Initial results

The simulations show that the reference AST (see table 2), operating in its acquisition mode, is capable of successful star identification for 9,934 out of 10,000 evenly spaced

FOV locations that cover the entire sky. The AST could not identify the stars at 64 locations (failed ID), while there were only two cases where the AST erroneously deemed the identification successful (false ID). The two false identifications were among the 213 cases where only 3 stars were identified. All cases where 4 or more stars were identified were successful. Using a Macintosh IIci computer, identification took 0.45 seconds on the average, with a minimum of 0.20 and a maximum of 0.81 s. The result is shown in the first row of table 3.

Table 3. Initial simulation results.

Case	AST Accu-racy, 1 σ (arcsec)	Mag. Tol.	Dist. Tol. (arcsec)	ID Results*			ID Compute Time**			Needed Array Space (Kbyte)
				Success	Failed	False	Mean (s)	Min. (s)	Max. (s)	
Reference AST	5	1.00	20	9,934	64	2	0.45	0.20	0.81	210
Tight Brightn. tol.	5	0.77	20	9,931	68	1	0.37	0.18	0.63	170
"No" Brightn. tol.	5	2.00	20	9,934	59	7	0.65	0.22	1.38	230
Less accurate AST	10	0.77	37	9,925	66	9	0.57	0.23	1.68	220

*) For a total of 10,000 evenly distributed FOV locations, covering the entire sky
**) Using a Macintosh IIci computer

The majority of the failures (56) are due to having less than 3 guide stars among the observed stars, clearly an indication that a more sophisticated guide star selection method is needed. Of the 9936 cases that were deemed successful by the star pattern recognition algorithm, there were 4730 cases with only one final match group and 5206 cases with multiple (up to 5) final match groups where the best match group was picked. In each of these groups there was 1 false identification. The minute probability of false identification can be eliminated by implementing some minor modifications to the algorithm.

The reference AST requires 210 Kbytes of data or work space to perform the mathematical operations associated with the identification process.

7.2. Sensitivity to brightness tolerance

In addition to picking the brightest observed stars for the identification process, the actual brightness of the stars is also used as one of the match criteria. Table 3 shows the effect of the brightness tolerance imposed on the guide stars. Reduction of this tolerance from 1 to 0.77 magnitude is beneficial in terms of (1) ID compute time, (2) amount of work space needed, and (3) robustness as may be seen by comparing row 2 with row 1. However, tightening of the tolerance decreases the success probability slightly from 99.34 to 99.31%.

In some applications (e.g., the ACRV) the AST may have been dormant for a long time prior to its use. Consequently, there may be a substantial error in the overall AST sensitivity. To determine the performance of an AST with unknown sensitivity, the star brightness was effectively eliminated from the recognition algorithm by opening up the tolerance to 2 magnitudes. The result, shown in row 3 of table 3, indicates that the success rate is hardly affected by elimination of the brightness criterion. False star identification probability, ID compute time, and work space do increase however.

7.3. Sensitivity to spatial accuracy

A comparison with the star trackers shown in table 1 shows that the 5 arc second (1 sigma) accuracy used for the reference AST, which translates into 0.06 pixels, should be attainable. A more pessimistic 10 arc second accuracy results in a moderate degradation of the success rate, an increase in false identification probability, and increases in ID compute time and work space as may be seen by comparing rows 4 and 2 in table 3.

7.4. Performance improvement

Failed and false star identifications were found to be caused by a lack of identifiable guide stars among the observed stars at certain locations of the sky. This problem can be alleviated by implementing one or more of the following changes: (1) adoption of a more sophisticated guide star selection algorithm, (2) increasing the number of observed stars used for identification, (3) allowing larger match groups, (4) permitting iterations in the identification algorithm, and (5) raising the total number of guide stars.

An improved guide star distribution was obtained by dividing the 110 main sky tiles into a total of 446 sub tiles and selecting the guide stars such that all main and sub tiles have their desired number of guide stars. This change increased the success rate from 99.34 to 99.44% as is indicated in Fig. 8 by the filled circle. If the reliability and completeness of the catalog from which the guide stars are extracted were 100% instead of the very conservative 95% assumed, the success rate would climb further to 99.72% (open circle).

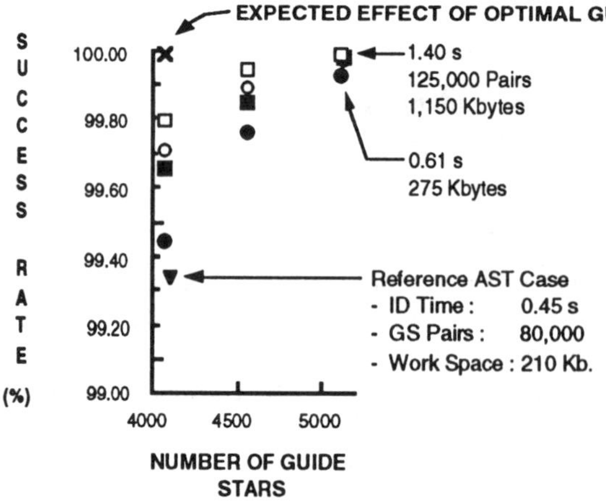

Symbol	Catalog Completeness	Catalog Reliability	Iteration OK ?	Numb. of ID Stars	Star Group Size Cap
□	1.00	1.00	Yes	12	11
■	0.95	0.95	Yes	12	11
○	1.00	1.00	No	8	6
●	0.95	0.95	No	8	6

Guide Star Selection :

- Ref. AST Case : Select reqd number of stars (brightest) in all 110 Sky Tiles.
- Other Cases : Add stars to obtain reqd number of stars in all 446 Sub Tiles

Fig. 8. Demonstrated improvement of star ID success rate and predicted benefit of optimal guide star selection.

By allowing the star pattern recognition algorithm to perform iterations for resolving identification failures and ambiguity, by increasing the number of observed stars used for identification from 8 to 12, and by raising the cap on the match group size from 6 to 11 members, a success rate of 99.66% is obtained (filled square in Fig. 8). If the catalog were perfect, the rate would go up to 99.79% (open square). The price for implementing these three changes consists of an increase in average ID compute time from 0.45 to 0.94 seconds and an increase in work space from 210 to 490 Kbytes.

Raising the number of guide stars further increases the success rate as is shown in Fig. 8. For the case where iterations are not allowed (circles), the success rate climbs to between 99.93 and 99.97%, depending upon the quality of the star catalog, when a total of 5116 guide stars is used. The larger number of guide stars raises the compute time to 0.61 s, increases the work space needed to 275 Kbytes, and necessitates the storage of 124,963 guide star pairs. Not using the guide star brightness information (tolerance was opened up to 2 magnitudes), had no effect on the success rate but did increase the compute time from 0.61 to 0.94 s and raised the work space to 300 Kbytes.

For the case where iterations are permitted (squares), a virtually perfect success rate of between 99.98 and 100% is realized when 5116 guide stars are used, at a cost of a 1.4 s ID compute time and 1,150 Kbyte of work space.

From the foregoing it may be concluded that (1) an identification success rate of better than 99.98% can be achieved, and (2) the crude method used to achieve this high rate is not very attractive. A better way to obtain a near-perfect success rate is to adopt an optimal guide star selection algorithm that ensures an adequate probability of having at least 4 identifiable guide stars among the observed stars everywhere in the sky. This can be accomplished by adapting the guide star density to local sky conditions.

Implementation of this optimal guide star selection algorithm is expected to result in a star identification success rate of at least 99.99% while reducing the amount of guide stars to less than 4100. In addition, it will not be necessary to perform iterations, 8 observed stars will suffice, and the match group size can be capped at 6. Consequently, the work space, ID compute time, and guide star database size of an AST with an 11.3 degree FOV diameter and optimally selected guide stars will be the same as those of the reference AST.

7.5. Memory requirements

The memory required for identification can be divided into: (1) memory for storing the identification code, (2) data or work space for performing the computations, and (3) memory for the guide star database. Storage of the identification code requires 17 Kbytes of ROM. An AST with an 11.3 degree FOV diameter and optimally selected guide stars requires a work space of 210 Kbytes of RAM as can be determined from simulations of the reference AST. Storage of the coordinates (12 bytes per star) and brightness (2 bytes per star) of the 4100 guide stars requires 57.4 Kbytes of ROM. A total of 485 Kbytes of memory is needed to store the some 80,500 guide star pairs (at 6 bytes per pair).

If the attitude acquisition function needs to be available instantaneously, the guide star pairs need to be stored in ROM. If a delay can be accepted, it may be preferable to generate the guide star pairs onboard and store them in RAM at the time when attitude acquisition is needed. An impression of the delay time may be obtained by noting that it takes 52 seconds to generate and sort 80,500 guide star pairs on a Macintosh IIci. The code performing this function would occupy 10 Kbytes of ROM.

7.6. Impact of FOV size

Using simulations, it was determined how the size of the FOV impacts: (1) the required star tracker sensitivity, (2) the number of guide star pairs, (3) the star ID compute time, (4) the required work space, and (5) the time required for generating and sorting all guide star pairs. The result for the first 4 parameters is shown in table 4 for FOV diameters ranging from 11.3 to 30 degrees. The time required for generating and sorting all guide star pairs ranges from 5 seconds for the 30 degree FOV diameter to 52 seconds for the 11.3 degree diameter. The average guide star density and relative spatial accuracy was 10 per FOV and 0.00012 FOV diameter (1 sigma) respectively for all 4 cases.

Table 4. Impact of AST field of view.

FOV Diameter (deg)	FOV Area (deg^2)	Reqd Sensitivity (magn.)	Number of Guide Stars*	Number of Guide Star Pairs	Spatial Accuracy, 1σ (arcsec)	Star ID Compute Time** (sec)	Work Space Reqd (Kbyte)
11.3	100	6.8	4000	80,000	5	0.45	190
16.0	200	6.3	2000	40,000	7	0.21	85
23.0	400	6.1	1000	20,000	10	0.11	45
30.0	700	5.2	600	12,000	15	0.07	25

*) 10 Guide Stars per FOV on average
**) Using Fortran 77 on Macintosh IIci

It may be seen that the number of guide stars, guide star pairs, the ID compute time, and the required work space are inversely proportional to the area of the AST FOV.

A preliminary analysis has shown that a high precision (1 arc second) AST with a 4 by 4 degree FOV is probably feasible. Using extrapolation, it follows that such a device would require a database of 25,000 guide stars and 500,000 guide star pairs, which would occupy 3.4 Mbytes of memory. A work space of 1.2 Mbytes would be required, while the ID compute time would be less than 3 s, assuming use of a Macintosh IIci equivalent flight computer. It should be noted that this compute time as well as the ones provided in table 4 would probably be reduced by a factor of 10 if a 33 Mhz R3000 compatible RISC processor were used.

8. AST DEVELOPMENT BY OTHER ORGANIZATIONS

JPL is testing a 15 by 15 degree FOV prototype AST equipped with a 1024 x 1024 pixel CCD. Their identification algorithm was developed in cooperation with the author. Initial star identification tests at the Table Mountain Observatory were successful.

JENA-Optronik in Germany is developing an advanced ASTRO 1M star tracker system comprising 3 trackers, each with a 5.3 by 8 degree FOV.[22] This system is capable of fully autonomous star identification as was demonstrated at the Karl Schwarzschild Observatorium in Tautenburg. The identification is however very slow, requiring up to 5 minutes.

OCA is developing an AST, referred to as the Advanced Star Tracker Assembly (ASTA). ASTA comprises dual-redundant star cameras and a R3000 compatible 33 Mhz LR33000 RISC microprocessor. Each camera incorporates a 384 x 576 CCD and has a lens that provides a 14 mm aperture and a FOV with a 23 degree diameter. It is designed to have a mass of only 1 kg, a power consumption of 4 W, and an accuracy of 7 arcsec (1 sigma). This accuracy corresponds to 0.03 pixel, which, if realized, would make it a factor 2 to 3 more accurate than the much narrower FOV CCD trackers in use today (see section 2.3.).

ASTA star identification is based on matching triplets using an algorithm due to IDI that is based on Groth's method. A database of 1024 guide stars is needed (density of 10 per FOV). Noting that A_{sky} / A_{fov} is 99.3, and using equation (1), it is found that 120,000 triangles need to be stored, requiring 1.4 Mbytes of memory without compression. This is to be compared with the need to store 20,000 guide star pairs, requiring 120 Kbytes of memory when using the method described in this paper. This illustrates that the star triplet matching scheme is incompatible with precision ASTs which have a moderate FOV.

9. CONCLUSIONS

Autonomous star trackers (ASTs) are currently being developed by various organizations in a number of countries. An AST is capable of (1) determining its attitude without having a-priori attitude knowledge, (2) autonomous attitude updating, and (3) providing its attitude at frequencies up to typically 40 Hz. These capabilities are needed for future advanced attitude control and navigation systems. They can also be exploited to improve spacecraft reliability, mass, power, and cost, and, in addition, to reduce the cost of spacecraft operation. Fabrication of miniature ASTs is made possible by the availability of accurate, anti-blooming capable CCDs, fast microprocessors, high density memory chips, and star pattern recognition algorithms.

An efficient, highly robust star pattern recognition algorithm was adapted and refined at Lockheed for use with ASTs. The algorithm finds the largest group of observed stars that matches a group of guide stars and maximizes its success rate by exploiting the full information contents of a star pattern. The database for this algorithm consists of the position and brightness of the guide stars plus a sorted list of all guide star pairs that fit in the FOV of the AST. The storage space required for this database is an order of magnitude smaller than that of an algorithm based on matching stored guide star triplets. Because of its memory efficiency, the algorithm makes it practical to build accurate, moderate-FOV ASTs.

Realistic computer simulations show that an AST with an 11.3 degree FOV diameter, a spatial accuracy of 5 arc seconds (1 σ), a brightness accuracy of 0.3 magnitude (1 σ), a database of 4100 guide stars (all brighter than m_V =6.8), a 25 MHz MC68030 class microprocessor, and 800 Kbytes of memory is capable of determining its attitude in 0.45 s while having no a-priori attitude knowledge. It can do so with a success probability of between 99.44 and 99.72%, depending on the quality of the catalog from which the guide stars are selected. Furthermore, it has been demonstrated that this success rate can be raised to between 99.98 and 100% by improving the guide star coverage of the sky. It will be possible to achieve this near-perfect success rate without penalizing the compute time or memory size through use of an optimal guide star selection technique.

While it reduces compute time and memory requirements, use of star brightness information has little or no effect on the star identification success rate. This implies that, although desirable, it is not necessary to know the overall sensitivity of an AST. Hence, an AST can be left in a dormant state for an extended period of time, which may be an advantage for certain applications (e.g., the Assured Crew Return Vehicle).
The compute time needed for identification, the amount of memory required, the number of guide stars, and the number of guide star pairs are all inversely proportional to the AST FOV area. This relationship holds for a fixed guide star density in terms of number of stars per FOV, and a fixed relative spatial accuracy. It is probably feasible to realize a high-precision AST with a FOV of only 4 by 4 degrees. The device would require 4.6 Mbytes of memory and should be capable of identifying the stars within its FOV in less than 3 seconds. This time could be reduced by an order of magnitude if a more advanced microprocessor were used.

Autonomous star identification has been tested on an experimental Brilliant Pebbles sub-orbital flight using a prototype miniature wide-FOV star tracker. In addition, other AST prototypes have been tested at observatories in the USA and Germany. The star identification algorithm used at Lockheed will be tested in 1993 using an AST prototype and the true sky.

10. ACKNOWLEDGEMENTS

The author gratefully acknowledges the Astronomical Data Center at the NASA Goddard Space Flight Center for providing the SKYMAP star catalog. He also would like to thank Paul Reshatoff and Ted Havas for their outstanding technical support.

11. REFERENCES

1. J. R. Janisick and S. T. Elliott, "History and Advancements of Large Area Array Scientific CCD Imagers," *Astronomical Society of Pacific Conference Series*, D. H. McNamara , editor, Vol. 23, *Astronomical CCD Observing & Reduction Techniques*, S. B. Howell, editor, BookCrafters Inc., 1992.

2. R. W. H. van Bezooijen, *Automated Star Pattern Recognition,* Ph. D. Dissertation, Stanford University, Stanford, California, June 1989.

3. R. W. H. van Bezooijen, "A Star Pattern Recognition Algorithm for Autonomous Attitude Determination," *Preprints of XIth IFAC Symposium on Automatic Control in Aerospace,* pp. 61-68, Tsukuba, Japan, July 1989.

4. R. W. H. van Bezooijen, "Success Potential of Automated Star Pattern Recognition," Paper AIAA-86-0254, *AIAA 24th Aerospace Sciences Meeting*, Reno, Nevada, Jan. 1986.

5. R. W. H. van Bezooijen, K. R. Lorell, and J. D. Powell, "Automated Star Pattern Recognition for use with the Space Infrared Telescope Facility (SIRTF)," *Preprints of 10th IFAC Symposium on Automated Control in Space*, pp. 59-67, Toulouse, France, June 1985.

6. F. A. Bogges et al., "The IUE Spacecraft and Instrumentation," *Nature*, Vol. 275, pp. 372-377, Oct. 5 1978.

7. W. J. Christis, "The Optical Sensors of the Netherlands Astronomical Satellite (ANS), the Star Sensor," *Philips Tech. Rev.*, Vol. 34, no. 8, pp. 218-224, 1974.

8. R. Holdaway, editor, "Infra-red Astronomical Satellite," *JBIS*, Vol. 36 no. 1 (special issue on IRAS), Jan. 1983.

9. R. W. H. van Bezooijen, "IRAS Attitude Control Subsystem," *Proceedings of AOCS Conference, held in Noordwijk, Netherlands, 3-6 October 1977, ESA SP-128*, pp. 21-27, Nov. 1977.

10. A. Pouw, "System Design and Operational Flight Experience of the IRAS Attitude Control Subsystem," Paper AIAA-84-0152, *AIAA 22th Aerospace Sciences Meeting*, Reno, Nevada, Jan. 1984.

11. R. R. Stephenson, "The Galileo Attitude and Articulation Control System," *Preprints of 10th IFAC Symposium on Automated Control in Space*, pp. 85-92, Toulouse, France, June 1985.

12. M. M. Birnbaum, R. L. Bunker, and J. T. Tavolacci, "A Radiation-Hardened Star Scanner for Spacecraft Guidance and Control," *J. Guidance*, Vol. 6, no. 1, pp. 39-46, 1983.

13. L. Merken and G. Singh, "Galileo Attitude Determination: Experiences with a Rotating Star Scanner," *Proceedings of the ESA Symposium on Spacecraft Flight Dynamics, held in Darmstadt, Germany, Sep. 30-Oct 4, 1991, ESA SP-326*, pp. 203-211, Dec. 1991.

14. I. Lewis, A. Ledebuhr, T. Axelrod, J. Kordas, and R. Hills, "WFOV Star Tracker Camera," *SPIE's International Symposium on Optical Engineering & Photonics in Aerospace Sensing*, Orlando, Florida, April 1991.

15. *Aviation Week & Space Technology*, p. 69, May 6, 1991.

16. J. Junkins and T. E. Strikwerda, "Autonomous Star Sensing and Attitude Estimation," Paper AAS 79-013, *Annual AAS Guidance and Control Conference*, Keystone, Colorado, 1979.

17. J. W. Bone, D. S. Parkinson, and J. J. Rodden, "Stellar Attitude Data Handling," *Guidance and Control 1992*, Vol. 78, *Advances in the Astronautical Sciences*, R. D. Culp and R.P. Zietz, editors, Univelt Inc, San Diego, California, 1992.

18. E. C. Wong and J. Y. Lai, "Autonomous Attitude Determination from Star Data for a Dual-Spin Planetary Spacecraft," *AGARD Spacecraft Pointing and Position Control, AGARD-AG-260*, P. Ph. van den Broeck and S. Z. Szirmay, editors, Neuilly sur Seine, France, Nov. 1981.

19. E. J. Groth, "A Pattern Matching Algorithm for Two-dimensional Coordinate Lists," *The Astronomical Journal*, Vol. 91, no. 5, pp. 1244-1248, 1986.

20. E. W. Dennison, R. H. Stanton, and K. Shimada, "The Development of a Charge-Coupled Device Tracker for Spacecraft," Paper AAS 87-007, *10th Annual AAS Guidance and Control Conference*, Keystone, Colorado, Feb. 1987.

21. L. W. Cassidy and R. Abreu, "Star Trackers for Spacecraft Applications," *Proceedings of SPIE Conference on Acquisition, Tracking, and Pointing IV, held in Orlando, FL, Apr. 19-20, 1990*, Vol. 1304, pp. 58-74, SPIE, Bellingham, WA, 1990.

22. Ch. Elstner, D. Ratzsch, and W. Skarus, "Star Pattern Recognition Sensor of the ASTRO Type," *Preprints of the 12th IFAC Symposium on Automatic Control in Aerospace*, pp. 557-561, Ottobrunn, Germany, Sep. 1992.

23. G. R. Hopkinson, R. A. Cockshott, D. J. Purll, M. D. Skipper, and B. Taylor, "Operation, performance, and reliability testing of charge-coupled devices for star trackers," *Optical Engineering*, Vol. 26, no. 8, pp. 725-733, Aug. 1987.

24. D. Purll, N. Gradmann, M. Bollner, "The ROSAT Star Tracker - Flight Experience," *Proc. First Internat. Conf. on 'Spacecraft guidance, navigation and control systems,'* ESTEC, Netherlands, June 1991, ESA SP-323, pp. 551-556, Paris, Dec. '91.

25. E. Shalom, J. W. Alexander, and R. H. Stanton, "Acquisition and Track Algorithms for the ASTROS Star Tracker," Paper AAS 85-050, *Annual Rocky Mountain Guidance and Control Conference*, Keystone, Colorado, Feb. 1985.

26. P. Eisenhardt and G. G. Fazio, "The SIRTIF concept and its realization," NASA Ames Research Center, Moffett Field, CA, Nov. 17 1988.

27. R. F. Draper, "The Mariner Mark II Program", Paper AIAA 88-0067, *AIAA 26th Aerospace Sciences Meeting*, Reno, Nevada, Jan 1988.

28. U. Renner and H. Meier, "Definition Study of a CCD Rate Sensor, Final Report," ESTEC Contract No. 7718/88/NL/DG, Institut fur Raumfahrtelectronik, Berlin, March 1989.

29. S. F. McLaughlin, "SKYMAP Star Catalog, version 3.5," National Space Science Data Center, Goddard Space Flight Center, Greenbelt, Maryland, 1989.

30. J. P. McQuerry, D. D. Wagner, M. L. Sullivan, R. A. Deters, M. A. Radovich, "A New Generation Stellar Attitude Sensor: The CT-601 Solid-State Star Tracker," *Proc. First Internat. Conf. on 'Spacecraft guidance, navigation and control systems,'* ESTEC, Netherlands, June 1991, ESA SP-323, pp. 607-611, Paris, Dec. '91.

Laser Remote Sensing in Atmospheric Sciences, Aviation Safety, Aeronautical Research, and Experiments from Space Platforms and High Flying Aircraft

Frank Allario, Philip Brockman, R. R. Antcliff, William B. Grant
NASA Langley Research Center
Hampton, Virginia 23681

Laser Remote Sensing in Atmospheric Sciences, Aviation Safety, Aeronautical Research, and Experiments from Space Platforms and High Flying Aircraft

Traditionally, the term laser remote sensing has been associated with active, optical measurements of the Earth's atmosphere, lands, and oceans. In this paper, we concentrate our overview of laser remote sensing upon the Earth's atmosphere in three disciplines: Atmospheric sciences, aviation safety, and aeronautical research. In atmospheric sciences, laser remote sensing has played a prominent role in the measurement of clouds, aerosols, the planetary boundary layer, chemical species, metals and ions, and in atmospheric dynamics in the temporal tracking of physical parameters and the direct measurement of atmospheric winds. Quite recently, laser remote sensing has been especially effective in correlative studies from ground and airborne platforms related to scientific studies in the eruption of Mount Pinatubo, and in following the dispersion of associated aerosols in the atmosphere, both in latitude and longitude. Laser remote sensing has also been very effective in studies related to the formation of the ozone hole, in the antarctic and arctic regions. Range-resolved measurements of atmospheric ozone have been made which track "in real time" the formation of the ozone hole and its subsequent dissipation. Laser measurements of the depolarization ratio of backscattering from particulates in the region of the ozone hole where Polar Stratospheric Clouds (PSC's) form have provided unique information on the physics of the PSC's and on the dynamics of the formation of the ozone hole phenomena. It is quite clear that laser remote sensing has proven to be an invaluable measurement technique for these types of chemistry investigations. The range-resolved measurement of atmospheric water vapor, correlated to the height of the planetary boundary layer and the distribution of aerosols over land and oceans, has also been demonstrated quite recently to be a unique measurement provided by laser remote sensing from aircraft. When this technique is developed from high flying aircraft and/or satellites, a major measurement technique will be available for the study of the hydrological cycle, globally. Soon, we should be seeing range-resolved measurements of atmospheric water vapor (50 meters) from a high-flying aircraft, with an accuracy better than 10 percent, as a routine measurement in atmospheric sciences. The historical evolution of laser remote sensing from the initial ground-based measurements of atmospheric aerosols in the early 1960's to the sophisticated measurements from aircraft of today represent a unique evolution of technology in lasers and electro-optics, coupled to persistent attention to sound engineering development of a unique technique. For this paper, I have asked Dr. William B. Grant, a pioneer in laser remote sensing of the atmosphere, to provide this section entitled "Laser Remote Sensing in Atmospheric Sciences."

The application of laser remote sensing to the measurement of wind velocities in the atmosphere has been the subject of a variety of research programs over the past 20 years. In atmospheric sciences, the direct measurement of winds in the Earth's lower atmosphere, to an accuracy of (+/-) 1 m/s, from space has been identified as one of the most important measurements in atmospheric sciences for weather forecasting and for improving understanding of the dynamics of the environment. This measurement need provided significant technological development of wind related laser sensing systems. As a result of the fatal accident at the Dallas-Fort Worth Airport in 1985 (Wingrove and Bach, 1987), there arose a national awareness of the role that active remote sensors, both radar and laser sensors, could play in the early detection of a windshear event from an airborne aircraft. As a result of this awareness and some earlier research from aircraft with carbon dioxide lasers, the NASA/FAA National Windshear Program was developed and in the summer of 1992, a series of measurements were conducted from the NASA Langley Research Center (LaRC) Boeing 737 aircraft in the vicinity of the Denver Stapleton Airport and the Orlando Florida Airport. The objective of these series of flights was to identify the ability of an airborne radar and a laser radar to detect microbursts in the atmosphere from an aircraft both in relatively dry and wet atmospheric conditions. The results are currently being thoroughly analyzed; however, preliminary evaluations indicate that indeed both the radar and the laser can detect microbursts with sufficient advanced warning for a pilot to take evasive action. The degree of accuracy for either a radar or laser radar instrument is dependent upon the meteorological conditions of the microburst (i.e., wet or dry microburst). A recent article summarized the details of the National Windshear Program and the preliminary results from the series of flight experiments was published in Aerospace America (Lewis, 1993). In addition to the use of laser remote sensing for detection of microbursts, NASA has explored the use of a laser remote sensor to provide range-resolved wind measurements in the vertical column ahead of a Shuttle, prior to the launch. Again, in this experiment, it was demonstrated that laser remote sensing is a powerful technique for the early detection of winds in the corridor of air, ahead of a launch vehicle, and especially in the measurement of high altitude winds and windshear. Laser remote sensing could provide unique information to a launch Director for the Shuttle or an Expendable Launch Vehicle. For this section of the paper, I have asked Mr. Philip Brockman, a laser researcher and one of the major technical consultants in the NASA/FAA Integrated Windshear Program to provide the overview of the section on Aviation Safety and Meteorology.

Long before there was public interest in atmospheric sciences and in aviation safety, there was national interest in the development of the airplane, for commercial and military purposes. Experimental testing of aeronautical vehicle designs in wind tunnels has been one of the major steps in taking a vehicle design from a concept to the reality of a commercial airplane. In fact, it is appropriate in this year, the 75th anniversary of the Langley Research Center, that a section summarizing the use of laser remote sensing for aeronautical research in wind tunnels be developed (Winds of Change, NASA Publication 130). With the planned resurrgence in aeronautical research for the National Aero-Space Plane (NASP), the High-Speed Research (HSR) aircraft, and the new initiatives in the development of technologies for the advanced subsonic aircraft, it is important to summarize in one place the unique measurements that laser techniques can make, and to recognize that with the emergence of powerful new analytical tools (Computational Fluid Dynamic (CFD) codes), laser measurements are indispensable for future wind tunnel investigations. I asked Dr. Richard R. Antcliff of the Langley Research Center to compile an overview of the evolution of laser sensing, focussing on wind tunnel experiments and to present the status of current, nonintrusive measurements using laser techniques. Dr. Antcliff is currently the Branch Chief of the Nonintrusive Diagnostics Branch, and is responsible for the advanced development and application of most of the laser techniques discussed in this section, and for ensuring they are quickly infused into wind tunnel operations at LaRC. It is interesting to note the wide range of laser techniques which have been explored and applied in wind tunnel experiments. Many of these experiments require complex electro-optical subsystems and very high accuracy, in signal to noise. However, especially in the very large wind tunnels where high flow rates at both high and low temperatures are encountered, the mechanical vibrations impinging upon an instrument are very high. The challenge which is being met by the laser instrumentation engineer is to obtain sophisticated measurements of fluid flow parameters, with high spatial and temporal resolution, in very short run times, and under extremely harsh conditions. In many cases, environmental conditions are more severe than either space or aircraft conditions. However, these engineering challenges are being met in wind tunnel experiments at LaRC and in other aeronautical research laboratories in the country (McKenzie, et al., 1992)

Ideally, laser remote sensing for the study of the Earth's atmosphere, the oceans, and land surfaces should be conducted from a space platform to obtain global measurements of Earth science parameters. Indeed, for the NASA Earth Observing System (EOS) program, at least four different categories of laser experiments were proposed by the scientific and engineering communities. Investigations were solicited and proposed in (1) Laser ranging and altimetry; (2) laser measurements of atmospheric winds; (3) laser measurements of tropospheric water vapor and associated aerosols and the planetary boundary layer; and (4) laser backscatter measurements leading to measurements of stratospheric and tropospheric aerosols, high and low altitude clouds and stratospheric temperatures. Only experiments in the first two categories were approved under the EOS program and even then, the engineering developments have been deferred into later generation EOS experiments. Between now and the year 2000, under the current planning cycle, only the Lidar In-space Technology Experiment (LITE) is scheduled to conduct experiments from space. LITE is a Space Shuttle experiment with the first flight scheduled for June 1994. The LITE experiment falls into the fourth category listed above. In 1994, a complex laser experiment to measure tropospheric water vapor to an accuracy better than 10 percent, with a vertical resolution better than 50 meters, should be operational from the high flying NASA aircraft, the ER-2. This totally autonomous experiment called the Laser Atmospheric Sensing Experiment (LASE), when fully operational and combined with the LITE engineering design, development, and testing database, should provide the confidence to sponsors to endorse a satellite experiment to measure atmospheric water vapor in the Earth's lower atmosphere through the stratosphere. Although the EOS experiment in the third category was not approved for the EOS program, primarily due to the high technical risk of its technological development, the combined LITE and LASE programs, which are close to flight, should finally provide the engineering information for a low risk, satellite experiment to be approved soon. Remarkably, a laser altimeter has been developed and integrated into an experiment currently headed for Mars. We, in the laser community, are proud to see this major accomplishment by the Goddard Space Flight Center. The experiment called the Mars Orbital Laser Altimeter (MOLA) is scheduled to begin collecting data in December 1993, and generically falls into the first category of laser experiments solicited and proposed for NASA's EOS program. Combining the design, development, and testing information of the LITE and MOLA experiments should also lead to an early definition for at least a cloud top lidar, from a low altitude microspacecraft or an Earth probe for Earth observing measurements. In the fourth section of this paper, we have summarized some of the major laser remote sensing experiments currently in one or more stages of advanced development. The listing is not all inclusive since we are aware of at least one space laser experiment planned for a fairly recent flight, utilizing an excimer laser. This experiment was verbally described to one of the authors during a recent visit to Russia (Anon., 1992b). However, this section we believe summarizes the major space experiments currently planned in the western world and Japan or major program developments, leading to a near-term space experiment for civilian applications. The material in this section of the paper is a combination of at least three review papers, recently published (Allario et al., 1991; Theon et al., 1991, Allario et al., 1992a), and represents a short summary and update of the status since the last papers.

In Allario et al., 1991, the major product of the paper was a series of recommendations for the development of new technology which should lead to more economical, lighter weight, and lower power lidar instruments from spacecraft. Again, not much has changed since these recommendations were made. Several initiatives in research have been initiated, and some accomplishments have been realized since the paper was published. In the last section of this paper, we will provide a short update on the status of new lidar technology, achieved or planned for applications to spacecraft. All elements of this new technology are highly beneficial to the other areas of laser remote sensing covered in this paper, and should continue to feed the evolutionary growth of laser remote sensing in atmospheric sciences, aviation safety, and aeronautical research. In the case of aeronautical research, however, these technological innovations should result in the rapid transition of some of the laser techniques currently being used in wind tunnel experiments into flight research, with high performance aircraft. In fact, although the near-term opportunities for additional lidar experiments from spacecraft currently appear slim, the need to measure flight parameters such as air speed, boundary layer transition, density, temperature, and velocities in and around the flight environment of a high performance aircraft are promising and should provide the lidar community with the technological "pull" to develop miniature laser transmitters and receivers comparable to what is required for microspacecraft. The challenge is up to the lidar technologist and instrumentation engineer to "make it happen."

Scientific Measurements with Laser Techniques

Lidar (LIght Detection And Ranging) experiments have been in use for the past 30 years and have provided useful and unique information concerning the physics and chemistry of the Earth's atmosphere. In this section, we present a brief review of the on-going contributions to the study of the Earth's atmosphere. The discussion begins with aerosols and clouds, proceeds to chemical species, and ends with atmospheric dynamics.

Aerosols have received the most attention in lidar studies since they are the easiest atmospheric component to measure using lidar, providing a scattering component in nearly all elastic-scattering lidar measurements of the atmosphere. Stratospheric aerosols were first measured using a ruby lidar in the early 1960's (Ligda, 1963 and Fiocco and Grams, 1964) shortly after Junge described the persistent layer in the middle stratosphere that bears his name (Junge and Manson, 1961). Stratospheric aerosols continue to be studied in the tropics where volcanic perturbations are of considerable interest since they affect ozone (Kinne et al., 1992; Grant et al., in press), in the mid-latitude regions where they affect climate (Lacis et al., 1992), possibly ozone (Hofmann and Solomon, 1989), and in the polar regions where polar stratospheric clouds (PSC's) are studied because of their role in destroying ozone. In addition to the vertical distributions and concentrations of PSC's in the polar regions, lidar measurements have also been used to differentiate between the various types of PSC's. Lidar measurements led to the discovery that the Type 1 PSC's (nitric acid trihydrate) should be divided into two distinct types (Types 1a and 1b), depending on their optical properties. In turn, these distinct types probably reflect different types of origin (Browell et al., 1990a). In the troposphere, aerosols play a significant role in affecting global climate directly through scattering (Charlson et al., 1992; Penner et al., 1992) and absorption, and indirectly by influencing the formation of clouds and their physical and chemical properties once formed (Twomey, 1977 and Albrecht, 1989). Lidar measurements have also played a role by tracking the dispersion and transport of various plumes (Reagan et al., 1989), and quantifying the sulfate mass loading in a plume from a power plant (Brock et al., 1990).

An understanding of the global aerosol backscatter distribution is of critical importance for designing Doppler lidar systems for measuring winds from a space-based platform. Both ground-based and airborne lidar systems have been used in an effort to provide data that can be used for this purpose (Menzies and Post, 1991). Clouds also play a major role in regulating the global climate. One of the major contributions to the study of clouds is that of the Experimental Cloud Lidar Pilot Study (ECLIPS) program, in which a number of lidar systems distributed around the Earth are being used to study the occurrence, vertical distribution, and optical depths of cirrus clouds during intensive measurement periods every year or so (Platt, 1991). Cirrus clouds play a major role in global climate by trapping longwave radiation and can easily be measured by spaceborne lidars. They are difficult to measure using passive satellite instruments because their size and optical depths are generally below detection threshold. In addition, lidar can study the formation of clouds due to the presence of additional Cloud Condensation Nuclei (CCN) in the atmosphere, such as volcanic aerosols which lead to the in situ formation of cirrus clouds (Sassen, 1992).

Turning now to chemical species, water vapor in the atmosphere is an extremely important element of the hydrological cycle (Chahine, 1992). Lidar has been demonstrated to be able to map water vapor distributions, using both Raman and Differential Absorption Lidar (DIAL) techniques (Grant, 1991). Raman lidar has been able to profile water vapor at night to altitudes of 10 km every 2 minutes during the passage of both warm and cold fronts (Melfi et al., 1989) and has been able to study atmospheric moisture structure (Melfi et al., 1991). Such measurements enable more detailed studies of frontal passages than are possible using infrequent balloon sondes. A high-pulsed repetition frequency (prf) Raman lidar system has been developed to study evaporation of water over land, providing useful data on the topic (Eichinger et al., 1992). DIAL systems, which can be airborne and operate during daytime, have been able to map inhomogeneities in water vapor distributions (Browell et al., 1991). It is in the measurement of ozone that the DIAL technique has been most useful to date. Stratospheric ozone is extremely important in providing the shield against the sun's ultraviolet radiation which is being affected by man-made chlorofluorocarbons (CFC's). While over 30 instruments have been developed for the measurement of stratospheric ozone (Grant, 1989), increased information is obtained through the synergism of several instruments opeated simultaneously. The DIAL technique, with its ability to measure vertical profiles of both aerosols and ozone simultaneously, can make important contributions to this on-going study. Ground-based ozone DIAL systems are being developed and demonstrated for use in the Network for Detection of Stratospheric Change (McDermid et al., 1990a,b, 1991). Eventually, there should be six such stations with a DIAL system serving as one of the key instruments. The airborne UV DIAL system (Browell, 1989) has played unique roles in studying both Antarctic and Arctic ozone decreases due to the presence of CFC's, as well as in studying ozone decreases near the Equator due to the presence of aerosols, especially following the eruption of Mt. Pinatubo (Grant et al., in press). Tropospheric ozone has also been well studied using the airborne UV DIAL system during deployment of NASA's on-going Global Tropospheric Experiment (GTE) program. In combination with aerosol measurements, the UV DIAL measurements of ozone have proven to be very useful in characterizing air masses encountered during the GTE airborne missions, especially in complementing the in situ instruments (e.g., Browell et al., 1990b). Stratospheric intrusions, as a source of tropospheric ozone, have also been studied on a number of occasions (Browell et al., 1987; Ancellet et al., 1991). Studies over Brazil have shown the differences in ozone mixing ratios from the wet to the dry seasons (Browell et al., 1990b), and made significant contributions to improving understanding of the effects of biomass burning.

UV DIAL systems have also been used to study other pollutants. Dispersion of plumes containing sulfur dioxide (SO_2) emissions from power plants have been studied a number of times (Beniston et al., 1990). Sources of mercury, such as from geothermal power plants, have also been investigated (Edner et al., 1991). The use of various lidar techniques to measure chemical species in urban plumes from ground-based platforms has reached a high degree of sophistication in several European laboratories and institutes. The ability to make quantitative measurements of chemical species, both range resolved and total column content in direct absorption, is resulting in some cases in the use of lidar as a measurement standard for other techniques. An excellent summary of recent measurements can be found in the proceedings of the SPIE Conference held in Berlin, the Federal Republic of Germany, June 22-26, 1992.

Fluorescence lidar systems from the ground have been used with great success to study metal ions and atoms in the mesopause from 83 to 110 km. Species such as sodium, iron, calcium, potassium, and magnesium have been studied with dye lasers ranging in wavelengths from the near UV to the near IR. Lidar data have been used to suggest that the metal ions and atoms arise from meteorite ablation in this region of the atmosphere (e.g., Batista et al., 1989). While that is probably the general source, Kane et al., 1991, argue that their ground-based and airborne data suggest that the appearance of the Na_s layer, which is linked closely with the E_s layer measured simultaneously using the Arecibo radar, is influenced by atmospheric tides. Kane et al., 1992, have studied the seasonal variations in the mesospheric-Fe layer and are making progress in relating the Fe layer changes to temperature-induced changes in the chemistry of the region. In atmospheric dynamics, lidar systems have also made unique contributions. The direct measurement of winds using a CO_2 laser-lidar system with heterodyne detection has enabled a number of unique case studies of wind phenomena, from a van containing the Doppler lidar to ranges of 10-20 km (Menzies and Hardesty, 1989; Hardesty, et al., 1991). Winds flowing down the eastern slopes of the Rocky Mountains and winds carrying sulfates in the Grand Canyon are two examples. Raman lidar measurements of water vapor, coupled with radiosonde measurements and meteorological data, have provided data on the dissipation of a low-altitude gust front (Koch et al., 1991). Rotational Raman/Rayleigh lidar systems have been used to study atmospheric density, temperature, and winds from the ground level to 90-100 km (Chanine et al., 1989; Adriani et al., 1991; Chanine, 1992). Rotational Raman scattering is used for measurements up to 30 km; the Rayleigh lidar system is used for higher altitudes (Chanine, 1992). As an example of the usefulness of the lidar, temperature measurements have complemented those from a satellite instrument, and demonstrated a temperature decrease of -0.10 ± 0.04 percent per year at 55 km at 44° N, which was expected from the observed increase in the atmospheric concentration of CO_2 (Aikin et al., 1991). A network of several such lidar systems around the Earth would provide a valuable mechanism to monitor the physical state of the upper atmosphere, with higher accuracy and vertical resolution than is possible using passive instruments from satellites. Finally, lidar systems have been able to provide data on gravity waves in the mesopause by using measurements obtained from laser induced fluorescence of sodium atoms. It was found that the frequency spectrum of density in the mesopause was significantly higher than in the stratosphere. The new, high-quality data helped establish a new theory for gravity waves in this region (Senft et al., 1992).

In the preceding section, a short summary of the application of lidar techniques to study the Earth's atmosphere has been presented. The interested reader is also referred to the proceedings of the 16th International Laser Radar Conference held in Boston, Massachusetts, July 20-22, 1992, for examples of some recent applications of lidar for Earth remote sensing including lands, oceans, and the atmosphere.

In the discipline of atmospheric sciences, lidar techniques have played a significant role in improving scientific understanding of the chemistry and dynamics of the Earth's atmosphere from the troposphere through the stratosphere, up through the mesosphere. Table 1 lists some of the existing laser systems currently being used by NASA scientists at the various field Centers to conduct some of the scientific investigations discussed in this section. In particular, the various performance characteristics of the laser transmitters are shown. Routine measurements have been conducted from a variety of ground and airborne platforms. The power of the lidar technique lies in its ability to provide range- resolved measurements, with fine spatial resolution of chemical species including molecules, atoms, and ions. The ability to directly measure atmospheric wind velocities remotely, with high spatial resolution, has to date provided detailed understanding of unique, regional atmospheric dynamics. Ultimately, when the measurement of winds is applied from space, the global profile of wind velocities will significantly improve our ability to predict weather, and our ability to study the transport of chemical species in the atmosphere. From the scientific contributions of lidar techniques demonstrated to date, it is clear that significant scientific contributions will continue in the future. As the technology of lidar systems continues to improve, lidar measurements will be routinely expanded to high flying aircraft and satellite platforms. In the next section, a summary of several lidar instruments under development for aviation metrology and safety will be presented.

Lasers in Aviation Metrology and Safety

Lasers Remote Sensing in Aircraft Flight Research

Nonintrusive measurement of true air speed, angle of attack, and conditions in free stream flow are important parameters in flight testing and flight operations of commercial and military aircraft. A recent paper (Bogue, 1992) contains a comprehensive review of the measurement problem, existing methods, limitations, and recent flight test results of optical techniques. In this section, we will summarize some of the laser remote sensing techniques which have been tested in research aircraft.

The basic measurement problem in aircraft flight measurements is that with in situ techniques, measurements can only be made at the surface of the aircraft; however with the new and powerful analytical codes, such as the Computational Fluid Dynamic Codes (CFD), aerodynamicists now require range-resolved measurements of fluid flow parameters from the surface of the vehicle to the free stream of the air flow. In addition, most conventional in situ sensors distort the flow and require careful interpretation of the data. For example, measurements with booms have limits in angle of attack and Mach number, and in addition, require extensive calibration. Current measurement techniques reviewed in the Bogue paper include the Flush Air Data System (FADS). FADS measures flow direction, mach number, and ambient pressure using multiple orifices on the vehicle nose. FADS applications have included measurements on the F-18 High Alpha Research Vehicle (HARV) (Moses and Whitmore, 1990). For mach numbers above 3, a system similar to FADS (called the Q-ball) was used on the X-15 research vehicle to measure hypersonic air data up to mach 6.7 (Wolowicz and Gossett, 1965). The Q-ball used a movable ported spherical nose in a pressure nulling concept. Laser remote sensing techniques can significantly improve these conventional measurement techniques.

Laser and other optical based air data sensors have the advantage of performing measurements in the free stream, without disturbing the airflow. Recent flight test results of two laser sensors were reviewed (Bogue, 1992).

A carbon dioxide (CO_2) laser doppler air speed sensor was tested on an F-16 aircraft on 18 flights (Silverman and Newman, 1990). A continuous wave laser was split into three beams. Each beam was focused 20 m from the aircraft, and backscattered signals from particles in the focal volume were mixed with the transmitted beam for heterodyne detection. The Doppler shift in the return signal was directly proportional to the velocity in the direction of the beam. Surface acoustic wave analyzers were used to analyze the signals. Tests were conducted to an altitude of 15.2 km, at flight velocities up to 283 m/sec. The angle of attack ranged from -10° to +16°, and the angle of sideslip covered the range from -10° to +10°. The system showed some signal to noise limitations at high altitudes, and as a result, cirrus clouds were required to provide for sufficient scattering due to the low concentration of atmospheric aerosols. System performance was improved by inclement weather, up to and including light rainfall.

A sheet-pairs air speed sensor was flight tested in a NASA F-104 aircraft on 16 flights to altitudes above 18.3 km and M = 2 (586 m/sec). The technique is described as follows: The sheet pairs sensor consists of pairs of concentrated light sheets projected from lasers located on the aircraft. Aluminum gallium arsenide (AlGaAs) semiconductor lasers, at a wavelength of 810 nm, were the active laser sources used to measure backscattering from particles located at a point approximately 1.2 m from the aircraft surface. Three laser sheet-pairs were used to measure air speed in three (nonorthogonal) directions. The velocity component of the aircraft, relative to the direction of the separation between the sheets, is inversely proportional to the time differentials between backscattering from the two sheets. The sheet pair separation was 1 cm. The natural particulate sizes and concentrations were sufficient at the time of these tests to satisfy system operation up to an altitude of 18.3 km, under night conditions. The sheet-pairs demonstrator system was tested in an F-16 aircraft, and results have been published (Smart and Abbiss, 1991). The results of the sheet-pairs sensor flight test on an F-104 clearly show that this measurement concept has promising potential for measuring vehicle air speed at ranges of 1 m from the surface of the flight vehicle. Further flight testing and assessment of these results will be used to suggest avenues for system modifications to improve performance over a broader range of flight conditions. It appears possible to extend the distance of the measurement volume from the surface of the flight vehicle to reduce the local flowfield effects, simply by increasing the average power of the laser transmitters.

Bouge, 1992, also reviewed optical measurements of ultraviolet molecular backscatter and oxygen fluorescence to measure atmospheric density and temperature, to flight altitudes of 15.2 km, during an F-16 flight test (Azzazy et al., 1990). The system acquired flight data by using a flashlamp as an optical source. The flight tests were conducted to demonstrate and assess the performance of the sensing concept under different lighting conditions (including day and night illumination). Eighteen flights were conducted during the program. Results indicate that a strong potential exists for measuring static density and temperature in flight; however, substantial improvement in the overall system performance is required to make the technique suitable for widespread use. The improvements can be achieved by using a narrow-bandwidth laser source, in place of the broadband flashlamp source.

Laser Remote Sensing for Aircraft Safety

Lasers provide the capability to measure wind velocity for the study of windshear, wake vortices, and clear-air turbulence. A recent article in Laser Focus World summarized lidar work on these topics (Allario et al., 1992b). As part of the NASA/FAA National Integrated Windshear Program, a measurable windshear hazard index has been defined as that which can be remotely sensed from an aircraft to give the pilot information about the wind conditions along the flight path. The current status of windshear detection is summarized in a recent conference proceeding "Airborne WindShear Detection and Warnings System" (Vicroy et al., 1992). Additionally, an analysis of the technology and the results of an end-to-end simulation identifying signal-to-noise ratio for a variety of tradeoff conditions and resulting wind velocity errors for different coherent laser radar (lidar) systems has been carried out (Targ et al., 1991). The results show that a Ho:YAG lidar (solid-state laser) at a wavelength of 2.1 μm and a CO_2 lidar (gas laser) at 10.6 μm can both provide a pilot with information on the line-of-sight component of a windshear threat, from a present position to a region extending 2-4 km in front of the aircraft. This constitutes a warning time of 20-40 seconds, even under conditions of moderately heavy precipitation. This study led to a program to install a Coherent Lidar Airborne Shear Sensor (CLASS) on a NASA B-737 aircraft as part of an airborne windshear detection program including a lidar, a Doppler radar, a passive infrared sensor, and an in-situ measurement system based on inertial and air data instruments. A 10.6-μm laser was chosen for the first lidar due to the technology readiness of CO_2 lasers. Future systems will utilize an "all-solid-state" diode laser pumped, 2-μm laser (Storm, 1991).

A radio frequency (RF) pumped injection seeded CO_2 Master Oscillator Power Amplifier was chosen for the laser, with the master oscillator continuously pumped to reduce chirp and the amplifier pulse pumped for efficiency. The output of the laser was 8 mJ at 100 Hz, in 2-μs pulses, with chirp less than 200 kHz. The data system was a poly-pulse pair processor and the beam was propagated through a 15-cm scanner mounted in the lower part of the aircraft. Dedicated windshear tests were conducted at Denver and Orlando during the summer of 1992. Preliminary results of these measurements have been reported (Robinson et al., 1992; Lewis 1993). The system showed acceptable operation in the "dry" environment at Denver. Under much higher humidity conditions at Orlando, performance was degraded with severe attenuation in heavy rain.

A 2-μm CLASS system upgrade is currently under development. The 2-μm system will have an increased pulse rate and shorter pulses than the 10-μm laser used on the CLASS experiment. The 2-μm laser should have improved performance for wet microbursts (Targ et al., 1991). A 2-μm system would also be a good candidate for measurements of trailing vortices. A comprehensive model for estimating performance of a 2-μm coherent lidar for wake vortex measurements has been published (Thomson et al., 1992).

Ground-based wind sensing is another application of laser remote sensing currently being studied within NASA programs. The Coherent LAunchsite Wind Sounder (CLAWS) is a technology demonstration which employs a Doppler laser radar (lidar) to provide accurate wind measurements in support of launch vehicle operations. Launch vehicles are susceptible to structural loadings due to wind variations above the launch site. Greatest loads occur as the booster passes through the maximum dynamic pressure regime, typically between 5 and 15 km Above Ground Level (AGL). Timely knowledge of wind variations up to approximately 20 km AGL is needed for efficient operation. In current operations, these measurements are made by radar tracking of balloons. Large margins are required to accommodate the uncertainties and time delays associated with this measurement technique.

During 1991, the CLAWS program constructed and tested a 1.06-μm lidar with a pulse energy of 900 mJ. This system was deployed at NASA Kennedy Space Center (KSC), and operated on a number of occasions during September and October 1991, including measurements taken 5 minutes after the launch of the Space Shuttle mission STS-48 (Hawley et al., in press). The system demonstrated a capability to provide useful measurements of horizontal winds up to an altitude of 26 km, in as little as 6 minutes. Wind measurements have also been demonstrated during the Space Shuttle (STS-2) landing, using a CO_2 laser remote sensing system (Huffaker and Hall, 1982).

CLAWS has demonstrated the feasibility of using a ground-based, short wavelength lidar system for measuring high altitude winds from a ground-based laser remote sensor. It has provided launch operations personnel with an opportunity to compare lidar wind data to data from existing measurement systems within the KSC meteorological environment. The program is now focussing upon instrument and system refinements, including an upgrade from the Nd:YAG laser to an eye-safe 2-μm laser, operating at 500 mJ or higher energy, and a systems capability for a low altitude, three-dimensional laser wind profiler, in support of Space Shuttle landings.

A Summary of Laser Remote Sensing to Wind Tunnel Applications

Before the Wright brothers flew at Kitty Hawk, fundamental questions had to be answered regarding an aircraft's lift, drag, propulsion system, and maneuverability. However, as aircraft have become more complex, a more fundamental understanding of these elements has been pursued to push the limit of aircraft performance. This understanding comes about through examining the air flow itself, rather than simply the effect of this air flow on the vehicle. The early work in this area was accomplished by inserting probes into the flow to sample a particular point in space. It was recognized early that these probes introduced their own disturbance to the flow being measured. Laser remote sensing of flow parameters was limited for many years to absorption and Schlieren type measurements with incoherent sources. The advent of moderate powered lasers allowed researchers for the first time to nonintrusively probe the airflow and obtain quantitative flow data. The following section will summarize the major laser remote sensing techniques currently being used to study the fundamental properties of fluids in wind tunnel research.

The wind tunnel data requirements today can be divided roughly into three categories. First, there is an examination of the **flow character**. This is the qualitative aspect of the fluid flow. This allows aeronautical researchers to see general trends, such as position and pattern of the flow, gross turbulent or laminar character, shock location and shape, general eddy size, interaction of flow with structure, and the generation and dissipation of vortices. The second category is **flow dynamics.** This involves the quantitative measurement of fluid properties so as to fully describe the fluid flow. This second category of data is often utilized in conjunction with Computational Fluid Dynamics (CFD). A list of requirements in this area always includes flow velocity, pressure distributions, temperature distributions, and often individual species densities. These data need to be spatially and temporally resolved. The final category is **flow physics**. This category includes the experiments and data examining the chemistry and physics producing the flow character under study.

The data are similar to that obtained for flow dynamics, but have the interaction between components and the fluctuating components are examined. In addition, the phenomena studied are very different including combustion systems, fluid mixing, vortex generation models, et al. These categories are not meant to be exclusive but only allow for a framework of discussion. The measurement systems to be discussed often fit into more than one of these categories depending upon the need and the constraints of the flow environment.

Seeding Based Techniques

The most straightforward application of lasers in ground facilities is the use of **laser light sheets** for visualization. Several institutions have installed dedicated systems typically based on argon ion lasers. These lasers are spread into a sheet across the flow, some form of condensate or smoke is added to the flow from which the laser can scatter, and the scattered light is typically detected by film or video cameras. These types of measurements have proven to be invaluable in visualizing previously unseen flow phenomena (Beiser et al., 1983; Rhodes et al., 1991; and Leighty et al., 1990).

Similarly, the argon ion laser has been the workhorse in **Laser Doppler Velocimetry** (LDV). In the typical implementation of this technique, a single laser beam is split and then crossed at the point to be measured. An interference pattern is set up from the crossed laser beams. Seed particles passing through this interaction zone scatter light which is typically detected on photomultiplier tubes and analyzed to obtain the velocity of the particle in the plane of the interference pattern. Often two and three components of velocity are obtained by replicating the system at a separate wavelength in a separate plane. This can readily be done with the multiwavelength output of the argon ion laser. Dedicated LDV systems are present at many ground facilities including supersonic wind tunnels (Hunter, Jr. et al., 1990). The utilization of diode lasers for LDV measurements continues to be studied for applications where physical size or instrument ruggedness are a concern (Franke et al., 1978).

Another single point particle, velocity measurement technique is **Laser Transit Anemometry** (LTA) (Smart and Mayo, 1984). With this technique, two laser beams are focused to two points placed very close in space. As the particle passes through each point, a scattered signal is generated and detected. With the laser spacing known and the detected time lag measured, the particle's velocity can be readily calculated. Again argon ion lasers were the primary laser source for these investigations. Although this technique has been utilized at large wind tunnel facilities (Humphreys, Jr. and Hunter, Jr., 1990), it has not yet had the impact of LDV.

One of the drawbacks of LDV and LTA is that data are collected at a single point. The optical system must be translated between data collection points to map out a full flowfield. This requires operation in a stable flow environment. One technique which has attempted to overcome this limitation is **Particle Image Velocimetry** (PIV) (Adrian, 1988). In this technique, two laser light sheets, coincident in space but separate in time, are used to track the particles. The two sheets are imaged onto a film camera to create a double exposure of the particles. The individual particle images will be shifted spatially on the film due to their movement during the period between the laser pulses. Therefore, by measuring the spatial separation and knowing the time delay between laser pulses, a calculation of the individual particle velocities is obtained. This technique requires pulsed lasers to produce time dependent laser sheets. The laser chosen for this task is typically a Nd:YAG laser. Normally, two lasers are slaved together with a precise timing circuit to produce the laser pulses. Although most of the early work with this technique was limited to water flows, this technique has been recently extended to a variety of flowfield situations (Humphreys et al., 1993). Several technological developments are extending the usefulness of this technique. Goss, et al., 1991, have implemented a two-color system which overcomes an inherent directional ambiguity (Post et al., 1991). This system has also been applied to simple flames. Other enhancements, such as holographic PIV, are currently under investigation to obtain three-dimensional measurements (Grant et al., 1991).

A recent development in the area of planer velocity measurements is a technique known as **Doppler Global Velocimetry** (DGV). This technique, developed jointly by the Northrup Corporation (Komine et al., 1991) and NASA Langley Research Center (Meyers, 1991), utilizes the properties of a molecular filter to convert velocity information to intensity information which can be readily detected. The light scattered from a field of particles, illuminated by a laser sheet, is shifted in frequency by an amount dependent upon the velocity (Doppler shifted). The unshifted light passes through the molecular filter with a known intensity; however, the intensity of the shifted light will be different because of the change in frequency and a corresponding change in molecular absorption. Most of the work with this technique has been done with argon ion lasers; however, laboratory tests with a system incorporating Nd:YAG lasers have been conducted. The use of a pulsed laser enables the researcher to obtain a "picture" from a much shorter timeframe. This relatively new technique has already been utilized in a wind tunnel and compared with standard LDV measurements.

Unseeded (Molecular) Measurement Techniques

One of the simplest measurement systems to conceive, utilizing laser systems on the molecular level, is **absorption spectroscopy**. However, well-known constraints of absorption spectroscopy limit its application in the diagnostic community. These constraints include the line of sight nature of the process producing a spatial averaging and the need to scan over spectral features to obtain useful information. Recently, Rea and Hanson (1993) have partially solved the second drawback with a clever modification of a ring dye laser. The argon ion pumped ring dye laser is modified by placing a rotating prism assembly within the cavity which allows a portion of the frequency spectrum to be swept out very rapidly. This instrument has been applied to several laboratory diagnostic applications. Recently, diode lasers have been investigated for utilization as diagnostic lasers in absorption (Stanton et al., 1989 and Trolinger, 1991).

Holographic remote sensing diagnostics of absorption and several other of the techniques to follow have recently been applied to practical flows (Philippe and Hanson, 1991, Tam et al.,

1991). A good review of this reemerging field has been presented by Trollinger and Havener, 1991.

Several standard spectroscopic techniques have been "rediscovered" in recent years with the advances in laser and detector capability. One of these techniques is **Raman spectroscopy**. This technique saw a meteoric rise in popularity with the introduction of the laser and is still being used in analytic laboratories as a standard diagnostic instrument. However, the weak nature of the interaction process has always limited its application as a diagnostic tool. Raman scattering is a process whereby light is scattered from molecules with its frequency shifted after interaction with the molecules; this shift is dependent upon the vibrational and rotational energy levels of the molecule being irradiated. This technique allows the researcher to study molecules without requiring laser frequencies at the hard-to-reach absorption bands. Although this technique still requires scanning over a frequency range, the data are spatially resolved. This technique has been applied utilizing standard Nd:YAG lasers at several facilities including nonpremixed methane flames and a Mach 6 wind tunnel (Shirinzadeh et al., 1991a). Recently, an excimer laser has been utilized to obtain Raman measurements in the exit of a turbulent diffusion flame (Cheng et al., 1991). The excimer laser allowed the researcher to take advantage of the ultraviolet enhancement in the Raman process by utilizing laser frequencies in this range.

Another "rediscovered" technique is **Rayleigh scattering**. This technique is a close cousin to Raman scattering since visible laser wavelengths can be utilized. It differs, however, since the scattering is not independent of a particular molecular energy level, but is only dependent on the total density of the flow. This nonspecificity is a drawback since the researcher cannot look at specific molecular constituents. However, it is also an advantage since the process is much stronger than the Raman process, thereby making it more generally applicable. Recent applications have utilized both argon ion lasers and excimer lasers. Seasholtz et al., (1991) have utilized argon ion lasers for Rayleigh measurements in rocket plumes. Shirinzadeh et al., (1991b), have utilized an excimer laser to make Rayleigh measurements in a supersonic wind tunnel.

A recent advance in Rayleigh scattering involves a technique very similar in principle to DGV, called **filtered Rayleigh scattering** (Miles, 1991). Like DGV, this technique utilizes a molecular filter to convert velocity to intensity information. However, unlike DGV, particulate seeding is not involved, as the technique depends upon scattering from molecular species in the flow. This technique is still in the early stages of development but could hold considerable promise.

Several other Raman relatives have received considerable attention over the last several years in the diagnostic remote sensing community. The most notable among these is **Coherent Anti-Stokes Raman Scattering** (CARS). This technique takes advantage of high laser powers to exploit nonlinear responses of molecular systems. The data are very similar to normal Raman data; however, the signal levels are several orders of magnitude greater and have several other important features. In addition, methods have been devised to overcome the need to scan the frequency, thus allowing the researcher to obtain information on temperature and density in a single laser firing. Typically, pulsed Nd:YAG lasers are utilized to make these measurements, although cw and excimer lasers have been used in laboratory tests. Several large operating facilities have utilized this technique including a supersonic wind tunnel (Antcliff et al., 1991), an MHD diffuser (Singh et al., 1991), a jet engine exhaust (Eckbreth et al., 1984), a bluff body combustor (Switzer et al., 1985), and inside an internal combustion engine (Lucht et al., 1987).

Another Raman relative is **Raman Doppler Velocimetry** (RDV). This technique detects the doppler shift of the molecules through a Raman process similar to CARS. The unique feature of this technique is that it is the only one that can provide simultaneous measurements of velocity, temperature, and pressure. It does have the drawback of requiring frequency scanning, but it has been demonstrated as a very sensitive measurement tool in stable facilities. The system, which was developed at NASA LaRC, has been demonstrated in the supersonic Unitary Plan Wind Tunnel (Exton and Hillard, 1986).

Another powerful technique is laser induced fluorescence. **Laser Induced Fluorescence** (LIF) has been explored by several researchers (Kychakoff et al., 1984; Seitzman et al., 1990; and Andresen et al., 1988). In this technique, light is absorbed by the molecule and then is reemitted at a different frequency as fluorescence radiation. The difference in frequency is dependent upon the energy levels of the molecules irradiated. This is a method of obtaining simultaneously, both temperature and density. In LIF, light must be absorbed. This is both an advantage and a disadvantage. It is a disadvantage since most absorption bands lie in the short wavelength range of the ultraviolet, which limits the number of accessible molecules. It has a great advantage since the signal to noise ratio (S/N) at these absorption frequencies is very large. This makes LIF a very sensitive diagnostic technique. The accuracy of this technique is still somewhat of a concern, to those attempting to apply it, because of the complexity of the interaction processes between the laser beam and the molecules. Most of the early work in this area used Nd:YAG lasers focussing the laser radiation to a point or dispersing it into a line. The development of high power excimer lasers has expanded this technique to planer measurements. Several groups have used this technique in the laboratory; however, only a few have applied it to operational ground facilities. A notable exception is the work by Fletcher and McKenzie, 1991. Recently, the LIF process has been applied to velocity measurements. Typically, this is done by photodissociating a molecule and subsequently probing the resulting species in the downstream flow, in a time delay measurement technique (Goss et al., 1991).

A similar flow tagging technique linked to Raman spectroscopy is termed **Raman Excitation and Laser Induced Excitation of Fluorescence** (RELIEF) (Miles et al., 1989). This technique utilizes the Raman process to "tag" the molecules and then uses the LIF technique to probe the tagged molecules and determine the molecular velocities. One and two dimensional figures have been "written" into flows to examine not only velocity but also vorticity.

It should also be mentioned that there have been several attempts to combine these techniques to obtain a wider array of information from the flow simultaneously. For instance, an early study by Hillard et al., 1974, combined Raman Scattering and LDV. A more recent study by Chen et al., 1988, combines the CARS technique with LDV to obtain densities, temperature, and velocity, simultaneously.

Lasers have made their presence known, in recent years, in remote sensing in aeronautical ground test facilities. A few basic laser systems have carried the bulk of the diagnostic work. They include ion lasers, Nd:YAG lasers (with and without dye systems), recently excimer lasers, and to a lesser extent diode lasers. The presence of diode lasers and new laser systems such as the Ti:sapphire have not yet had a large impact on the diagnostic community in ground facilities. However, preliminary investigations with these systems look very promising especially when anticipating the needs of flight systems in carrying the measurement tools developed on the ground to aircraft flight research.

Experiments from Space Platforms and High Flying Aircraft

Mars Observer Laser Altimeter (MOLA)

The Mars Observer Laser Altimeter (MOLA) developed by the Goddard Space Flight Center (GSFC) is currently on route toward Mars. The instrument, along with the rest of the Mars observer instrument complement, has been described by Komro and Hujber, 1991. The Mars Observer will be inserted into Mars orbit in August 1993 and will reach its mapping orbit in December 1993. Approximately 2 years of mapping operations are planned.

Measurements, with the MOLA instrument (Zuber et al., 1992), will determine the global topography of Mars at a horizontal and vertical resolution that allows investigations in the areas of geology, geophysics, and dynamic meteorology. Science goals include: Measurements of the internal structure and the global tectonics of Mars, the stress fields within the outermost layers of the planet, surface processes (e.g., erosion), volcanic flow volumes and gradients, impact cratering, and the effects of the topography, on atmospheric circulation, local topography and surface roughness. The MOLA will also contribute to the characterization of the surface composition by measuring the way the surface backscatters the transmitted 1.06 μm signal. The MOLA instrument includes a 50-cm diameter receiver telescope; a diode-pumped 10-Hz; neodymium:yttrium aluminum garnet (Nd:YAG) laser operating at 1.06 μm, with a pulse energy of 45 mJ; a pulse width of 10 ns at half maximum; a detector package; and an electronics box.

ATmospheric LIDar (ATLID)

ATmospheric LIDar (ATLID) (Endemann et al., 1990) is a proposed European Space Agency instrument planned for a flight in polar orbit. ATLID goals include measurements of cloud top heights, the height of the planetary boundary layer, the optical depth of thin clouds, the extent of clouds, and cloud depolarization. The laser transmitter is planned to be a diode laser pumped Nd:YAG laser, with 10-watt average power, operating at 95 Hz. The baseline measurement will be a measurement of the cloud top height at 1.064 μm, using an avalance photodiode (APD) detector. ATLID uses a unique receiver with four rectangular scanning telescopes. Fiber optics are designed to connect the telescopes to the four detectors. A recent status of the ATLID instrument was presented at the SPIE Conference 1714, Berlin, Federal Republic of Germany, June 22-26, 1992.

Geoscience Laser Altimeter System (GLAS)

The Geoscience Laser Altimeter System (GLAS) is an approved instrument investigation in NASA's Earth Observing System (EOS) program. A description of the scientific objectives and the proposed instrument have been described in detail (Cohen et al., 1987). GLAS is being developed by the Goddard Spaceflight Center (GSFC), essentially as an in-house development. The instrument description provided below was supplied by the NASA Earth Science Support Office located at GSFC.

GLAS is a laser altimeter designed to measure ice-sheet topography and associated temporal changes, as well as cloud and atmospheric aerosol properties. In addition, operation of GLAS over land and water will provide along-track topography. For ice-sheet applications, the laser altimeter will measure height from the spacecraft to the ice sheet, with an intrinsic precision better than 10 cm with a 70 m surface spot size. The height measurement, coupled with knowledge of the radial orbit position, will provide the determination of topography. Characteristics of the return pulse will be used to determine surface roughness. Changes in ice-sheet thickness at a level of a few tens of cm (anticipated to occur on a subdecadal time scale) will provide information about ice-sheet mass balance and will support prediction analyses of cryospheric response to future climatic changes. The ice-sheet mass balance and contribution to sea level change will also be determined. The accuracy of height determinations over land will be assessed using ground slope and roughness. The height distribution will be digitized over a total dynamic range of several tens of meters.

Along-track cloud and aerosol height distributions will be determined with a vertical resolution of 75 to 200 m, and a horizontal resolution from 150 m for dense clouds to 50 km for aerosol structure and height of the planetary boundary layer. Unambiguous measurements of cloud height and the vertical structure of thin clouds will support studies on the influence of clouds for radiation balance and climate feedbacks. Polar clouds and haze will be detected and sampled with much greater reliability and accuracy than can be achieved by passive sensors. Planetary boundary layer height will be directly and accurately measured for input into surface flux and air-sea and air-land interaction models. Direct measurements of aerosol vertical profiles will contribute to understanding of aerosol-climate effects and aerosol transport.

The GLAS laser is a frequency-doubled, cavity-dumped, solid-state Nd:YAG laser with energy levels of 120 mJ (1064 nm) and 60 mJ (532 nm). The laser transmitter is designed to use semiconductor laser arrays as optical pumping sources. The pulse repetition rate is 40 Hz, and the beam divergence is approximately 0.1 mrad. The infrared pulse is used for surface altimetry, and the green pulse is used for atmosphere measurements. The altimeter uses a 90-cm diameter telescope. The detector for 1064 nm will be a silicon avalanche photo detector. The detector at 532 nm will be either a photomultiplier or a Geiger mode, silicon avalanche photodetector.

Laser Atmospheric Wind Sounder (LAWS)

One of the instruments under NASA's EOS program is LAWS (Baker, 1992; Baker and Curran, 1985). Currently envisioned as a free-flyer in sun-synchronous orbit, at an altitude of 525 kilometers, the instrument would utilize a lidar instrument essentially consisting of a pulsed CO_2 laser transmitter, a scanning telescope of 1.5 meters diameter, and a coherent heterodyne receiver. Horizontal winds are inferred from the Doppler-shifted backscatter from aerosol particles in the troposphere and lower stratosphere. The use of a symmetrical conical scan around nadir, with a cone half-angle of approximately 45°, allows a given region of the atmosphere to be observed successively from forward and rearward viewing directions. This provides two linearly independent velocity measurements of the aerosols in the region. Neglecting vertical air motion and compensating for satellite motion, three independent measurements of horizontal atmospheric wind can be obtained in each range-resolved bin for each 100 km by 100 km square horizontal region. These range bins determined from the pulse length are averaged to obtain a representative measurement for each 1-km vertical region of the atmosphere.

The LAWS concept has been studied extensively since the mid-1970's (Huffaker, 1978; Menzies, 1986) and, in its latest form, LAWS has been carried through extensive system studies and laser breadboard tests (Petheram et al., 1989; Anon. 1992a). In its original configuration, LAWS would provide line-of-sight wind measurements to an accuracy of approximately 1 m/s in regions of strong aerosol scattering and approximately 5 m/s in weaker regions. The system could be flown in the interval 2002-2005 and satisfy a required 5-year, 10^9 shot lifetime. Pertinent system parameters are shown in Table 2 (Anon., 1992a).

Lidar In-Space Technology Experiment (LITE)

The LITE experiment (Allario et al., 1992) is a manifested Space Shuttle experiment which has several important objectives toward improving our scientific understanding of the Earth's atmosphere. The first flight of the LITE experiment will focus upon technology, and will determine the dynamic range of backscatter from aerosols, clouds, the ground, and molecules in the atmosphere. Scientists in the physics of radiative transfer will obtain, for the first time, fundamental measurements from the top of the atmosphere to the Earth's surface in three laser wavelengths. This data set will be invaluable in validating radiative transfer codes from space-to-ground. The LITE experiment is planned to have several flights on the Space Shuttle. In LITE II, the focus will be on scientific objectives. LITE III and beyond will refocus upon technology upgrades with emphasis on scientific experiments related to needs of "Mission to Planet Earth." The LITE payload is being viewed as a lidar testbed, and is guided by an international scientific team (Couch et al., 1991).

Details of the design, development, and testing of the LITE Laser Transmitter Module (LTM) were presented previously (Allario et al., 1991). In summary, the LTM design consists of a three-color, flashlamp pumped, Q-switched Nd:YAG laser. The configuration consists of an oscillator followed by two amplifier stages. A CD*A doubler and KD*P tripler are used to convert the fundamental frequency at 1.06 μm to 0.532 and 0.355 μm, respectively. The LITE laser has demonstrated 550, 600, and 206 mJ at these wavelengths. There are two laser transmitters mounted on each side of an I-beam mechanical structure. The two lasers are redundant with the exception of a common thermal control system. The LTM is housed in a sealed, pressurized canister filled with dry nitrogen gas. The LTM completed qualification testing and integration into the Space Shuttle Pallet in January 1993. The rest of the LITE experimental hardware and software is progressing to schedule. A series of atmospheric tests will be performed to verify end-to-end performance by March 1993. The LITE experiment is scheduled to fly on the Space Shuttle Endeavor in June 1994. In Table 3, a summary of the key performance parameters for the LITE laser is given. The first column of numbers represents the initial requirements; the second column shows the measured performance prior to integration into the Space Shuttle pallet.

Lidar Atmospheric Sensing Experiment (LASE)

The LASE experiment (Allario et al., 1992a) is a scientific experiment which uses the Differential Absorption Lidar (DIAL) technique to obtain measurements of tropospheric water vapor to an accuracy better than 10 percent, with vertical resolutions of 200 m for night and 300 m for daytime measurements. The LASE experiment will be flown on the NASA ER-2 aircraft at an altitude of 16-21 km, providing extended lateral coverage up to 4.8 km. The instrument is a "hands-off" self-contained system. As such, this lidar system is completely autonomous, and will make extensive use of microprocessors and control systems. The major subsystems include the laser transmitter, a thermal control system, telescope receiver, signal processing, data storage, and data transmission system.

A conceptual design of the titanium-doped sapphire laser used as the dual laser transmitter for the LASE experiment was presented in the previous SPIE publication (Allario et al., 1991). Original requirements for the titanium-doped sapphire have been met or exceeded in a breadboard model and assembly of the flight laser is underway. Table 4 summarizes major performance requirements of the laser transmitter. The second column summarizes performance measured on a breadboard unit in the laboratory. In Figure 1, a schematic diagram of the end-to-end Ti:sapphire laser is shown. The numbers in the schematic indicate the actual energy values that have been achieved. In the top of the schematic diagram, a double-pulsed, Nd:YAG pump laser is shown, frequency doubled to 0.532 μm. At this point in the optical train, the laser energy is 600 mJ per pulse at a repetition rate of 5 Hz, doubled-pulsed. The double-pulsed output pumps a Ti:sapphire power oscillator, which produces output pulses in excess of 150 mJ at 0.815 μm. The Ti:sapphire power oscillator is injection seeded with an AlGaAs semiconductor laser to achieve the narrow linewidth and proper wavelength required for the DIAL measurements of atmospheric water vapor.

Summary of Lidar Flight Experiments

In this section, we have summarized some of the major scientific investigations under development using lidar techniques. In Table 5, a summary of the major programs under development are listed with the performance parameters for the laser transmitters delineated. It is to be noted that with the exception of the LAWS experiment, all laser experiments are based upon Nd:YAG, a well-developed solid-state laser technology. Satellite experiments, with long lifetime, and requirements of high electrical-to-optical conversion efficiencies plan to utilize semiconductor laser arrays as pump sources, highlighting the maturity of this technology. Flashlamp pumping is restricted to the Shuttle experiment LITE and the ER-2 experiment LASE. The average power for the experiments in Table 5 range in the 10-100 watts range, which requires careful attention to design of the optical components, in order to avoid optical damage, as a function of lifetime. In Table 5, in addition to the major experiments discussed in this section, we have also listed an experiment with acronym EAGLE. EAGLE stands for EOS Atmospheric Global Lidar Experiment. EAGLE was a proposal originally submitted by the Langley Research Center to the EOS program, in response to the original Request for Proposals (RFP). The EAGLE laser transmitter represents state-of-the-art technology in 1989, for an all solid-state laser transmitter, with a Ti:sapphire laser as the tunable laser source. The scientific objective of the experiment was to measure vertical profiles of atmospheric water vapor similar to the LASE experiment, except from satellite altitudes.

Since the original proposal was completed, technology in the laser transmitter, especially the Ti:sapphire laser, and associated subsystems has improved significantly to the point where a new EAGLE instrument could be accommodated on a much smaller spacecraft. In the next and final section of this paper, we will summarize some of the more significant improvements that can be achieved by improvements in laser remote sensing subsystems.

Advanced Technological Developments for Laser Remote Sensing Instruments

As described in the previous sections of this paper, laser remote sensing is being effectively applied in a variety of scientific investigations, and is providing unique measurements in a variety of scientific and engineering disciplines. Commercial products in lasers; electro-optic components and devices; and lidar subsystems in telescopes, detectors, and signal processing subsystems are very effective in meeting today's measurement needs for laser remote sensors. However, to expand the application of laser remote sensors to routine use in space and advanced research aircraft, unique requirements must be addressed. For both space and aircraft investigations, three principles, guiding the development of technology must be kept in mind. First, conversion efficiencies from electrical to optical should be in excess of 5 percent and preferably closer to 10 percent. This should ensure compatibility with spacecraft and aircraft power systems, in the one kilowatt class. Second, within this envelope of efficiency, lasers should have the ability to tune across the electromagnetic spectrum from the the near ultraviolet (0.28 μm) to the far infrared (5.0 μm), although there are preferred wavelengths for near-term experiments from space for atmospheric sciences. Third, the technology must have the capability to be qualified for long lifetime, in a space environment, and sometimes in the harsher environments of aircraft. This requirement for long lifetime in the device and component technology includes optical components where high, internal optical fluences exceed commercial design requirements. Therefore, if we accept that tighter standards must be developed for aerospace lidar systems, then we must ensure that optical damage criteria are understood and standards developed as guiding principles for development of aerospace instruments.

In the area of laser remote sensing for atmospheric sciences from space, the two most important measurements include winds and water vapor. The measurement of vertically distributed ozone in the atmosphere is equally important, but difficult to realize because of the large concentration of ozone in the stratosphere. With the DIAL technique, this leads to saturation of the absorption signal and precludes measurements in the lower atmosphere. The possibility of using a combination of weak and strong absorption lines, in a clever measurement scheme, could lead to meaningful measurements of vertically resolved ozone concentrations. This could be a possibility with wavelengths in the near infrared range of the EM spectrum, 1-2 μm. However, the technical difficulties of realizing this measurement suggests a longer term opportunity than the measurement of winds and water vapor.

In measuring winds, CO_2 lasers represent a lower technological risk for development when compared to solid-state lasers, with the possible exception of an isotopic sensitive catalyst for closed cycle operation (Allario et al., 1992). Issues of lifetime and systems design for the laser transmitter have been adequately adressed. The risk, as in all space experiments, is in the development and calibration of the total system prior to launch, to ensure a 4-year mission lifetime. The experience with the LITE experiment has shown that the highest technical risk was in the engineering integration of technologically developed susbsystems. Therefore, for the measurement of winds from space, technology development is not required. However, development of engineering models or precursor missions with reasonably planned mission lifetimes should be developed, as a risk reduction mechanism.

In Allario et. al., 1991, the status of an "all" solid-state laser technology at 2 μm was presented. Progress since then has been achieved in demonstrating the advantage of Nd:YLF over Nd:YAG, especially in achieving higher efficiency for relatively high powered laser transmitters, in the 10-100 watt class. AlGaAs semiconductor laser array technology as a viable pump source for Nd:YLF has also been demonstrated and device yields appear to be high. The technology for an "all" solid state laser transmitter for 2 μm is planned for demonstration at the 0.75 joule level, in a NASA Langley Research Center program, scheduled for September 1993. In Targ et. al., 1992, an "all" solid-state laser transmitter for aviation safety (detection of microbursts) is currently under development. All these developments, point to the high potential for an "all" solid-state laser transmitter as a leading technology for the measurement of atmospheric winds in the next generation wind sensor (Anon. c).

In measuring water vapor, the titanium doped sapphire laser (Ti:sapphire) should provide the technology for a near-term measurement from space, especially a 1-year mission. The LASE experiment described in the previous section should provide the confidence in the maturity of the engineering design when engineering flights from the ER-2 are completed. These engineering flights are planned for December 1993. The areas of technological development that are recommended for improving water vapor measurements from space include a more reliable injection seeding source than the current semiconductor lasers being used, the engineering development of optically precise, composite structures for optical benches to reduce weight, and the utilization of semiconductor laser arrays as optical pump sources. Light weight optical telescope technology should also be included and all "light weight" materials technology should be incorporated for a space experiment. In the longer term, for measurements of water vapor from space, new solid-state laser crystal technology should be developed which can be directly pumped by semiconductor laser arrays, removlng the need for the Nd:YAG. Again, solid-state laser crystal technology with electrical to optical efficiencies in excess of 5 percent should be targetted. An ideal situation would be the development of semiconductor laser arrays with the engineering performance parameters for the LASE experiment, and a 1 joule per pulse capability, at 10 Hz.

One other atmospheric molecule that begs to be measured from space, with laser remote sensing is the carbon monoxide (CO) molecule (Allario et al., 1992). In this SPIE Conference, the Plenary speaker, Paul J. Crutzen, "Problems in Global Atmospheric Chemistry and the Role of Remote Sensing to their Solution," identified global measurements of carbon monoxide as one of the highest priority remote sensing measurements from space. Middle and upper tropospheric mixing ratios of CO have been made with a passive remote sensor from space (H. G. Reichle, Jr., et al, 1986) and (H. G. Reichle, Jr., et al., 1989); however, the passive Gas Filter Correlation Radiometer (GFCR) could not measure the CO in the biosphere, 0-3 km. A laser remote sensor has the capability to make this measurement in the biosphere, with a vertical resolution in principle, better than 100 m. What is required is a laser source which tunes into either the fundamental absorption band of CO (4.7 μm) or the first overtone (2.3 μm). This requires tunable mid-infrared laser sources, again with efficiency and lifetimes, compatible with the middle infrared wavelength. In N. P. Barnes, 1991, a summary of non- linear materials which might be used for measurements of atmospheric molecules with absorption bands in the middle infrared range of the spectrum is presented. In combination with the rapid developments in 2 μm laser technology, the measurement of CO from space is not far away from realizing engineering development. Ultimately, laser remote sensors will, at a minimum, measure 2 major atmospheric molecules and one important atmospheric parameter uniquely: Water vapor and carbon monoxide and atmospheric winds to an accuracy of (+/-) 1 m/s.

This paper is dedicated to the memory of Clayton H. Bair (7/21/47 - 10/28/92), who devoted his career to the study of atmospheric spectroscopy, lasers for atmospheric remote sensing and the training of young researchers in these disciplines.

EXISTING LASER SYSTEMS WITHIN NASA FIELD CENTERS: AIRBORNE AND GROUND LIDARS

PROGRAM	λ(μm)	PULSE LENGTH	REP RATE	ENERGY	PLATFORM
Stratospheric Aerosol Program (LaRC)	0:6943	17 ns	0.1 Hz	2.0J	Ground Based, 1.2m
	1.06/.532	7 ns	10 Hz	2.0/1.1J	Ground Based, 1.2m
	1.06/.532	7 ns	10 Hz	900/450mJ	Airborne, 35 cm
Cloud/Aerosols (GSFC)	1.06/.532	5 ns	20 Hz	200/90mJ	Airborne, 17 cm
UV-DIAL/Ozone (LaRC)	1.06/.532	12 ns	10 Hz	1.0/.50J	Airborne, 35 cm
	.290	12 ns	10 Hz	30 mJ	
UV-DIAL/Ozone (JPL)	.308/.353	35 ns	150 Hz	670 mJ	Ground Based, 90 cm
GLOBE/Aerosol (GSFC)	1.06/.532	5 ns	50 Hz	300/50mJ	Airborne, 40 cm
	2.1			50 mJ	
GLOBE/Winds (JPL)	9.25	2 μs	5 Hz	1 J	Airborne, 15 cm
CLASS (LaRC)	10.6	2 μs	100 Hz	10 mJ	Airborne, 15 cm
CLAWS (LaRC)	10.6	2 μs	10 Hz	900 mJ	Ground Based, 90 cm

Table 1

CHARACTERISTICS OF LAWS INSTRUMENT

OPTICAL SUBSYSTEM	
Telescope Aperture	1.5 meters
Nadir Angle	Fixed at 45 degrees
Rotation Rate	12 RPM
Lag Angle Compensation	Variable - Support
Asynchronous PRF	
LASER SUBSYSTEM	
Wavelength	9.1 micrometers
Pulse Energy	> 15 Joules in Far Field
Pulse Repetition Frequency	Asynchronous, <= 20 Hz, Avg. 10 Hz
Chirp	< 200 KHz over 3 microseconds
Beam Quality	< 1.1 x Diffraction Limit
RECEIVER SUBSYSTEM	
Detection Method	Heterodyne
Detector	5-Element HgCdTe
Bandwidth	1.2 GHz

Table 2

Lidar In-Space Technology Experiment (LITE): Significant Performance Parameters Required Versus Measured Performance

Objective: Development of elastic backscatter lidar system for global monitoring of aerosols, clouds, atmospheric density and temperature and evaluation of lidar system operations in Space.

Laser subsystem: Space qualified, redundant flash lamp pumped Nd:YAG doubled (CD*A) and tripled (KD*P)

	Requirement	Performance
Nd:YAG energy		
(1.06, .532, .355 μm)	200, 400, 150 mJ	550, 600, 206 mJ
pulsewidth	< 100 nsec	18, 20, 9 nsec
beam divergance	0.5 - 1.5 mrad	0.9, 0.6, 0.6 mrad
pulse repetition frequency	10 Hz	10 Hz
weight	<500 lbs (227 kg)	495 lbs (225 kg)
power	<2200 watts	1800 watts
volume	<0.54 m^3	0.42 m^3
lifetime	10^6 shots	>10^6 shots

Table 3

Lidar Atmospheric Sensing Experiment (LASE): Significant Performance Parameters Required Versus Measured Performance

Objective: Development of a Different Absorption Lidar (DIAL) system for regional monitoring of tropospheric water vapor and aerosols with 10% accuracy and 10-20 km horizontal and 200 m vertical resolution.

Laser subsystem: Ti:Sapphire laser with doubled Nd:YAG pump and tunable laser diode injection seeding

	Requirement	Performance
Nd:YAG energy 1.06 μm	1.3 J	>1.3 J
0.532 μm	560 m J	>650 m J
Ti:Sapphire energy	150 m J	>150 m J
wavelength	813-818 nm	720 - 830 nm
pulse repetition frequency	5 Hz, double pulse	5 Hz, double pulsed (400 μ sec separation)
line width	< 1 pm	0.5 pm
bandwidth control	>99% of energy in 1 pm	>99.7%
wavelength stability	0.25 pm	0.18 pm
energy stability	± 15%	± 5%
pointing stability	25 μrad	25 μrad

Table 4

NASA AND ESA ADVANCED LASER SYSTEMS UNDER DEVELOPMENT OR PLANNED

PROGRAM	λ(μm)	PULSE LENGTH	REP RATE	ENERGY	EXPECTED FLIGHT
MOLA	1.06 (Nd:YAG)	7 ns	10 Hz	30 mJ	1992
GLAS	1.06/.532/.355 (Nd:YAG)	75 ps	40 Hz	120 mJ/60 mJ	2000+
ATLID (ESA)	1.06 (Nd:YAG)	20 ns	95 Hz	100 mJ	<2000
LAWS	9.1 ($C^{12}O_z^{18}$)	1 μs	10 Hz	15 J	2000+
LITE	1.06/.532/.355 (Nd:YAG)	20 ns	10 Hz	1 J	1994
LASE	1.06/0.53 (Nd:YAG .82 Ti:Sapphire)	20 ns	10 Hz	1.30/0.65 J 150 mJ	1994
EAGLE	0.72-0.95 (Nd:YAG Pumped Ti:Sapphire)	<100 ns	10 Hz	1 J	2000+

Table 5

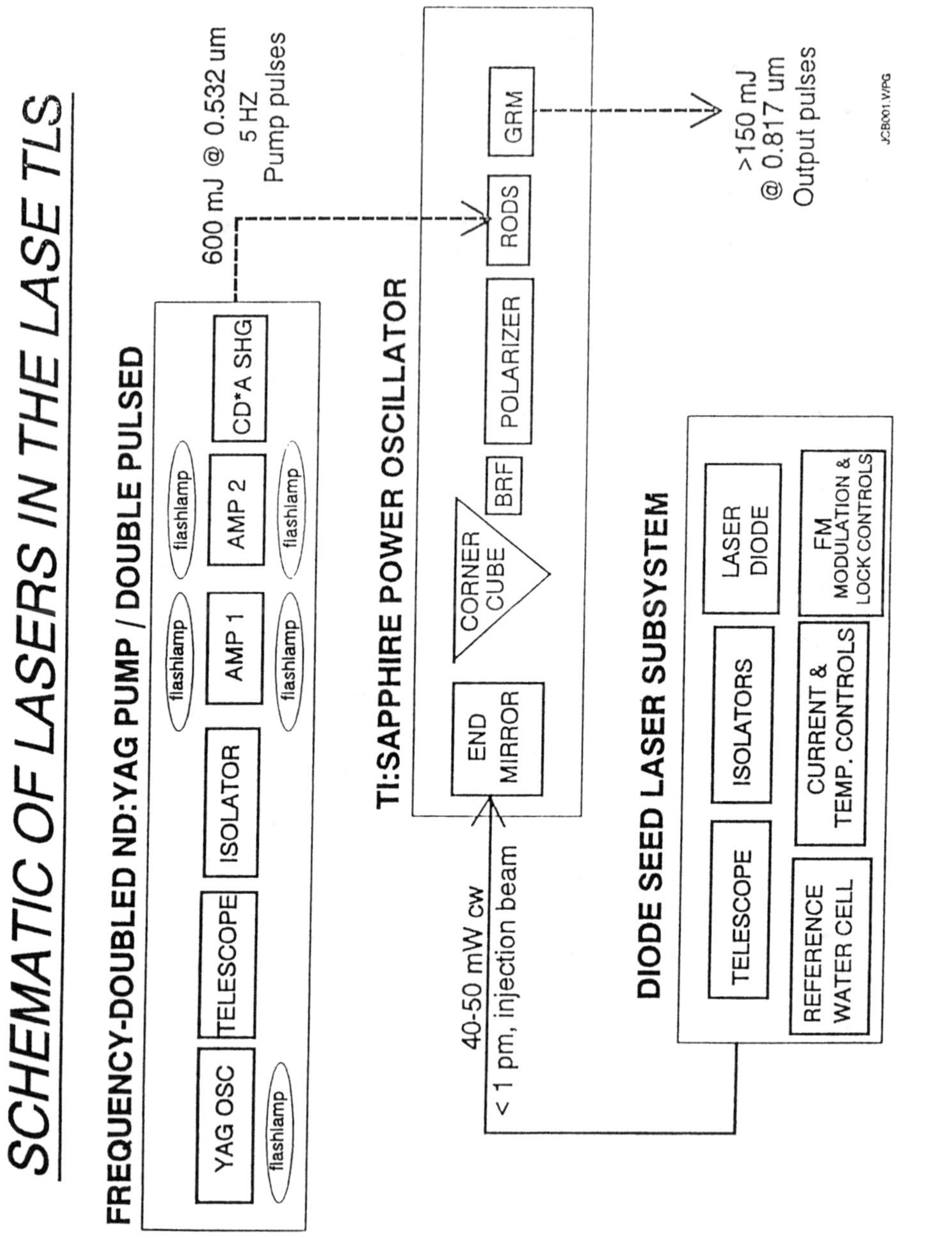
SCHEMATIC OF LASERS IN THE LASE TLS
FREQUENCY-DOUBLED ND:YAG PUMP / DOUBLE PULSED
YAG OSC
flashlamp
TELESCOPE
ISOLATOR
flashlamp
AMP 1
flashlamp
flashlamp
AMP 2
flashlamp
CD*A SHG
600 mJ @ 0.532 um
5 HZ
Pump pulses
TI:SAPPHIRE POWER OSCILLATOR
END MIRROR
CORNER CUBE
BRF
POLARIZER
RODS
GRM
>150 mJ
@ 0.817 um
Output pulses
40-50 mW cw
< 1 pm, injection beam
DIODE SEED LASER SUBSYSTEM
TELESCOPE
ISOLATORS
LASER DIODE
REFERENCE WATER CELL
CURRENT & TEMP. CONTROLS
FM MODULATION & LOCK CONTROLS
JCB001.VPG

Figure 1

Bibliography

The information contained in this paper was developed from research conducted by the authors and excerpts from the reports and journal articles listed in the bibliography. The authors wish to acknowledge the many researchers who have contributed to the field of laser remote sensing and give special thanks to those who have contributed materials for this paper.

In particular, the authors would like to express their thanks to Ms. Jane Leffel for assembling the original text from the four authors. Our special thanks to Ms. Tammy Spellman for the long hours she spent in the final editing of this paper, coordinating across four authors.

Adrian, R. J.: Particle Image Displacement Velocimetry," von Karman Institute for Fluid Dynamics, Lecture Series 1988-06, March 1988.

Adriani, A.; G. P. Gobbi; F. Congeduti; and G. Di Donfrancesco: "Lidar Observations of Stratospheric and Mesospheric Temperature," November 1988-November 1989, Ann. Geophysicae, 9, pp 252-258, 1991.

Aikin, A. C.; M.-L. Chanin; J. Nash; and D. J. Kendig: "Temperature Trends in the Lower Mesosphere, Geophys. Res. Lett., 18, pp 416-419, 1991.

Albrecht, B. A.: "Aerosols, Cloud Microphysics, and Fractional Cloudiness," Science, 245, pp 1227-1230, 1989.

Allario, F.; N. P. Barnes; and M. E. Storm: "Future Technologies for Lidar/DIAL Remote Sensing, International Symposium on Optical Engineering and Photonics in Aerospace Sensing, SPIE 1492, pp 92-110, 1991.

Allario, F.; W. E. Miller; L. D. Staton; and M. M. Sokoloski: "NASA Laser Remote Sensing Program," International Symposium on Environmental Sensing, SPIE 1714, pp 4-20, 1992a

Allario, F.; R. R. Antcliff; P. Brockman; and M. E. Thomas: "Lasers for Remote Sensing," Laser Focus World, pp 59-68, April 1992b.

Ancellet, G.; J. Pelon; M. Beekmann; A. Papayannis; and G. Megie: "Ground-based Lidar Studies of Ozone Exchanges Between the Stratosphere and the Troposphere," Journal Geophysical Research, 96, pp 22,401-22,421, 1991.

Andresen, P.; A. Bath; W. Gröger; W. Lüff; G. Meijer; and J. J. ter Meulen: "Laser-based Flurescence with Tunable Excimer Lasers as a Possible Method for Instantaneous Temperature Field Measurements at High Pressures, Checks with an Atmospheric Flame," Applied Optics 27, pp 365-378, 1988.

Anon. a: Definition and Preliminary Design of the LAWS, Final Study Report Phase II, NASA MSFC Contract No. NAS8-37589, June 1992.

Anon. b: "OACT Laser Sensing Technology Team's Report of Visit to Selected Russian Labs, October 18-23, 1992. (Available by request at the Langley Research Center.)

Anon. c: "2 μm Solid-State Laser Technology Review, Findings and Recommendations," Washington, D.C., May 18-19, 1992.

Antcliff, R.; M. Smith; O. Jarrett, Jr.; A. Cutler; G. Northam; and D. Taylor: "A Hardened CARS System Utilized for Temperature Measurements in a Supersonic Combustor," American Institute of Aeronautics and Astronautics (AIAA)-91-0457, 1991.

Azzazy, M.; J. B. Abbiss; and R. W. McCullough: "Determination of Aircraft Pressure Attitude and Ambient Temperature using Fluorescence and Rayleigh Scattered Radiation from a Continuum Source," Spectron Development Laboratories, Costa Mesa, California. NATO Advanced Study Institute, Algarve, Portugal, 1990.

Baker, W. E. and R. J. Curran: "Report of the NASA Workshop on Global Wind Measurements," NASA CR-190376, December 1985.

Baker, W. E.: "Science Goals and Mission Objectives of NASA's Laser Atmospheric Wind Sounder Program," Sixteenth International Laser Radar Conference, July 1992.

Barnes, N. P.: "Tunable Mid-Infrared Sources Using Second-Order Nonlinearities," Invited Paper in the International Journal of Nonlinear Optical Physics, 1, pp 639-672, 1991.

Batista, P. P.; B. R. Clemesha; I. S. Batista; and D. M. Simonich: "Characteristics of the Sporadic Sodium Layers Observed at 23°S," J. of Geophysical Research, 94, pp 15,349-15,358, 1989.

Beiser, L., ed.: "Advances in Laser Scanning and Recording," Proceedings of SPIE--The International Society for Optical Engineering, 396, 1983.

Beniston, M., et al.: "Use of Lidar Measurements and Numerical Models in Air Pollution Research, J. of Geophysical Research, 95, pp 9879-9894, 1990.

Bogue, R. K.: "Recent Flight-test Results of Optical Airdata Techniques," SAE Technical Paper Series, 921997, Aerotech '92, Anaheim, California, October 1992.

Brock, C. A.; L. F. Radke; and P. V. Hobbs: "Sulfur in Particles in Arctic Hazes Derived from Airborne in situ and Lidar Measurements, J. of Geophysical Research, 95, pp 22,369-22,387, 1990.

Browell, E. V.; E. F. Danielsen; S. Ismail; G. L. Gregory; and S. M. Beck: Tropopause Fold Structure Determined from Airborne Lidar and in situ Measurements, J. of Geophysical Research, 92, pp 2112-2120, 1987.

Browell, E. V.: "Differential Absorption Lidar Sensing of Ozone," Proceedings of IEEE, 77, pp 419-432, 1989.

Browell, E. V., et al.: "Airborne Lidar Observations in the Wintertime Arctic Stratosphere: Polar Stratospheric Clouds, Geophys. Res. Lett.,17, pp 385-388, 1990a.

Browell, E. V.; G. L. Gregory; R. C. Harriss; and V. W. J. H. Kirchhoff: "Ozone and Aerosol Distributions over the Amazon Basin During the Wet Season, J. of Geophysical Research, 95, pp 16,887-16,901, 1990b.

Browell, E. V.; S. Ismail; and B. Grossmann: "Temperature Sensitivity of Absorption Lines for H_2O DIAL Measurements, Applied Optics, 30, pp 1517-1524, 1991.

Chahine, M. T.: "The Hydrological Cycle and its Influence on Climate," Nature, 359, pp 373-380, 1992.

Chanine, M.-L.; A. Garnier; A. Hauchecorne; and J. Porteneuve: "A Doppler Lidar for Measuring Winds in the Middle Atmosphere, Geophys. Res. Lett., 16, pp 1273-1276, 1989.

Chanine, M.-L.: "Rayleigh/Raman Lidars: Intercomparisons and Validation," Abstracts, Sixteenth International Laser Radar Conference, MIT, NASA Conference Pubulication 3158, pp 337-340, July 20-24, 1992.

Charlson, R. J., et al.: "Climate Forcing by Anthropogenic Aerosols," Science, 255, pp 423-430, 1992.

Chen, T.; L. Goss; D. Trump; and W. Schmoll: "Studies of a Turbulent Premixed Flame using CARS-LDV Diagnostics," AIAA-88-3194, 1988.

Cheng, T.; J. Wehrmeyer; and R. Pitz: "Concentration and Temperature Measurement in a Turbulent Diffusion Flame by a Narrowband Tunable Excimer Laser, AIAA-91-0181, 1991.

Cohen, S. C.; J.J. Degnan, III; J. L. Bufton,; J. B. Garvin; and J. B. Abshire: "The Geoscience Laser Altimetry/Ranging System," IEEE Trans. Geoscience Remote Sensing, Vol. GE-25, pp 581-592, September 1987.

Couch, R. H., et al.: "Lidar In-Space Technology Experiment (LITE): NASA's First In-space Lidar System for Atmospheric Research," Optical Engineering, 30, 1, pp 88-95, January 1991.

Eckbreth, A. C.; G. M. Dobbs; J. H. Stufflebeam; and P. A. Tellex: "CARS Temperature and Species Measurements in Augmented Jet Engine Exhausts," Applied Optics, 23, pp 1328-1338, 1984.

Edner, H., et al.: "Lidar Search for Atmospheric Atomic Mercury in Icelandic Geothermal Fields," J. of Geophysical Research, 96, pp 2977-2986, 1991.

Eichinger, W. E., et al.: "Lidar Analysis Techniques for Use in the Atmospheric Boundary Layer," Abstracts, Sixteenth International Laser Radar Conference, MIT, NASA Conference Publication 3158, pp 593-596, July 20-24, 1992.

Endemann, M.; R. Lange; R. Krawczcyk; T. Banos; A. Dupuy; B. Hofer; H. Strebel; N. Limbourg; and E. Zanzottera; "ATLID-Phase A Study," Executive Summary, Final Report, Part 1, May 1990.

Exton, R. J. and M. E. Hillard: "Raman Doppler Velocimetry: A Unified Approach for Measuring Molecular Flow Velocity, Temperature, and Pressure," Applied Optics, 25, pp 14-21, 1986.

Fiocco, G. and G. Grams: "Observation of the Aerosol Layer at 20 km by Optical Radar," J. of Atmospheric Sciences, 21, pp 323-324, 1964.

Fletcher, D. and R. McKenzie: "Simultaneous Measurements of Temperature and Density in Air Flows Using UV Laser Spectroscopy," AIAA-91-0458, 1991.

Franke, J.; S. Ocheltree; W. Hunter, Jr.; and J. Meyers: "Laser Velocimetry with Semiconductor Lasers, Proceedings of the Electro-Optics/Laser '78 Conference, pp 374-380, September 1978.

Goss, L.; T. Chen; D. Trump; B. Sarka; and A. Nejad: "Flow-tagging Velocimetry Using UV-photodissociation of Water Vapor, AIAA-91-0355, 1991.

Grant, I.; Y. Zhao, et al.: "Three Component Flow Mapping: Experiences in Stereoscopic PIV and Holographic Velocimetry, Proceedings of Laser Anemometry - Advances and Applications, Cleveland Ohio, pg 365, 1991.

Grant, W. B., ed.: "Ozone Measuring Instruments for the Stratosphere," Optics, Vol. 1, Optical Society of America, pg 438, 1989.

Grant, W. B.: "Differential Absorption and Raman Lidar for Water Vapor Profile Measurements: A Review, Optical Engineering, 30, pp 40-48, 1991.

Grant, W. B., et al.: "Volcanic-aerosol-induced Changes in Stratospheric Ozone following the Eruption of Mt. Pinatubo," Proceedings of Quadrennial Ozone Symposium Charlottesville, Virginia, June 4-13, 1992, in press.

Hardesty, R. M.; M. J. Post; and R. M. Banta: "Observing Atmospheric Winds with a Doppler Lidar, Optics and Photonics News," 2 (10), pp 12-15, 1991.

Hawley, J. G.; R. Targ; S. Henderson; M. Kavaya; and D. Moerder: "Coherent Launch-site Atmospheric Wind Sounder (CLAWS) Theory and Experiment, Applied Optics, in press.

Hillard, M., Jr.; W. Hunter, Jr.; J. Meyers; and W. Feller: "Simultaneous Raman and Laser Velocimeter Measurements," AIAA Journal, 12 (10), pp 1445-1446, October 1974.

Hofmann, D. J. and S. Solomon: "Ozone Destruction through Heterogeneous Chemistry following the Eruption of El Chichón," J. of Geophysical Research, 94, pp 5029-5041, 1989.

Huffaker, R. M., ed., "Feasibility Study of Satellite-Borne Lidar Global Wind Monitoring System," NOAA Technical Memorandum ERL WPL-37, 1978.

Huffaker, R. M. and Hall, F. F., Jr.: "Wind Comparison Measurements Between a Pulsed Doppler Lidar and Other Wind Sensors," Eleventh International Laser Radar Conference, NASA Conference Publication 2228, 1982.

Humphreys, Jr., W. M. and W. W. Hunter, Jr.: "Laser Transit Anemometer Measurements on a Slender Cone in the Langley Unitary Plan Wind Tunnel," NASA Technical Memorandum 102775, December 1990.

Humphreys, Jr., W. M.; S. M. Bartram; and J. L. Blackshire: "A Survey of Particle Image Velocimetry Applications in Langley Aerospace Facilities," 31st Aerospace Science Meeting and Exhibit, Reno, Nevada, January 1993.

Hunter, Jr., W. W.; L. R. Gartrell; C. E. Nichols, Jr.; and D. W. Witte: "Investigation and Development of a Laser Velocimeter Capability for Langley's 20-inch Mach 6 Wind Tunnel," Ninth NASP Technology Symposium, Orlando, Florida, November 1990.

Junge, C. E. and J. E. Manson: Stratospheric Aerosol Studies," J. Geophysical Research, 60, pp 2163-2182, 1961.

Kane, T. J.; C. A. Hostetler; and C. S. Gardner: "Horizontal and Vertical Structure of the Major Sporadic Sodium Layer Events Observed During ALOHA-90, Geophys. Res. Lett., 18, pp 1365-1368, 1991.

Kane, T. J.; P. H. Mui; and C. S. Gardner: "Evidence for Substantial Seasonal Variations in the Structure of the Mesospheric Fe Layer," Geophys. Res. Lett., 19, pp 405-408, 1992.

Kinne, S.; O. B. Toon; and M. J. Prather: "Buffering of Stratospheric Circulation by Changing Amounts of Tropical Ozone: A Pinatubo Case Study," Geophys. Res. Lett., 19, pp 1927-1930, 1992.

Koch, S. E., et al.: "Structure of an Internal Bore and Dissipating Gravity Current as Revealed by Raman Lidar, Mon. Weather Rev., 119, pp 857-887, 1991.

Komine, H.; S. Brosnan; A. Litton; and E. Stappaerts: "Real-time, Doppler Global Velocimetry," AIAA-91-0337, 1991.

Komro, F. G. and F. N. Hujber: "Mars Observer Instrument Complement," Journal of Spacecraft and Rockets, 28, 5, pp 501-506, September - October 1991.

Kychakoff, G.; R. Howe; and R. Hanson: Applied Optics, 23, pp 704-712, 1984.

Lacis, A.; J. Hansen; and M. Sato: "Climate Forcing by Stratospheric Aerosols," Geophys. Res. Lett., 19, pp 1607-1610, 1992.

Leighty, B.; J. Franke; J. Jones; and D. Rhodes: "A Two-mirrored Galvanometer Laser Light Sheet Generator," AIAA-88-4680, 1990.

Lewis, Michael S.: "Sensing a Change in the Wind," Aerospace America, pp 20-30, January 1993

Ligda, M. G. H.: "Meteorological Observations with Pulsed Laser Radar," Proceedings of the First Conference on Laser Technology, U. S. Navy, pp 63-72, 1963.

Lucht, R. P.; R. E. Teets; R. M. Green; R. E. Palmer; and C. R. Ferguson: "Unburned Gas Temperatures in an Internal Combustion Engine," Combustion Science and Technology, 55, pg 41, 1987.

McDermid, I. S., et al.: "Comparison of Ozone Profiles from Ground-based Lidar, Electrochemical Concentration Cell Balloon Sonde, ROCOZ-A Rocket Ozonesonde, and Stratospheric Aerosol and Gas Experiment Satellite Measurements," Journal of Geophysical Research, 95, pp 10,037-10,042, 1990a.

McDermid, I. S.; S. M. Godin; and L. O. Lindqvist: "Ground-based Laser DIAL System for Long-term Measurements of Stratospheric Ozone," Applied Optics, 29, pp 3603-3612, 1990b.

McKenzie, R. L.; Hanson, R. K.; and Eckbreth, A. C.: "Shedding New Light on Gas Dynamics," Aerospace America, pp 20-25, November 1992.

McDermid, I. S.; D. A. Haner; M. M. Kleiman,;T. D. Walsh; and M. L. White: "Differential Absorption Lidar Systems for Tropospheric and Stratospheric Ozone Measurements," Optical Engineering, 30, pp 22-30, 1991.

Melfi, S.H.; D. Whiteman; and R. Ferrare: "Observation of Atmospheric Fronts using Raman Lidar Moisture Measurements," Journal of Applied Meteorology, 28, pp 789-806, 1989.

Melfi, S. H.; B. N. Whiteman; and R. Ferrare: "Atmospheric Moisture Structure Revealed by Raman Lidar, Optics and Photonics News," 2 (10), pp 16-18, 1991.

Menzies, R. T.: "Doppler Lidar Atmospheric Wind Sensors: A Comparative Performance Evaluation for Global Measurement Applications from Earth Orbit," Applied Optics, 25, 15, pp 2546-2552, 1986.

Menzies, R. T. and R. M. Hardesty: "Coherent Doppler Lidar for Measurements of Wind Fields," Proceedings of IEEE, 77, pp 449-462, 1989.

Menzies, R. T. and M. J. Post: "GLOBE Backscatter: Climatologies and Mission Results," Proceedings of SPIE, 1416, pp 139-146, 1991.

Meyers, J. F.: "Doppler Global Velocimetry - A New Way to Look at Velocity," Fourth International Conference on Laser Anemometry - Advances and Applications, Cleveland, Ohio, August 1991.

Miles, R.; J. Coners; E. Markovitz; P. Howard; and G. Roth: "Instantaneous Profiles and Turbulence Statistics of Supersonic Free Shear Layers by Raman Excitation + Laser-induced Electronic Fluorescence (RELIEF) Velocity Tagging of Oxygen," Exp. in Fluids 8, pg 17, 1989.

Miles, R.: "Instantaneous Velocity Fields and Background Suppression by Filtered Rayleigh Scattering, AIAA-91-0357, 1991.

Moses, Timothy R. and Stephen A. Whitmore: "A Preliminary Look at Techniques Used to Obtain Air Data from Flight at High Angles of Attack, NASA TM-101729, 1990.

Penner, J. E.; R. E. Dickinson; and C. A. O'Neill: "Effects of Aerosol from Biomass Burning on the Global Radiation Budget," Science, 256, pp 1432-1434, 1992.

Petheram, J.C., et. al.: "The Laser Atmospheric Wind Sounder," SPIE, 1062, pp 274-286, 1989.

Philippe, L. and R. Hanson: "Tunable Diode Laser Absorption Sensor for Temperature and Velocity Measurements of O_2 in Air Flows," AIAA-91-0360, 1991.

Platt, C. M. R.: "The Experimental Cloud Lidar Pilot Study (ECLIPS) Program," Technical Digest, Optical Remote Sensing of the Atmosphere, Williamsburg, Virginia, Optical Society America, pp 163-165, November 18-21, 1991.

Post, M.; L. Goss; L. Brainard; J. Fridell; B. Cohrs; C. Murawski; and C. MacArthur: "Comparison of Two-color Particle Imaging Velocimetry Data and Calculated Velocity Field in a Turbine Cascade," AIAA-91-0274, 1991.

Rea, Jr., E. C.; and R. K. Hanson: "Rapid, Extended Range Tuning of Single-mode Ring Dye Lasers," Applied Optics 22, pp 518-520, 1983.

Reagan, J. A.; M. P. McCormick; and J. D. Spinhirne: "Lidar Sensing of Aerosols and Clouds in the Troposphere and Stratosphere," Proc. of IEEE, 77, pp 433-448, 1989.

Reichle, H. G., Jr., et. al.: "Middle and Upper Tropospheric Mixing Ratios as Measured by a Satelite-Borne Remote Sensor During 1981," J. Geophysical Res., 91, pp 10865-10887, 1986.

Reichle, H. G., Jr., et al.: "Feasibility of Determining the Vertical Profile of Carbon Monoxide from a Space Platform," Applied Optics, 28, No. 11, pp. 2105-2110, June 1989.

Rhodes, D.; J. Franke; S. Jones; and B. Leighty: "Laser Light Sheet Generation Techniques for Flow Visualization at Langley Research Center," ISA 91-120, May 1991.

Robinson, P.; R. Bowles; and T. Targ: "Flight Test Results of the NASA/LMSC Coherent Lidar Airborne Shear Sensor," International Conference on Lasers '92, Houston, Texas, December 1992.

Sassen, K.: "Evidence for Liquid-phase Cirrus Cloud Formation from Volcanic Aerosols: Climatic Implications, Science, 257, pp 516-519, 1992.

Seasholtz, R.; S. Schnieder; and F. Zupac: "Spectrally Resolved Rayleigh Scattering Diagnostic for Hydrogen Oxygen Rocket Plume Studies," AIAA-91-0462, 1991.

Seitzman, J.; P. Paul; and R. Hanson: AIAA-90-0160, 1990.

Senft, D. C.; C. A. Hostetler; and C. S. Gardner: "Characteristics of Gravity Wave Activity and Spectra in the Upper Stratosphere and Upper Mesosphere at Arecibo during Early April 1989," Journal of Atmospheric Terr. Phys., in press, 1992.

Shirinzadeh, B.; M. Hillard; A. Blair; and R. Exton: "Study of Cluster Formation and its Effects on Rayleigh and Raman Scattering Measurements in a Mach 6 Tunnel," AIAA-91-1496, 1991a.

Shirinzadeh, B.; R. Jeffrey Balla; M. E. Hillard; J. B. Anders; R. J. Exton; and I. A. Waitz: "Study of Rayleigh Scattering for Visualization of Helium-air Mixing at Mach 6," 14th International Congress on Instrumentation in Aerospace Simulation Facilities, Rockville, Maryland, October 1991b.

Silverman, B. B. and L. A. Newman: "Flight Testing of a Three-axis CO_2 Laser Doppler Velocimeter," United Technologies Optical Systems, Inc., West Palm Beach, Florida, 1990.

Singh, J.; F. Yueh; and R. Cook: "CARS Temperature Measurements in an MHD Diffuser," AIAA-91-0211, 1991.

Smart, A. E. and W. T. Mayo: "Technique for the Measurement of Turbulence and High-speed Flow with a Laser Transit Anemometer," Arnold Engineering Development Center Technical Report, AEDC-TR-84-13, I, May 1984.

Smart, A. E. and John B. Abbiss: "NASA F-104 Flight Testing of the Titan Spectron Optical Air Data Sensor (OADS) System," Titan Spectron Report Number 91-2877-24, 1991.

Stanton, A.; D. Bosme; and J. Silver: "Nonintrusive Fast Response Oxygen Monitoring System for High Temperature Flows," SBIR Phase I Final Report, NASA Contract No. NAS1-18829, 1989.

Storm, M. E.: "Coherent 2-Micrometer Sources Burst into Wind-Shear Detection," Laser Focus World, pp 117-120, April 1991.

Switzer, G. L.; L. P. Goss; D. D. Trump; C. M. Reeves; J. S. Strutrud; R. P. Bradley; and W. M. Roquemore: "CARS Measurements in the Near-wake Region of an Axisymmetric Bluff-body Combustor," AIAA-85-1106, July 1985.

Tam, T.; N. Brock; J. Cavolowsky; and A. Strawa: "Holographic Interferometry at the NASA-Ames Ballistic Range," AIAA-91-0568, 1991.

Targ, R.; M. J. Kavaya; R. M. Huffaker; and R. Bowles: "Coherent Lidar Airborne Windshear Sensor: Performance Evaluation," Applied Optics, 30, 15, pp 2013-2026, May 1991.

Theon, J.; W. W. Vaughan; E. V. Browell; W. D. Jones; M. P. McCormick;S. H. Melfi; R. T. Menzies;G. K. Schwemmer; and J. D. Spinhirne: "NASA's Program in Lidar Remote Sensing," Conference on Earth and Atmospheric Remote Sensing Conference, SPIE 1442, pp 2-23, 1991.

Thomson, J. Alex; R. M. Huffaker; and R. D. Richmond: "Two-micron Coherent Laser Radar for On-Board Aircraft Vortex Wake Detection: Pattern Recognition Techniques and Technology Assessment," Proceedings of the Aircraft Wake Vortices Conference, Vol. 2, June 1992.

Trolinger, J. and G. Havener: "A Renaissance in Holographic Flowfield Diagnostics," AIAA-91-0566, 1991a.

Trolinger, J.: "A New Multiplex Holographic Interferometer for Hypersonic Flow, AIAA-91-0566, 1991b.

Twomey, S. A.: "The Influence of Pollution on the Short-wave Albedo of Clouds," J. Atmospheric Science, 34, pp 1149-1152, 1977.

Vicroy, D.; R. Bowles; and R. H. Passman: "Airborne WindShear Detection and Warning Systems," Fourth Combined Manufacturers' and Technologists' Conference, NASA Conference Publication 10105, Part 1, September 1992.

Wingrove, R. C. and R. E. Bach, Jr.: "Severe Winds in the DEW Microburst Measured from Two Aircraft," AIAA Guidance, Navigation, and Control Conference, Monterey, California, Vol. 1, August 17-19, 1987.

Wolowicz, Chester M. and Terrence D. Gossett: "Operational and Performance Characteristics of the X-15 Spherical Hypersonic Flow-direction Sensor," NASA TN D-3070, 1965.

Zuber, M. T.; Smith, D. E.; Solomon, S. C.; Muhleman, D. O.; Head, J. W.; Garvin, J. B.; Abshire, J. B. ;and Bufton, J. L.: "The Mars Observer Laser Altimeter Investigation, Journal of Geophysical Research, Vol. 97, no. E5, pp 7781-7797, May 25, 1992.

Photonics for Aerospace Sensors

John Pellegrino, Eric Adler, Andree Filipov, Lorna Harrison, Joe van der Gracht, Dale Smith, Tristan Tayag, Ed Viveiros

Army Research Laboratory
2800 Powder Mill Rd., Adelphi, MD 20783

1 Introduction

1.1 Objective

The maturation in the state-of-the-art of optical components is enabling increased applications for the technology. Most notable is the ever-expanding market for fiber optic data and communications links, familiar in both commercial and military markets. The inherent properties of optics and photonics, however, have suggested that components and processors may be designed that offer advantages over more commonly considered digital approaches for a variety of airborne sensor and signal processing applications. Various academic, industrial, and governmental research groups have been actively investigating and exploiting these properties of high bandwidth, large degree of parallelism in computation (e.g., processing in parallel over a two-dimensional field), and interconnectivity, and have succeeded in advancing the technology to the stage of systems demonstration. Such advantages as computational throughput and low operating power consumption are highly attractive for many computationally intensive problems. This review covers the key devices necessary for optical signal and image processors, some of the system application demonstration programs currently in progress, and active research directions for the implementation of next-generation architectures.

1.2 Applications

Optical processing technology most readily lends itself to the implementation of such transformations as the Fourier transform, convolution, and correlation. When problems can be formulated in these terms, optical modules can be incorporated as application-specific processors into signal and image processing systems. Under the DARPA-led program for Transition of Optical Processors into Systems (TOPS), for example, such optical processors are being incorporated for demonstration across a variety of applications. These include pulse compression processors for radar applications [1, 2], signal channelization/Fourier transformer and direction location processing [3], optical control of phased-array antennas [4, 5], optical null-steering [6], optical correlation for target recognition [7], real-time synthetic aperture radar image formation [8, 9], and optical data-base processing [10]. These projects are cooperative in execution between industry groups and the armed forces laboratories, with demonstrations planned by early 1995.

Acousto-optic processing offers the greatest maturity of the available photonic technologies for processing, hence, the majority of the above processors are acousto-optic based. All optical processing applications, however, have the same basic requirement for sources, modulators, and detectors as active elements of their architectures. Source considerations are not be treated here, as lasers are (and are becoming) available that have been developed for many different purposes. Modulators and detectors, however, have more processing-specific requirements imposed upon them and are discussed in the following section.

2 Devices for Optical Processing

2.1 Bragg Cells

At the heart of many real-time optical processors is an AO device or Bragg cell that performs the electrical to optical interface. The Bragg cell consists of an optical crystal with a transducer bonded to one end and an acoustic absorber at the other end. The transducer converts an input electrical signal into an acoustic pressure wave that modulates the index of refraction of the crystal by the photoelastic effect. This creates a diffraction grating with spacings that are proportional to the signal frequencies. This grating diffracts an incident light beam at angles proportional to the temporal frequencies of the signal and at amplitudes proportional to the signal strength. This diffracted beam is also frequency shifted by the grating from the incident optical frequency ν to $\nu + f$.

When the interaction length of the acoustic wave L is small the AO cell is operating in the Raman-Nath regime. In this mode the incident beam is normal to the acoustic wave and diffracted by the grating into multiple orders. The AO cell operates in the Bragg regime when the interaction length is relatively large. In this mode, with proper phase matching of the input and output wave vectors, the light is primarily diffracted into the first order. This phase matching occurs at the Bragg angle $\theta_b = \lambda_l/\lambda_a$ where λ_l is the optical wavelength and λ_a is the acoustic wavelength of the center frequency of the device bandwidth. The output beam is diffracted at an angle of 2 θ_b for a particular acoustic wavelength. The combination of phase matching and large interaction length leads to fairly high diffraction efficiencies in the Bragg mode which are generally required in most signal processing applications.

A key parameter in signal processing is the time-bandwidth product, which is a measure of processing capability. The time-bandwidth product of the Bragg cell is determined by the time aperture and bandwidth and is associated with the number of resolvable angles produced by the device. Time-bandwidth products for AO devices vary widely depending on the application with maximum values typically between 1000 and 2000. The limitation on time aperture is generally due to acoustic attenuation that varies greatly with material choice and generally increases with the square of the acoustic frequency. This leads to a tradeoff between time aperture and bandwidth since in order to achieve higher bandwidths the center frequency of the device must increase to satisfy the octave bandwidth criterion. This criterion is satisfied when the frequency range is less than one octave. The counteracting effect of these factors tends to keep the upper limit on the time-bandwidth product fairly constant.

The requirements on Bragg cells for AO processing include high diffraction efficiencies, large spurious-free dynamic ranges, and large time-bandwidth products. Acoustic beam steering and tangential phase-matching can be used to provide better phase matching over a wider range of frequencies and thereby improve efficiency and increase bandwidth. Improvements in efficiency lead to a reduction of the intermodulation products because acoustic drive power is reduced, which in turn leads to an increase in the spurious-free dynamic range, i.e., a range that is free from two-tone intermodulation products.

Because they can provide longer time apertures high-frequency-resolution devices are typically made from materials that have slow acoustic velocities and low acoustic attenuation. Devices have been reported that have time-bandwidth products of greater than 2000 and spurious-free dynamic ranges of greater than 50 dB [11, 12]. Bandwidths for these devices are as high as 60 MHz, and time apertures are as long as 100 μs.

Wideband Bragg cells that have intermodulation products less than 50 dB have been reported [13, 14, 15]. The reduction in intermodulation products is achieved by using phased-array transducer designs. The devices have bandwidths between 1 and 2.5 GHz. These devices tend to have small time apertures of 1 μs or less because of acoustic attenuation. The largest time-bandwidth products of such wideband cells are typically about 1000.

2.2 SLM's

Although high-performance AOcells are critical for optical processingof one-dimensional(1-D)signals, the future success of optical processing of two-dimensional(2-D) signals is highly dependent on the development of 2-D spatial light modulators (SLM's). Holographic film is often used to demonstrate the ability of optics to process large images. In a standard VanderLugt correlator architecture, film is used at the input plane to store input imagery and at the pupil plane to store Fourier filter information. The film spatially modulates the transmittance characteristics of the impinging optical wavefield. For real-time optical processing to be performed, film must be replaced by optically or electrically addressable SLM's. The ideal SLM would have a large number of effective pixels, continuous phase and amplitude modulation capability, and high-speed modulation characteristics. Currently available SLM's have a limited number of pixels, limited modulation capability, and limited speed of modulation. We believe that an improvement in SLM technology is necessary in order for optical processing to make a significant impact in the processing of 2-D signals. In this section, we briefly describe some of the most promising electrically addressable SLM technologies. Most of the discussion is directed toward the use of the SLM's in a VanderLugt correlator architecture, which can be applied to target recognition of imagery gathered from aerospace sensors.

Magneto-optic spatial light modulators [16] (MOSLM's) have received widespread use in the demonstration of 2-D optical correlators [17, 18]. MOSLM's modulate the polarization of the incident light through the magneto-optic effect. When linearly polarized light is incident on a given pixel, the output polarization is rotated by an angle of θ either clockwise or counterclockwise, depending on the signal sent to the particular pixel. Appropriate orientation of input and output polarizers can be chosen to give either binary amplitude operation or binary phase modulation [16]. In a MOSLM-based VanderLugt correlator architecture, binary amplitude modulation is typically employed for the input device, and binary phase modulation is employed at the filter plane [18]. Researchers have also exploited a third modulation state to design ternary filters in which each pixel assumes one of three values: $+1$, 0, or -1 [19]. Operation of a 128 by 128 pixel device has been reported at rates as high as 2 kHz, and 256 by 256 pixel devices currently under development are anticipated to operate at speeds up to 10 kHz [20]. We view the possibility of high frame rates as the most advantageous feature of the MOSLM, but relatively low light throughput and the limited number of modulation values are significant limitations.

Deformable mirror devices (DMD's) [21] have also been used to implement 2-D coherent correlators [22]. DMD's are pixel devices that consist of electrostatically controlled mirrors. Different pixel architectures are used to produce amplitude-dominant or phase-dominant modulation characteristics [23]. In a correlator architecture, an amplitude-dominant DMD is placed at the input, and a phase-dominant DMD is used at the filter plane. The devices can be operated in either an analog or binary modulation mode. A frame-addressed device having 128 by 128 pixels can be operated at speeds as high as 7.8 kHz [23]. Future plans call for the development of larger arrays for optical processing applications having 512 by 512 pixels. The application of high-definition television displays has spurred work on devices with lower frame rates and higher pixel counts. The DMD offers the potential of high-speed operation, continous phase or amplitude modulation, and relatively high pixel counts. In addition, the devices can be operated in a coupled phase and amplitude mode [22]. Despite the great promise of these devices, evaluation devices are not widely available at this time because problems with fabrication procedures.

The ferroelectric crystal (FLC) SLM is another promising 2-D device for optical processing applications [24, 25]. FLC's exhibit a large birefringence and can be operated as switchable half-wave plates. Appropriate application of an external electric field reorients the FLC optic axis. In order to maintain high switching speeds, the devices are typically designed to support binary modulation. As with the MOSLM's, appropriate orientation of input and output polarizers allows the device to be operated effectively as either a binary amplitude SLM or a binary phase SLM. Operation of a 64 by 64 pixel binary modulation device has been reported at rates as high as 4.5 kHz [24]. New FLC materials and techniques that can provide continuous modulation are currently under investigation [26, 27]. FLC's offer great potential in terms of modulation capability and speed. In addition, FLC's are compatible with standard integrated circuit fabrication procedures that enable the construction of "smart" SLM's with on-chip circuitry [24]. As with the DMD's, evaluation devices are not currently available in significant numbers.

Twisted nematic liquid-crystal display panels for projection television have found widespread use as SLM's in laboratory investigations of optical correlators [28, 29]. Such liquid crystals (LCTV's) can be operated as continuous polarization rotators or as continuous phase modulation devices [30, 31]. When used in the continuous polarization mode, input and output polarizers can be configured to provide continuous amplitude modulation. This capability can be used at the input plane SLM to provide grey-scale input imagery. At the pupil plane SLM, polarizers can be used to produce binary phase with or without amplitude filters. Alternatively, adjusting the bias level of the analog drive signal allows continuous analog phase modulation for the filter plane SLM through the electro-optic effect. Coupled amplitude and phase modulation modes have also been exploited [29, 32]. Lack of phase flatness may prove to be a problem, particularly for the input plane SLM [33]. LCTV's with 440 by 480 pixels are available at relatively modest prices. Because devices are designed to operate at video rates, extension to the much higher speeds necessary in optical processing is needed before they can be considered legitimate candidates for realistic optical processing applications.

2.3 Detectors

A critical component in the interfacing between the optics and subsequent processing elements is the detector array. Until recently, the detector has been the critical component that has limited the performance of optical processors [34]. Most commercially available detector arrays were developed for imaging applications where high speed and high dynamic range are not required. Significant effort has been put into the development of both 1-D and 2-D sensors aimed at satisfying the performance parameters required of optical processing. Improvements have been reported in device dynamic range, image lag, crosstalk, and readout rates for devices with large numbers of elements [35, 36].

For many 1-D signal processing applications, the detection requirements are for a device with at least 1000 detector elements that can be read out at microsecond rates that has optical detection dynamic range capabilities in excess of 50 dB. Fast readout rates for large devices can be obtained by dividing the detector array into sections that are read out in parallel. Factors that limit the dynamic range performance of CCD detectors include the charge handling capability of the device and the background noise level. Other factors, such as the analog to digital (A/D) conversion after detection, make it desirable to have a detector that implements some method of voltage compression on the output signal. Voltage compression methods have been realized that provide either a logarithmic response [37] or a piecewise linear approximation of a logarithmic response [38, 39].

Performance specifications for high-speed linear detector arrays that were developed specifically for optical processing have more than 1000 elements, dynamic ranges >60 dB with no image lag, and readout rates of less than 10 μs [36, 37, 39].

Two-dimensional signal and information processing pose similar requirements on area arrays for large number of elements, typically 1000 by 1000, to provide high resolution imaging with dynamic ranges of greater than 30 dB. Several high-frame-rate 2-D imagers that have been developed to satisfy these requirements use frame-transfer and interline-transfer readout schemes [40, 41]. The frame-transfer device is a high-fill-factor, back-side-illuminated device that requires shuttering of the light during readout to avoid smearing of the image. The imager dead time is minimized by the use of a storage region next to the imaging region that holds the charge during readout. The interline-transfer device has an interline readout structure in the imaging section that reduces the fill-factor to approximately 50%. These devices have the advantages of high resolution for the frame-transfer device and high probability of intercept for the interline-transfer device. Another device in development is a frame-interline-transfer device [42]. This includes a storage section next to the imaging region where the charge is quickly transferred. This architecture is used to reduce image smear, that can result from substrate crosstalk to the interline transfer registers, by minimizing the time that the charge resides in that area.

The performance specifications of several multi-output high-speed 2-D detector arrays that are suitable for signal and image processing applications have been reported. These arrays have 1024 × 1024 elements with dynamic ranges of >30 dB and commensurate crosstalk performance. The frame rates of these devices are at least 100 frames per second [40, 42].

An alternative approach to the development of CCD detectors for 1-D AO processors has been the development of a fiber-optic detector system. A joint program between ARL and Lockheed has produced a detector system with both heterodyne detection and intensity detection capabilities. This approach uses an array of 512 optical fibers to transfer the signal at the focal plane of the processor to individual photodetectors. Each detector is a photodetector/logarithmic amplifier hybrid with a PIN (positive-intrinsic-negative) photodiode operating in the photovoltaic mode. The output voltage is a logarithmic function of the incident light intensity allowing for the compression of a large optical dynamic range.

When operated at room temperature, this system has an optical detection dynamic range of >100 dB. The dynamic range increases to >125 dB when the system temperature is lowered to −15 degrees C. In addition, this system can be used for heterodyne detection at frequencies of up to 150 kHz.

3 Optical Processors

3.1 Channelizers

A channelized receiver is a spectrum analyzer that implements a large number of narrowband filters or channels. The channelized receiver has the advantage over a scanning super-heterodyne receiver of being able to achieve a high probability of intercept in a dense signal environment. Further, an AO channelizer can provide hundreds of channels in a small, rugged package.

Two types of AO spectrum analyzers are commonly used: the power spectrum analyzer and the interferometric spectrum analyzer. The power spectrum analyzer, shown in figure 1, provides an output that is proportional to the light intensity incident on the photodetector array that is a measurement of the input signal power. The interferometric spectrum analyzer, shown in figure 2, uses a reference beam to heterodyne with the signal beam so that the output is proportional to the input signal amplitude. The ability to measure amplitude, as opposed to intensity, allows for the detection of complex signals, as well as providing for increased dynamic range performance. The power spectrum analyzer has the advantage of simpler post-detection processing and is more stable mechanically.

Each processor consists of essentially the same components: a laser. AO Bragg cell, and a linear array of photodetectors. The collimated laser illuminates the Bragg cell that contains a transducer that converts the electrical signal to an acoustic pressure wave. This pressure wave diffracts light incident upon the cell at an angle proportional to the temporal frequency of the electrical drive signal. A lens placed a focal length behind the Bragg cell converts the angular frequency spectrum to a linear spatial spectrum that is sampled by the detector array. The detector also serves to channelize the frequency spectrum. With currently available devices, AO spectrum analyzers are capable of processing between 1000 to 2000 channels.

A channelized receiver has the ability to provide frequency, amplitude, and time-of-arrival information on multiple simultaneous signals. The TOPS EW channelizer is a small, lightweight, low-power, wide instantaneous bandwidth channelizer which will be integrated into an electronic support measures (ESM) system for demonstration and evaluation [3]. It is intended to provide increased capability over conventional IFM (intermediate frequency modulation) receivers by having a 100 % probability of intercept in dense signal environments.

An extension to the 1-D channelized receiver is a 2-D processor that uses a multi-channel AO device with a phased-array antenna to provide accurate direction-of-arrival (DOA) information on multiple simultaneous signals. The TOPS precision direction finder (PDF) program will demonstrate a wideband channelized receiver in a low size, weight, and power configuration that has the added feature of DOA capability [3]. Within each Bragg cell channel, the system acts as a spectrum analyzer or single channelized receiver that provides emitter frequency information in one dimension. The other dimension consists of a phase interferometer where each separate Bragg cell channel is fed by an element from a phased-array antenna. The interferometer uses the phase difference in the antenna elements related to the DOA of the signal to form an input grating. The Fourier transform produced by the lens then provides a spatial location, which depends on the emitter location angle.

3.2 Correlators

Optical correlators have been proposed for use in target detection and classification. A 1-D correlator produces an output that is proportional to the correlation between two input signals:

$$i\left(\frac{z}{v}\right) = \int s_1(\nu)s_2\left(\nu + \frac{z}{v}\right)\, d\nu \tag{1}$$

where z is the position variable in the AO device and v is the acoustic velocity. If $s_2(\nu)$ is a shifted replica of $s_1(\nu)$, a correlation peak occurs at the lag τ position, where $\tau = z/v$.

The integration in equation (1) can be performed either spatially or temporally. In the space-integrating approach, the light that is used to illuminate the Bragg cell is collected onto a single detector so that the time-varying output of the detector contains the desired correlation. Since the correlation is realized by the spatial integration of all parts of the signal contained in the acoustic device simultaneously, this technique is limited by the time-bandwidth product of the AO device. In the time-integrating approach, the integration is performed temporally by a linear detector array. The time-bandwidth product of the time-integrating approach is then determined by the detector integration time and the Bragg cell bandwidth [43]. An advantage of the time-integrating technique over space-integrating is its ability to operate on large time-bandwidth products signals. Due to the increased processing gain that time-integration offers this paper focuses on time-integrating correlators as opposed to space-integrating correlators.

Correlation can be obtained through systems based on either additive or multiplicative architectures. An example of an additive architecture is a Mach-Zender interferometer used with a Bragg cell in each arm to intensity modulate the light. The two beams are then recombined and detected. The simplest example of a multiplicative architecture, shown in figure 3, consists of an intensity-modulated source that illuminates a Bragg cell which is imaged onto a detector array. A portion of the light beam may be split off in a number of ways and used as a reference for coherent processing [44].

Since correlations and convolutions represent the basis of most conventional signal processing, correlators are found in such diverse applications as broadband signal detection, code synchronization in communication systems, filtering, radar ranging, and pulse compression. There currently exist several programs for demonstrating the capabilities of optical processing in detection and classification of signals for both ESM and radar applications. Part of the TOPS pulse compressor program consists of the development of a 1-D time-integrating correlator for a wideband ESM system [2]. The correlator is being developed for detecting and analyzing wideband spread-spectrum signals. The goal of the program is to provide an increased capability in wideband signal detection to an existing ESM system.

Correlators are not limited to processing only two signals: 2-D correlators are capable of processing three or more signals by producing a 2-D surface. Because of their capacity for increased processing power over current digital capabilities, 2-D AO processors are attractive in applications such as communications intercept, radar, and image processing/target recognition. The increased processing power is achieved through the massively parallel processing structure of 2-D optical architectures, which provides large throughput and high bandwidth capabilities.

For example, the 2-D triple-product correlator is designed to perform the three-product correlation given by

$$c(\tau_1, \tau_2) = \int_T A(t)B(t-\tau_1)C(t-\tau_2))dt \tag{2}$$

where $A(t)$, $B(t)$, and $C(t)$ are signals input into the correlator and T is the integration time of the detector. Depending on the form of the input signals, different signal properties can be represented in the two dimensions. Many operations such as ambiguity function generation, high-resolution spectrum analysis, cyclo-stationary processing, computation of the bispectrum, and Wigner-Ville transform generation can be performed optically using the 2-D correlator. Architectures for 2-D processing have been developed that make use of three Bragg cells [45], two Bragg cells and a modulated source [45], two counter-propagating surface-acoustic-wave (SAW) delay lines [46], and one four-transducer SAW delay line with a dove prism that provides rotation of the orthogonal dimension [47].

In applications such as radar systems where Doppler frequency shifts occur between the reference signal and the return signal, use of a 1-D correlator becomes difficult. A 2-D correlator can provide Doppler shift information, as well as time difference of arrival (TDOA) information, provided the correlator is configured as an ambiguity function generator. The ambiguity function is given by

$$A_{xy}(\tau, f_d) = \int_T x(t)y^*(t-\tau)e^{j2\pi f_d t}dt, \tag{3}$$

where τ is the time delay and f_d is the Doppler shift between the two signals $x(t)$ and $y(t)$. If y is a time-shifted version of x the correlator produces a 2-D correlation peak at coordinates proportional to the time delay and Doppler shift.

A range-Doppler processor designed to perform the ambiguity function is under development as part of the TOPS program for insertion into a radar testbed. The processor, shown in figure 4 [1], uses a 2-D time-integrating AO correlator in an additive architecture. To produce the ambiguity function, the laser diode is intensity modulated with the reference signal on a chirped carrier, a chirp signal is input into AO cell 1, and the radar return signal is input into AO cell 2. The output is a 2-D range-Doppler image on a spatial carrier. The goal of this demonstration is to demonstrate a system that is capable of wideband processing at a reduction in size and weight that can be incorporated into future wideband radar systems.

3.3 SAR

Due to the natural fit of optical processing primitives to sythetic aperture radar (SAR) image formation and due to the lack of computers having sufficient computational horsepower, SAR imagery has been typically processed using optical techniques. In fact, coherent optical processors for SAR image formation represent one of the most significant success stories for optical processing. Even with advances in analog and digital processing, optical processors are still used to process SAR imagery stored on film. Only recently have computers become powerful enough to tackle the real-time formation of SAR imagery (this capability still requires a sizeable computer that occupies significant space). However, optical processors are also being developed to implement real time image formation algorithms; these optical processors will not only equal the performance of digital processors, they will typically do so in smaller, lighter packages and consume significantly less power.

3.3.1 SAR Background

A SAR is an imaging sensor that finds many applications in the remote sensing arena. When a remote object is sensed by a radar, image resolution is primarily determined by two parameters: range resolution, and azimuthal or cross-range resolution. Range resolution for both SARs and conventional radars is determined by the use of short pulses or, equivalently, any waveform having a high time-bandwidth product (typically a chirp). On the other hand the azimuthal resolution problem is solved differently for SAR than for conventional radars.

In a conventional radar, azimuthal resolution is limited by the antenna characteristics; the larger the antenna, the better the resolution. The 3 dB beamwidth γ of an antenna is approximated by

$$\gamma = \frac{\lambda}{D}, \tag{4}$$

where λ is the wavelength and D is the antenna diameter (or length in the azimuth direction in the case of a rectangular array). The resolution on the ground δY is

$$\delta Y = R\gamma, \tag{5}$$

where R is the range to the target area. At typical operating frequencies up to tens of GHz it is impractical to construct antennas sufficiently large to achieve the resolutions demnaded by modern applications.

However by using a standard antenna and coherently processing radar return data from many locations along a collection path, we synthesize the equivalent of a long aperture antenna. Azimuthal information is deduced from a Doppler analyses of the returned signal. The Doppler shifts arise from phase changes in the radar return as the collection platform moves from one point to another. Image formation is not affected by the velocity of the collection, but by the density of data collection. Since SAR exploits Doppler generated by platform motion for image formation, there is an implicit assumption that the objects in the area to be imaged are stationary. Any motion of targets in the imaging area will cause blur or ambiguity in target location. Two points on the ground at the same range are resolvable if the frequency difference in their Doppler shifts is resolvable. As might be expected, this condition is satisfied by

$$\delta Y = R \times \frac{\lambda}{L} \tag{6}$$

where L is the length of synthetic aperture (i.e., it is the same as if one had a physical antenna of length L).

3.3.2 SAR Processing

SAR image formation from phase history data consists of two steps. The data must first be compressed in range. This is achieved by correlating the radar return with a copy of the transmitted waveform. For a transmitted chirp waveform the signal can be dechirped by mixing the radar return with a copy of the transmitted chirp, so that range data is transformed into frequency and the range compression step can be accomplished with a Fourier transform. The range compressed data must then be compressed in azimuth. If the SAR data was focused (that is, compensated for platform motion relative to the center of the target area), then azimuth compression can also be accomplished by a Fourier transform operation. In this case the Fourier transform is performed on all the data from the collection path that lies within a particular range bin. This must be done in sequence for each of the range bins and is shown schematically in figure 5[48].

The nature of this azimuth compression processing requires digital processors to store all of the return data in a corner turning memory either befire or during processing. In particular, all the data must be collected before azimuth compression can begin. Alternatively it is possibble to perform a 2-D Fourier transform on the entire data set, which is one of the reasons that optical processors were originally used to form SAR imagery. If the data recorded in a raster format on film, illuminated by coherent light, and followed by a spherical lens, the 2-D Fourier transform is computed and observed at one focal length past the lens. This type of processor is classified as a space integrating architecture, since both range and cross-range information is encoded and processed spatially.

Although the space integrating architecture is an elegant solution, it is not real time. Due to their limited dynamic range, rapid developing films are not sufficient. If spatial light modulators, as discussed in section 2.2, of sufficient density and performance were available, it would be trivial to construct a real time SAR image formation processor, however these devices are still under development. The versatility and potential performance of AO processors has prompted their inclusion in the TOPS program. Two different processors are to be designed, constructed and tested using real sensor data. These two processors are described briefly in the following paragraphs.

From the description above the SAR imaging problem can be decomposed into two sequential 1-D operations. Each return pulse is processed in one dimension to determine range and processed in the orthogonal direction to compute cross-range. The lack of real-time film or spatial light modulators has promptedthe development of a time space integrating (TSI) processor for SAR image formation [49, 50, 51, 52, 53, 54, 55]. This architecture uses the spatial dimension for range compression and temporal integration for azimuth compression. The focused image is formed in time on a 2-D imaging sensor, e.g., a vidicon camera or a CCD detector array. The image is integrated as a function of time, and exibits an improvement in resolution as more radar pulses are integrated. An operator watching the output of this 2-D detector sees at first a range only image that slowly comes into focus.

The architecture of the TSI SAR processor is shown in figure 6[49] consists primarily of a pair of orthogonal Bragg cells. The first Bragg cell performs the range compression function. The range return and a reference function input into this cell interfe at the detector plane, which allows for subsequent temporal processing of target phase information. To understand how this processor functions consider the return from a single radar scatterer within the field of view. For this case each radar return produces on the output detector plane a narrow stripe along the target range which is processed coherently to determine its phase. As the collection platform moves, the phase of the single point scatterer changes due to changes in its range to the collection platform. This change in phase must be detected to perform the cross-range compression function. In the TSI architecture cross range compression is performed by correlating in a temporal fasion the range compressed data with a signal in the second crossed Bragg cell. As discussed previously, cross-range is determined by Fourier transforming the time history for a particular range gate. However Fourier transformation can be considered as a crosscorrelation of the signal with a series of sinusoids at each of the frequencies to be analyzed. Thus by inputting the appropriate signal into the second Bragg cell a different frequency can be presented at each cross range stripe on the 2-D detector. Through the process of time integration a cross correlation of the range stripe is achieved with every possible Doppler frequncy (the cross range stripes) which yields a correlation peak only at the coordinates that correspond to the point scatterer on the 2-D detector.

An other architecture for SAR image formation such as the time time integrating (TTI) processor which is a generalized triple product processor [56, 57, 58]. In this processor both range and cross-range compression are achieved using time integration. The TTI architecture also requires a pair of crossed Bragg cells, however here they serve a different function than in the TSI architecture. The TTI processor represneted in figure (figure 7) [59] is a time integrating Fourier transform processor. The laser diode is modulated with the radar return data and the two Bragg cells are used to scan the signal beam into a modified Koster interferometer (MKI). The MKI is used to generate a pair of light beams that are subsequently interfered to form a sinusoidal set of fringes. A set of fringes is formed for every target return from every radar pulse which are coherently integrated on the detector array. It can be shown [59] that this process yields the 2-D Fourier transform of the reformatted input signal and hence the desired formed image.

The above simplistic explanations are presented to give the reader an intuitive feel for the processing necessary for SAR image formation. For more in depth information, the reader is referred to the cited references. Also (as usual) not everything is as simple as it seems: dynamic range must be considered ,as well as techniques for providing finely focused operation. These issues, together with system interfaces, are being addressed as part of the TOPS program. We expect to demonstrate AO based SAR image formation processors in 1995 that are small, lightweight, and consume significantly less power than equivalent digital processors.

3.4 ATR

Optical processing could play an important role in providing automatic target recognition (ATR) of SAR imagery gathered by airborne sensors. Most optical architectures proposed for ATR applications are based on coherent optical correlators that employ electrically addressable SLM's at the input plane and the filter plane. The input imagery generated from the phase history is preprocessed digitally and presented to the input plane SLM, and pattern recognition filters are written to the pupil plane SLM. Ideally, the presence of a target will be indicated by a bright spot of light at the output plane of the correlator. Important issues that need to be addressed include filter design, filter management, and the limited performance characteristics of currently available SLM's.

The design of the best pattern-recognition filter for a given application is a nontrivial problem. The matched filter is said to be the optimum filter for detection in the presence of additive white noise, but background noise and clutter in typical scenes do not fit the additive white noise model. Despite the preponderance of articles on binary phase filters, they are only optimum in terms of light efficiency. The problem of determining the optimum filter has received some attention, but further research is necessary. An excellent review of optical pattern-recognition filters has been prepared by Kumar [60].

Another critical aspect of system design is filter management: that is, how the system searches through the set of filters. The speed of the correlator will have a strong effect on the filter management strategy. If the correlator can operate at a high rate, the filter designer may elect to prepare a large number of highly discriminating filters and allow the correlator to search over the whole bank of filters for each input scene. If the correlator cannot operate fast enough to access all filters, the designer is forced to use composite filters that are designed to account for variations of the target, such as, scale or rotation. If we take the composite filter approach to the extreme, we can envision a single filter to cover all possible variations of the target. Such a general composite filter is not practical, since the probability of false alarm will rise as the filter is designed to handle more and more variations. The designer must reach a compromise between using a large number of highly discriminating filters and using a small number of composite filters. The best solution may be to start with highly composite filters and then perform a smart search over filter space that ends with the use of highly discriminating filters that perform more exact identification or classification. The approach chosen for a particular application will be determined by the speed of the correlator and the amount of target variation that is anticipated.

As discussed in section 2.2, the limited performance characteristics of currently available SLM's adversely affects the performance of optical correlators. An optical correlator is an analog processor that is limited by device performance and optical system noise. The key devices are the input and pupil plane SLM's, the laser source, and the two-dimensional detector array. The performance of the SLM's represents the largest stumbling block [61]. The ideal correlator would have full grey-scale modulation at the input SLM and full complex modulation at the filter plane SLM. In practice, a limited approximation to the ideal correlator may be sufficient to solve some ATR problems. The issue really is how close an approximation is sufficient for a particular application.

While there is general agreement that SLM's with faster modulation and higher pixel count are needed, the importance of an improvement in modulation capability is subject to some debate. The limitations of simple binary phase filtering approaches are beginning to receive some attention in the literature [61, 62], but some optical pattern recognition experts still feel that binary or ternary modulation is sufficient. Filter design algorithms that take the SLM modulation characteristics into account may compensate in part for the device limitations [63, 64]. In fact, diffractive optical element design techniques can be used to synthesize virtually arbitrary correlation templates even when restricted to binary phase modulation, provided that the SLM has a very high pixel count [63]. An alternative solution to the SLM performance problem is the use of incoherent AO image correlators [65] that rely on more mature device technology.

The issue of limited modulation at the input plane is also important. Binary modulation SLM's at the input plane are only sufficient if a robust segmenter is available to binarize the original greyscale imagery. Many researchers in the digital ATR community feel that segmentation is often the operation that limits the effectiveness of ATR systems. Further investigation is needed to determine whether available segmentation approaches are robust enough to handle realistic imagery.

A number of programs are under way to demonstrate the potential of optical correlators. The most visible of these programs is part of DARPA's TOPS program [66, 67]. This program is directed toward imagery gathered in the visible spectrum, but lessons learned should be useful for future systems designed to operate on SAR-generated imagery. The aggressive time scale of the TOPS program forces the use of SLM's having limited modulation (binary or ternary), limited pixel counts (256 by 256 pixels or smaller), and limited modulation speed. The degree of success that the TOPS program encounters should be very useful in quantifying the amount of SLM performance improvement necessary to tackle more challenging ATR problems.

Optical correlators are facing competition from digital correlators because of the advent of custom high-speed FFT chips and parallel processing architectures [68, 69]. The development of SLM's with higher pixel counts and speeds is necessary to provide a clear advantage over rapidly improving electronic architectures. If major breakthroughs in SLM technology do occur, optical correlators should be in an excellent position to tackle challenging ATR problems. While the speed and compactness of projected digital correlators are impressive, it is generally acknowledged that optical implementations should enjoy a significant advantage with respect to power consumption.

The direct competition of digital correlators is overshadowed by the indirect competition represented by highly flexible digital ATR systems. Typical digital ATR systems do not rely on simple correlation schemes for target identification and classification. Since digital systems are not inherently limited to Fourier-based operations, they can implement complex nonlinear pattern-recognition algorithms. Despite the flexibility, digital ATR approaches have yet to meet the difficult challenges that real ATR problems pose. The complementary strengths of digital and optical approaches suggest that a hybrid optical digital approach to ATR should be explored. As an example of a hybrid approach, the inherent parallelism of optics could be used to perform a broad-area search where the task is to identify regions of interest (ROI's) for subsequent processing by a digital processor [61]. AO incoherent correlators [65] and SLM-based incoherent correlators [70] are particularly suited for the ROI task. Optical implementations of morphological algorithms [71, 72] could also play a role in ROI operations. As another example of a hybrid approach, optical interconnects could be used to enhance the performance of digital ATR hardware. The key is to exploit the strengths of both optics and electronics.

4 Optical Control for Phased-Array Antennas

Next-generation communications and radar systems will employ phased-array antenna systems to provide greater operational flexibility. These antenna systems consist of a large number of antenna elements and transmit/receive (T/R) modules to effectively control the shape and direction of the radiation pattern. The T/R modules use phase shifters and variable attenuators to alter the beam shape and direction. Because the larger number of elements and subsequent size of the antenna, the signal distribution networks are bulky and lossy, and vary appreciably with temperature. For this reason fiber optic systems have been considered for use in such networks, which are small and light weight.

In phased array systems optical fibers can be used effectively as time delay units, which provides an increase in system bandwidth over those that use electronic phase shifters. Optical fibers are currently being used to provide long time delays in electronic warfare systems to allow sufficient time to preprocess the received signals. Fiber optics delay lines are also being used to simulate radar target return signals for calibration and testing of radar target generators. The advantages of fiber optic technologies over conventional microwave technologies include wider instantaneous bandwidth, immunity from electromagnetic interference, small size, and light weight [73].

4.1 Phased-Array Antenna Background

High performance communication and radar systems often require antennas with high gain and narrow beamwidth to adequately perform their mission. Since antenna gain is given by

$$G = \frac{4\,\pi\,\rho_a\,A}{\lambda^2} \tag{7}$$

increase gain can be realized by increasing the physical size of the antenna, where ρ_a is the antenna aperture efficiency, A is the physical area of the antenna and λ is the wavelength. Because beamwidth is inversely proportional to the gain, increasing the area of the antenna also decreases the beamwidth.

However the gain can also be increased by forming an array of individual elements provided that coupling between elements is minimized. The radiation pattern of such an array is effectively the vector addition of the radiated fields of each element. The individual elements must be excited so that their radiated fields constructively interfere in the desired direction and destructively interfere in all other directions. The shape and direction of the radiated beam can be controlled by proper selection of the following: the configuration of the array (linear, circular, rectangular, etc.), the spacing between elements, the signal amplitude at the individual elements, the signal phase at the individual elements, and the radiation pattern of the individual elements [74]. The radiated field of the array is given by

$$E(total) = (arrayfactor)(elementfactor) \tag{8}$$

where the array factor is a function of the array configuration and the element factor is the radiated field of an individual element.

Phased arrays can be steered by changes in transmit frequency or by electronic variation of the relative phases of the signals at the elements. Electronically scanned phased arrays use electronic phase shifters to dynamically steer the radiated beam. We can steer the beam by adjusting the relative phases of the signals to the elements to attain the desired beam direction. The direction of the radiated beam is given by

$$\theta_0 = \sin^{-1}\left(\frac{\phi\, c}{2\,\pi\, d\, f}\right), \tag{9}$$

where ϕ is the relative phase shift between adjacent elements, c is the speed of light, d is the spacing between elements, and f is the frequency. These arrays can generate multiple beams to search and also track several targets.

Phase shifter technology has evolved over the years from the first mechanically actuated phase shifters to the current electronically controlled ones. Ferrite phase shifters use a metal-oxide material whose permeability can be altered electronically to change the velocity within the medium and hence produce a phase shift. Digitally switched phase shifters use constant phase shifters that can be switched in or out electronically in a series or parallel configuration to effectively alter phase. Many other types of phase shifters are also described in the literature [75].

One problem to using phase shifters to steer the radiated beam is a phenomenon called beam squint, which occurs when the array is operated at frequencies other than its center frequency [76]. As seen in equation (9), a constant phase shift causes the radiated beam to change direction, or squint, if there is any variation in the frequency. This is a significant problem in high-resolution radar systems, which must transmit wide-bandwidth signals to achieve the required fine range resolution. The broad range of input frequencies causes the different frequency components to be radiated in different directions which results in a smeared beam.

This can be corrected if time delay elements are used instead of phase shifters. Photonics techniques being investigated to realize time delay elements include several novel AO architectures[77, 78, 79]. Although these techniques may have future applications, advances in fiber optics appear to be the most promising to realize a true time delay element.

4.2 System Applications

High-speed fiber-optic links have been demonstrated in remote signal distribution applications where the transmit, receive, reference oscillator, and control signals are transferred from the system to a remote antenna via fiber optic cables. In a coherent aperture, the T/R modules at each element must be synchronized to a master oscillator. We can accomplish this synchronization either by directly distributing the master oscillator via fiber-optic lines or by placing an oscillator at each T/R module and locking optically them via fiber.

One drawback of the fiber optic technique is the inefficient conversion of RF energy to optical and vice versa. The complexity of these conversions tends to degrade performance with respect to dynamic range, signal-to-noise ratio, and bandwidth[80].

As mentioned above, other applications of fiber optics for phase array antenna include true time delay beam steering, as opposed to the traditional microwave phase shifter approach. As discussed earlier, the microwave phase shifters used for beam steering cause beam smear in broadband systems which has led to the replacement of phase shifters by time delay elements. For a length l of transmission line the relative phase shift between elements is given by

$$\phi = \frac{2\,\pi\,l\,f}{v} \tag{10}$$

where f is the frequency, and v is the velocity in the transmission line. The velocity is given by $v = c/\sqrt{\epsilon_r}$, where c is the speed of light, and ϵ_r is the dielectric constant of the medium. Substitution of equation (10) into equation (9) yields

$$\theta_0 = \sin^{-1}\left(\frac{l\sqrt{\epsilon_r}}{d}\right). \tag{11}$$

The direction of the steered beam is now independent of frequency, and broadband signals can be transmitted without smearing.

Because the effective dielectric constant of microstrip of waveguide transmission lines is a function of frequency, dispersions results when signal propagate in non-TEM modes. For wideband signals dispersion also leads to squinting[81]. Because the modulation bandwidth is smaller than the center frequency inside a fiber optic cable, dispersion in a fiber optic system is minimal and demonstrations have verified the lack of beam squint over large frequency changes in an optical time delay phased array antenna [73].

Two TOPS programs currently under way seek to exploit the advantages of fiber-optic and photonic components by demonstrating true time-delay phased-array systems. The first program will use rf and digital fiber-optic signal distribution with 4 to 6-bit switched fiber-optic time-delay units in a 96-element array. This atenna array is designed for an airborne surveillance radar and will operate at L band with a 50 percent bandwidth (850-1400 MHz) and a ± 60 deg azimuth scan angle. System analysis is currently under way to develop a low-noise-figure receive path given the high insertion loss of the time-delay units. Also under investigation are ways to reduce the amplitude and phase noise of the fiber-optic links and ways to provide better amplitude and phase matching between channels. System analysis suggests that a 60-dB dynamic range and a 5-dB noise figure can be achieved with current state-of-the-art photonic components [82].

The second program seeks to use an integrated optical delay line to achieve a broad-band phased array. The proposed system is for an airborne early warning radar to provide a 380 nmi detection range against 0.1 m^2 targets at S band with over 7000 elements in an elliptical array and up to 60 deg scan angle [5]. The delay line consists of micromechanical integrated optic switches and low-loss phosphosilicate glass waveguides to provide the desired delay. These networks will be fabricated on silicon substrates using VLSI technology, which permits the use of a single off-chip emitter/detector to provide the conversion from optical to rf and back for each delay unit. Two types of switches are currently under development to provide a minimum of 20-dB isolation. In addition, system analysis is being performed to minimize noise figure and maximize dynamic range. Analysis of several cascaded networks indicates an 8-dB noise figure and a 56-dB dynamic range for high-loss (64 dB) time-delay networks [5].

Improvements to the fiber-optic approach for true time-delay phased-array antennas can be achieved in the future using integrated optics to fabricate the optical modulator and variable delay components on a single optical chip [83, 84].

5 Integrated Optics

Many of the optical processing architectures discussed in the previous sections conform to the planar geometry of integrated optics. In addition, due to the relative maturity of the AO modulator [85, 86], AO signal processing is a particularly active area of research. This section focusses on wide-band integrated optic (IO) signal processors, that are based on SAW modulators.

IO processing systems offer a number of desirable features for aerospace applications. Airborne and spaceborne sensing platforms capable of real-time signal processing greatly reduce the downlink and data storage requirements of the platform. Additionally, applications such as image formation aboard airborne and missile platforms necessitate high-speed real-time data manipulation. In section 3, bulk optical processing architectures are described that perform real-time signal and image processing for specified aerospace applications. IO implementations of these architectures offer real-time operation plus reductions in the size, weight, and power consumption of the processor.

Other inherent advantages of IO signal processors accrue from the integration of the individual components onto a common substrate. The critical optical alignment required of bulk coherent optical processing architectures increases both its fabrication cost and susceptibility to environmental effects. IO processors, however, can take advantage of the mature and highly generic fabrication processes of the semiconductor industry.

The potential for batch fabrication of IO processors translates to reduced fabrication costs. However, innovative fabrication technologies such as epitaxial liftoff [87], flip-chip bonding [88], and patterned epitaxial regrowth [89] must be developed before completely integrated optical processors are realized.

Airborne and spaceborne platforms will certainly expose the processor to harsh environmental conditions. A compact optical system that is integrated onto a common substrate forms a rugged package capable of maintaining the critial alignment of the optical components. This inherent feature of IO is invaluable in taking optical signal processing architectures from the lab bench into fieldable systems for aerospace sensing applications.

Given the above stated attributes of IO processors, numerous SAW-based architectures for rf signal processing have been proposed and investigated. These architectures primarily have been those that perform 1-D signal processing tasks, since they are directly conformable to the planar geometry of guided-wave optics. Examples of 1-D IO processing tasks include spectrum analysis [90], chirp pulse compression [91], and convolution and correlation of wideband rf signals [92, 93, 94].

Laboratory demonstrations of these IO processors have achieved limited success. The simplest of these architectures, which understandably has received the most attention, is the rf spectrum analyzer. One possible implementation of an IO spectrum analyzer is depicted in figure 8. This architecture is based on a GaAs/AlGaAs waveguide structure. The lenses are negative embedded waveguide lenses [95] and the SAW transducer incorporates a piezoelectric overlay film for efficient electric to acoustic transduction [96].

Since it contains all the essential building blocks of each of the other 1-D IO systems, let us consider the performance of this fundamental architecture. Important performance parameters of the IO spectrum analyzer include dynamic range, frequency bandwidth, frequency resolution, and response time [90]. Achievement of the desired specifications of each of these parameters depends upon the individual performances of the individual components of the system and the interplay among them.

Practical systems for rf spectrum analysis require a dynamic range in excess of 40 dB. The dynamic range of current IO systems is limited to about 20 dB. This is due to saturation of the detector at the high signal end and by the presence of a fixed pattern background signal at the low signal end [97, 98, 99]. As described in section 2.3, significant progress has been made in extending the dynamic range of linear CCD arrays to greater than 60 dB. Thus, efforts to achieve wide dynamic range IO processors should concentrate on reducing the noise floor seen by the detector array. The background light signal originates from three principle sources: in-plane waveguide scattered light [100, 101], lens boundary scattered light [102, 103, 104], and substrate coupled light (this results from inefficient source/waveguide coupling) [99]. The scattered and substrate-coupled light remains trapped in the guiding portion of the waveguide structure, where it contributes to the high noise floor. To achieve a targeted dynamic range specification of 50 dB, waveguide power attenuation on the order of 0.5 dB/cm and a per lens loss of about 1 dB should be maintained [105].

Efficient source/waveguide coupling is significant not only for improved processor dynamic range, but also for practical component integration. Currently, laboratory demonstrations of spectrum analyzers have either used butt-coupling techniques to inject the laser light into the waveguide [98, 106] or they have avoided the problem by using bulk optics to couple the light into the waveguide. $LiNbO_3$ is an attractive material system for IO processors, since it has both a high piezoelectric coupling constant and a very low acoustic propagation loss (about 1 dB/cm at 1 GHz [107]). However, a hybrid configuration is necessary, since neither sources nor detectors can be fabricated in $LiNbO_3$. New processing technologies such as epitaxial liftoff [87] or flip-chip bonding [88] might provide manufacturable and robust processes to couple sources and detectors to $LiNbO_3$-based waveguides.

In contrast, GaAs-based waveguides have the potential for total integration of sources, waveguides, modulators, lenses, detectors, and even electronics. Some progress has been made in integrating multiple IO components on GaAs substrates [99, 108, 109]. The integration of laser sources with low-loss passive waveguides on GaAs is particularly difficult because of the different epitaxially grown structures. Patterned epitaxial regrowth [89] is a particularly challenging research area which attempts to overcome this problem. It must be realized, however, that whatever material system is chosen, the fabrication process must be manufacturable for minimum cost and feasibility of implementation.

Ambitious design specifications of an IO spectrum analyzer are a 1-GHz center frequency, 500-MHz bandwidth, and 4-MHz rf frequency resolution. SAW Bragg cells operating at gigahertz frequencies have been realized in both $LiNbO_3$ [110] and GaAs [96]. The objective is to design a waveguide Bragg cell for the chosen material system with sufficient light diffraction over the specified bandwidth for adequate dynamic range and sensitivity. The bandwidth of the SAW Bragg cell is mainly determined by the frequency dependence of three factors: the transducer conversion efficiency, the Bragg condition (phase matching), and the overlap between the guided optical wave and the surface acoustic wave. A number of wideband interdigitated transducer configurations have been proposed and investigated [110]. These include multiple tilted transducers, phased-array transducers, multiple tilted phased-array transducers, tilted finger chirp transducers, and parallel finger chirp transducers. Interdigitated transducer designs have been extensively investigated; however, the system bandwidth is dependent on the individual waveguide components acting in concert. For example, the Fourier transform lens must be diffraction limited over a field of view corresponding to the AO bandwidth. For a bandwidth of 1-GHz, this is about 6 deg.

To achieve the highest frequency resolution possible, the photodetector periodicity must be less than the diffraction limited spot size. For a waveguide lens to be diffraction-limited, it must be corrected for the salient third-order Seidel aberrations (spherical, coma, and Petzval curvature) over the transducer's entire field of view. Waveguide lenses have been fabricated in a variety of substrates including $LiNbO_3$, glass, oxidized Si, and GaAs. These lensing techniques include Luneburg [111], geodesic [112], chirped grating [113], Fresnel [114], grating/Fresnel hybrid [104], and embedded waveguide [95]. As we have seen, the lens must be diffraction limited, have low scatter, and have high throughput. Furthermore, the fabrication process must be highly manufacturable to reduce costs [115].

So far, the system performance specifications for an IO spectrum analyzer have been discussed. Required research efforts in waveguide propagation characteristics, component integration (particularly, source/waveguide coupling), and waveguide lenses have been addressed. Progress in these research areas would similarly benefit the other 1-D IO architectures mentioned earlier.

A strikingly novel approach to the IO architecture attempts to process two-dimensional data fields by communicating the guided wave from the plane of the substrate vertically into the third dimension. Two-dimensional signal processing problems typically require nonplanar optical implementations. Waveguide gratings, however, are used to vertically outcouple the 2-D information. An IO architecture which forms the 2-D image from the received signal of a SAR system has been proposed [116, 117] (see Figure 9). This system uses both time and space integration in an IO format to form the 2-D SAR image.

Components of this system include wide-aperture, low-f-number waveguide lenses, wide-band SAW transducers, low cross-talk rib or channel waveguides, vertical waveguide out-coupling gratings, a fixed 2-D mask, and an area CCD array. Of particular interest is the large-area selective out-coupling grating [118, 119]. The outcoupling grating is not shown in figure 9, but is incorporated in the rib waveguide array. This grating must provide uniform out-coupling over the length of the interaction region (approximately 1 cm), near vertical diffraction angle, and high diffraction efficiency [120, 121].

IO processors promise numerous advantages for aerospace sensing applications that involve both 1-D and 2-D data fields. Areas of research emphasis include waveguide lenses, source/waveguide coupling, and vertical out-coupling gratings. Of primary importance in designing IO processors is material system selection, compatibility of component fabrication processes with respect to component integration and manufacturability, and overall component interaction for peak system performance.

6 Conclusion

The research and development community is beginning to make significant strides in bringing optical processing technology into useful systems applications. As the various demonstrations prove the effectiveness and reliability of the technology, opportunities for additional applications will lead to further interest in the technology, additional avenues of research, and the possibility for more powerful, compact signal and image processing systems.

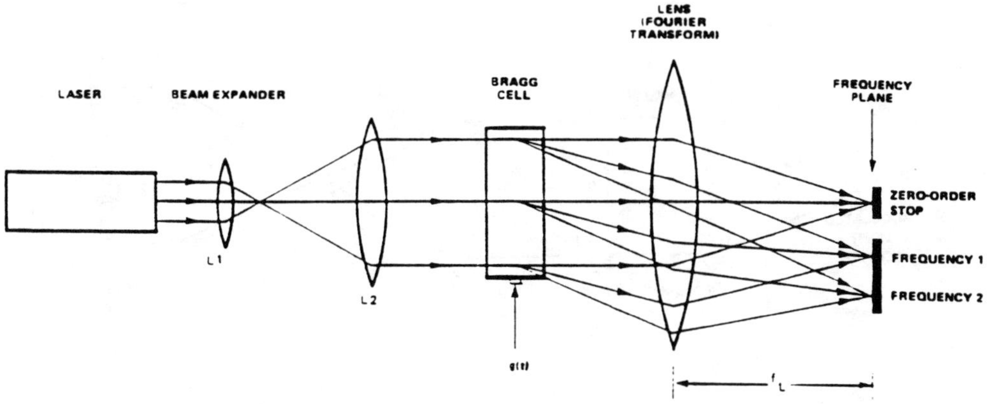

Figure 1: AO power spectrum analyzer.

[122]

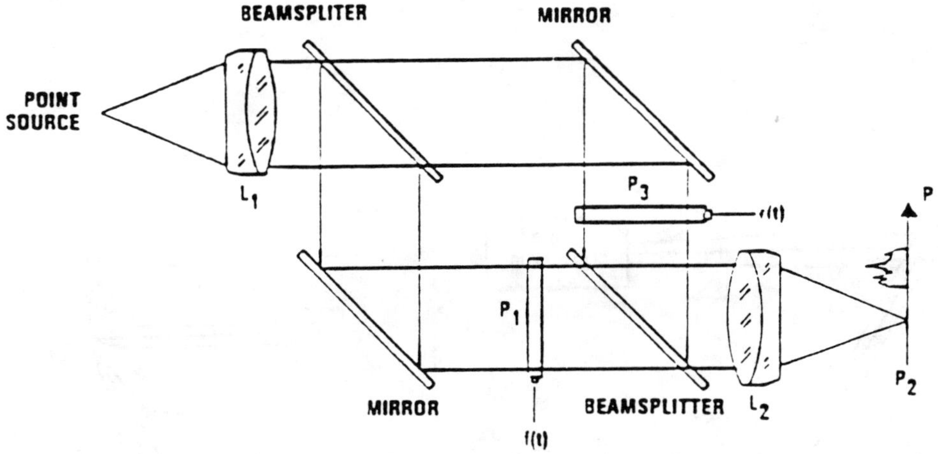

Figure 2: AO Interferometric spectrum analyzer.

[123]

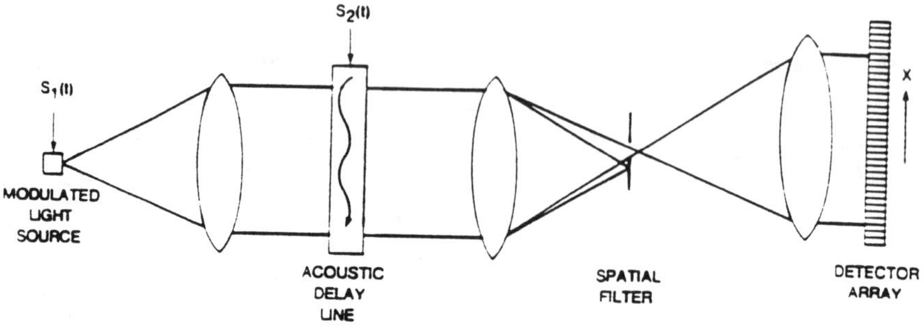

Figure 3: Incoherent time-integrating correlator.

[122]

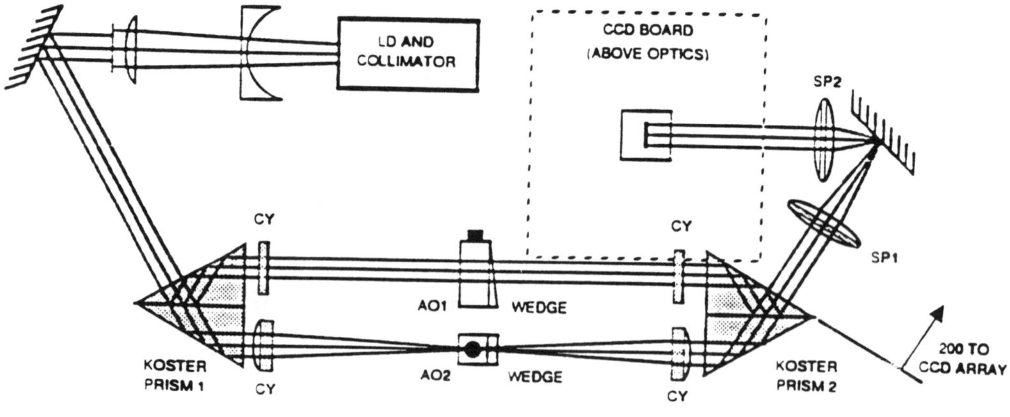

Figure 4: AO range-Doppler architecture.

[1]

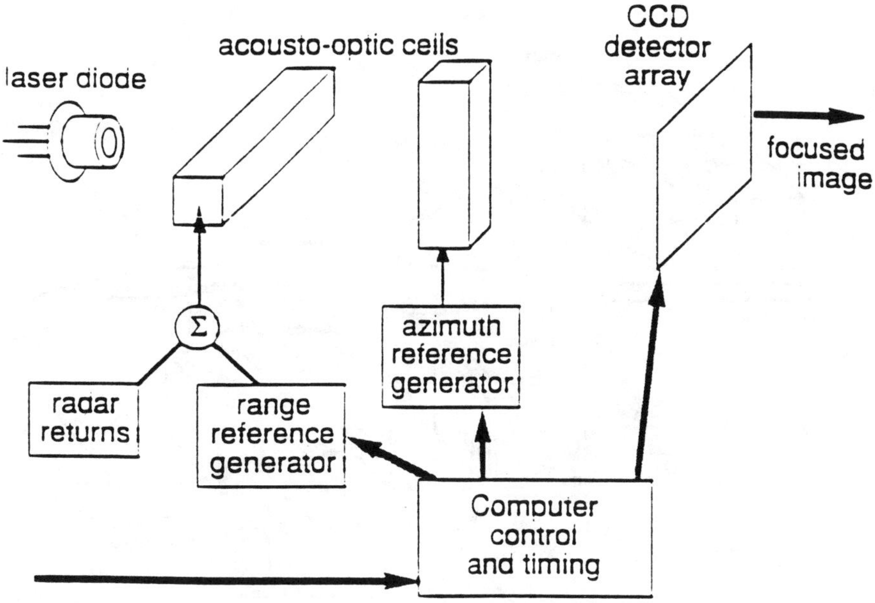

Figure 5: Schematic of TSI architecture.
[49]

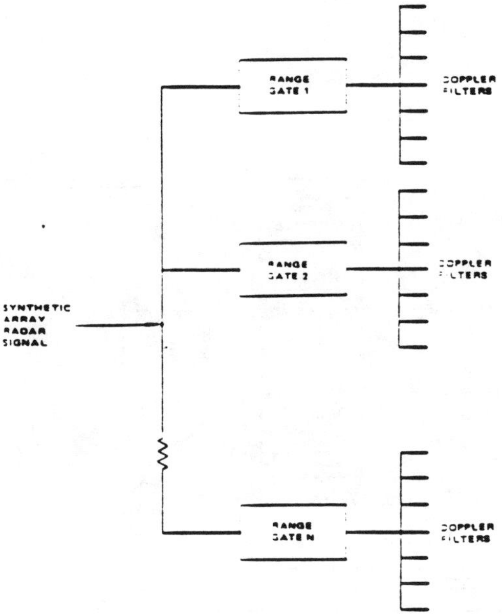

Figure 6: Range gates followed by Doppler filters for SAR image formation.
[48]

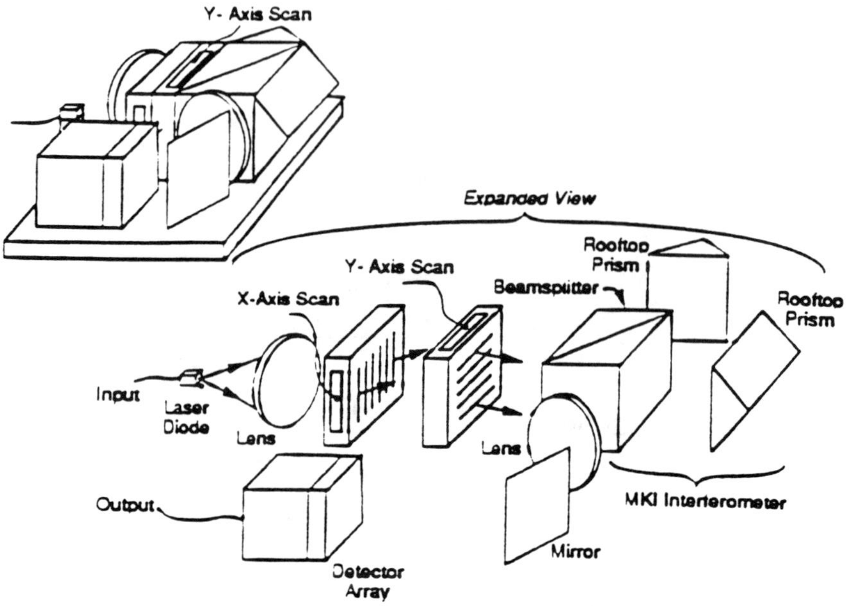

Figure 7: Schematic for TTI architecture.

[59]

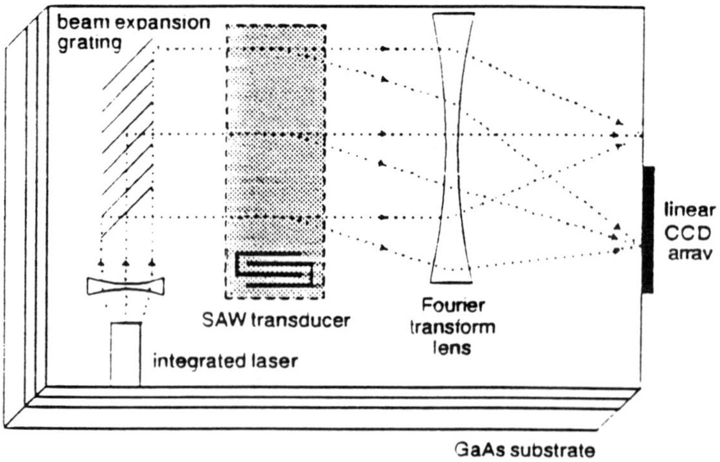

Figure 8: Integrated optical rf spectrum analyzer on a GaAs substrate.

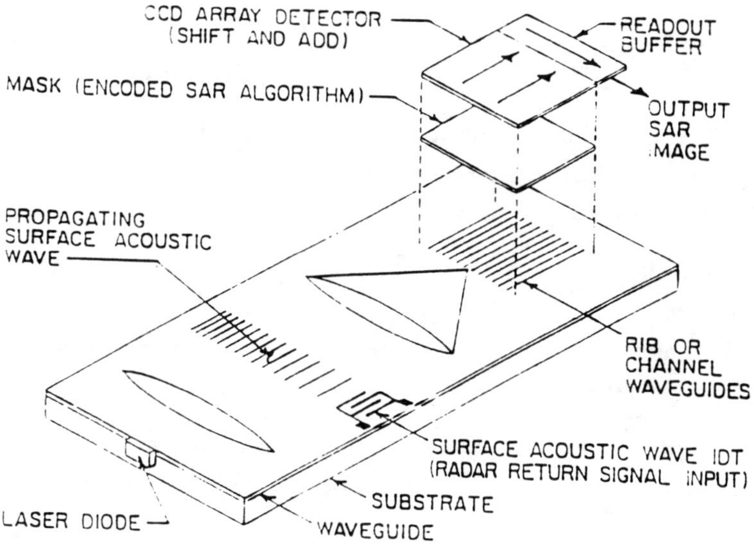

Figure 9: Integrated optical synthetic aperture radar processor.

[116]

References

[1] E. A. Viveiros, L. J. Harrison, M. S. Patterson, R. J. Berinato, K. W. Williams, V. R. Riasati, and R. A. Durrett. Acousto-optic range-doppler processor design for radar insertion. *SPIE Proceedings, Advances in Optical Information Processing V*, 1704:19–30, 1992.

[2] A. F. Zwilling, C. S. Anderson, R. A. Durrett, E. D. Adler, J. M. Pellegrino, and A. N. Filipov. Wideband acousto-optic processor for ESM applications. *SPIE Proceedings, Advances in Optical Information Processing V*, 1704:10–18, 1992.

[3] J. Wood and M. Koontz. Transition of acousto-optics in ELINT and ESM systems. *SPIE Proceedings, Advances in Optical Information Processing V*, 1704:2–9, 1992.

[4] J. N. Lee and A. VanderLugt. Acousto-optic signal processing and computing. *Proceedings of the IEEE*, 77(10):1528–1557, October 1989.

[5] C.O. Hemmi and C. E. Takle. Optically controlled phase array beamforming using time delay. *Proc. SPIE*, 1703, 1992.

[6] Dan Friedman. Design requirements development for the AONSP acousto-optic adaptive processor. *Proc. SPIE, Microwave optical and phased array processing*, 1703, 1992.

[7] C.F. Hester, M.G. Temmen, and D.A. Gregory. Operational considerations for pattern recognition demonstration for transition of optical processing to systems (TOPS). *Proc. SPIE, Optical Pattern Recognition IV*, 1701, 1992.

[8] M. W. Haney, J. J. Levy, and M. P. Christensen. Time and space integrating acousto-optic SAR image-formation processor. *SPIE Proceedings, Advances in Optical Information Processing V*, 1704:31–42, 1992.

[9] C. C. Aleksoff, N. S. Subotic, I. Cindrich, A. Klooster, J. O. Absheir, E. I. LeBaron, and K. K. Ellis. Time-integrating compact hybrid optical processor for SAR image formation. *SPIE Proceedings, Advances in Optical Information Processing V*, 1704:43–58, 1992.

[10] W.B. Beaudet and L.M. Ralston. Optical disk associative memory performance simulation. *Proc. SPIE, Optical Information Processing V*, 1704:59–66, 1992.

[11] E. H. Young, H. C. Ho, and L. J. Harrison. Optically rotated long time aperture TeO_2 Bragg cell. *SPIE Proceedings, Advances in Optical Information Processing IV*, 1296:304–315, 1990.

[12] M. L. Shah and D. R. Pape. Intermodulation suppression in slow shear TeO_2 Bragg cells using a phased array transducer. *SPIE Proceedings, Advances in Optical Information Processing IV*, 1296:316–322, 1990.

[13] S. B. Chinn and D. K. Henderson. Design, fabrication and characterization of wideband high diffraction efficiency phased array Bragg cells. *IEEE Ultrasonics Symposium Proceedings*, 2:679–684, 1990.

[14] I. C. Chang. Acousto-optic and bulk acoustic wave devices for EW applications. *SPIE Proceedings, Optical Technology for Microwave Applications IV*, 1102:189–192, 1989.

[15] J. I. Soos, R. G. Rosemeier, T. P. McFerrin, and R. L. Scheerer. 2.5 GHz bandwidth $LiNbO_3$ shear Bragg cells. *SPIE Proceedings, Advances in Optical Information Processing III*, 936:85–94, 1988.

[16] W.E. Ross, D. Psaltis, and R.H. Anderson. Two-dimensional magneto-optic spatial light modulator for signal processing. *Opt. Eng.*, 22:485–490, 1983.

[17] D. Psaltis, E.G. Paek, and S.S. Venkatesh. Optical image correlation with a binary spatial light modulator. *Opt. Eng.*, 23:698–704, 1984.

[18] D.L. Flannery, A.M. Biernacki, J.M. Loomis, and S.L. Cartwright. Real-time coherent correlator using binary magnetooptic spatial light modulators at input fourier planes. *Appl. Opt.*, 25:466, 1986.

[19] B.A. Kast, M.K. Giles, , S.D. Lindell, and D.L. Flannery. Implementation of ternary phase amplitude filters using a magneto-optic spatial light modulator. *Appl. Opt.*, 28:1044–1046, 1989.

[20] W.E. Ross and D.N. Lambeth. Advanced magneto-optic spatial light modulator device development. *Proc. SPIE*, 1562:93–102, 1991.

[21] D.R. Pape and L.J. Hornbeck. Characterization of the deformable mirror device for optical information processing. *Opt. Eng.*, 22:675, 1983.

[22] J.M. Florence and R.O. Gale. Coherent optical correlation using a deformable mirror device spatial light modulator in the fourier plane. *Appl. Opt.*, 27:2091, 1988.

[23] L.J. Hornbeck. Deformable-mirror spatial light modulators. *Proc. SPIE*, 1150:86–102, 1989.

[24] L.K. Cotter, T.J. Drabik, R.J. Dillon, , and M.J. Handschy. Silicon VLSI/ferroelectric liquid-crystal silicon integrated circuit spatial light modulator. *Optics Letters*, 15:291–293, 1990.

[25] D.A. Jared and K.M. Johnson. Ferroelectric liquid crystal spatial light modulators. *Proc. SPIE*, 1150:46–60, 1989.

[26] G.D. Sharp and K.M. Johnson. Chiral smectic liquid crystal modulator. In *Technical Digest of the Optical Society of America 1992 annual meeting*, volume 23, 1992. (Abstract only).

[27] Y. Toyoda, T. Mori, N. Mikami, N. Nunoshita, and T. Masumi. Ferro-electric liquid crystal having gray scale for high speed spatial light modulator. *Proc. SPIE*, 1665:338–344, 1992.

[28] N. Clark, K.J. Powell, and M.K. Giles. Using liquid crystal devices as input and filter slms. *Proc. SPIE*, 1704:237–247, 1992.

[29] S.S. Cotariu, S.E. Monroe, and J. Knapp. A live input, live filter liquid crystal correlator. *Proc. SPIE*, 1704:248–256, 1992.

[30] N. Konforti, E. Marom, and S.T. Wu. Phase-only modulation with twisted nematic liquid crystal spatial light modulators. *Optics Letters*, 13:251–253, 1988.

[31] J. Kirxch, D. Gregory, M. Thie, and B. Jones. Modulation characteristics of the epson liquid crystal television. *Opt. Eng.*, 31:963–969, 1992.

[32] K. Lu and B.E. Saleh. Complex amplitude reflectance of the liquid crystal light valve. *Appl. Opt.*, 30:2354–2362, 1991.

[33] J.D. Downie, B.P. Hine, and M.B. Reid. Effects and correction of magneto-optic spatial light modulator phase errors in an optical correlator. *Appl. Opt.*, 31:636–643, 1992.

[34] G. W. Anderson, B. D. Guenther, J. A. Hynecek, R. J. Keyes, and A. Vanderlugt. Role of photodetectors in optical signal processing. *Applied Optics*, 27:2871–2886, 1988.

[35] D. J. Smith, L. J. Harrison, J. M. Pellegrino, and D. R. Simon. Performance of high dynamic range CCD arrays with various epi-layer structures. *SPIE Proceedings, Devices for Optical Processing*, 1562:242–250, 1991.

[36] S. J. Strunk, J. F. Coward, and L. J. Newel. Electro-optical characterization of CCD imagers designed for use in acousto-optic channelized receivers. *SPIE Proceedings*, 1308:345–355, 1990.

[37] W. D. Washkurak, S. G. Chamberlain, and N. D. Prince. Wide dynamic range tapped linear array image sensor. *SPIE Proceedings, Advances in Optical Information Processing III*, 936:14–21, 1988.

[38] J. D. Swarthout, L. J. Harrison, J. M. Pellegrino, and P. P. Suni. High dynamic range CCD detectors for acousto-optic applications. *SPIE Proceedings, Optical Technology for Microwave Applications IV*, 1102:61–68, 1989.

[39] L. J. Harrison, D. C. Reaves, D. Vu, and P. P. Suni. Performance of 1024 element high dynamic range CCD detector array. *SPIE Proceedings, Advances in Optical Information Processing IV*, 1296:296–303, 1990.

[40] P. A. Levine, D. J. Sauer, F. L. Hsueh, F. V. Shallcross, G. M. Meray, G. C. Taylor, G. W. Hughes, J. M. Pellegrino, D. R. Simon, L. J. Harrison, and W. Lawler. Performance of a high frame-rate CCD imager. *SPIE Proceedings, Surveillance Technologies II*, 1693, 1992.

[41] Loral Fairchild Imaging Systems. personal communication.

[42] R. A. Bredthauer. 64 output 1024x1024 pixel imager for high frame rate applications. *Proceedings of the IEEE*, 1155:89–93, 1990.

[43] R. A. Sprague and C. L. Koliopoulos. Time-integrating acousto-optic correlator. *Applied Optics*, 15(1):89–92, January 1976.

[44] P. Kellman. *Time-Integrating Optical Signal Processing*. PhD thesis, Stanford University, 1979.

[45] J. D. Cohen. Incoherent-light time-integrating processors. In N. J. Berg and J. N. Lee, editors, *Acousto-Optic Signal Processing: Theory and Implementation*, pages 225–288. Marcel Decker, Inc., New York, 1983.

[46] I. J. Abramovitz, N. J. Berg, and M. W. Casseday. Interferometric surface-wave acousto-optic time integrating corrleators. *IEEE Ultrasonic Symposium*, pages 483–486, November 1980.

[47] N. J. Berg, M. W. Casseday, and C. Garvin. A new single AO element two-dimensional signal processor. *Proceedings Ultrasonics Symposium*, 1:482–486, October 1983.

[48] S. A. Hovanessian. *Introduction to Synthetic Array and Imaging Radars.* Airtech House Inc., 1980.

[49] Michael W. Haney. Time and space integrating acousto-optic SAR image-formation-processor. *Proc. SPIE, Advances in Optical Information Processing V*, 1704:31–42, 1992.

[50] D. Psaltis and K. Wagner. Real-time optical synthetic aperture radar (SAR) processor. *Opt. Eng.*, 21(5), 1982.

[51] D. Psaltis, M. Haney, and K. Wagner. Real-time synthetic aperture radar processing. *Proc. of the NASA Optical Information Processing Conference*, 1983.

[52] M. Haney and D. Psaltis. Acousto-optic techniques for real-time SAR processing. *Proc. SPIE*, 545, 1985.

[53] D. Psaltis and M. Haney. Book chapter: Real-time synthetic aperture radar processors. *Optical Signal Processing*, 1987.

[54] M. Haney and D. Psaltis. Real-time programmable acousto-optic synthetic aperture radar processor. *Appl. Opt.*, 27(9), 1988.

[55] M. W. Haney. Real-time acousto-optic processor for spotlight mode synthetic aperture radar. *Proc. SPIE*, 1291, 1990.

[56] C.C. Aleksoff. Optical-hybrid backprojection processing. *Proc. SPIE*, 422:89–95, 1983.

[57] M.A. Kraenak and D.E. Brown. Interferometric triple product processor (almost common path). *Appl. Opt.*, 24:1385, 1985.

[58] C.C. Aleksoff and N.S. Subotic. Compact real-time interferometer fourier transform processors. *Proc. SPIE*, 1291:174–184, 1990.

[59] C.C. Aleksoff, N.S. Subotic, and I. Cindrich. Time-integrating compact hybrid optical processor for SAR image formation. *Proc. SPIE, Advances in Optical Information Processing V*, 1704:43–58, 1992.

[60] B. Kumar. Tutorial survey of composite filter designs for optical correlators. *Appl. Opt.*, 31:4773–4801, 1992.

[61] G. Gheen, E. Washwell, and C. Huang. Problems facing optical correlators. *Proc. SPIE*, 1772, 1992.

[62] D. Wunsch, R. Marks, T. Caudell, and C. Capps. Limitations of a class of binary phase-only filters. *Appl. Opt.*, 31:5681–5687, 1992.

[63] J. Mait, J. van der Gracht, and S. Sarama. Diffractive filter design for SLM's in optical processing and pattern recognition. *Proc. SPIE*, 1732, 1992. (to appear).

[64] R. Juday. Globally optimal filters. *Technical Digest of the Optical Society of America 1992 Annual Meeting*, 23:162, 1992.

[65] P. Molley and B. Kast. Automatic target recognition and tracking using an acousto-optic image correlator. *Opt. Eng.*, 31:956–962, 1992.

[66] C. Hester, M. Temmen, J. Brasher, J. Kinser, and J. DeWitt. Operational considerations for pattern recognition demonstration for transition of optical processing to systems (TOPS). *Proc. SPIE*, 1701:11–16, 1992.

[67] S. Lindell and W. Hahn. Overview of the Martin Marietta transfer of optical processing to systems (TOPS) program. *Proc. SPIE*, 1701:21–30, 1992.

[68] M. Hose. FFT chips for transform-based image processing. *Adv. Imaging*, 7(6):56–59, June 1992.

[69] C. Huang, G. Gheen, and E. Washwell. Throughput comparison of optical and digital correlators for automatic target recognition. *Proc. SPIE*, 1564, 1991.

[70] J. van der Gracht and J. Mait. Incoherent pattern recognition with phase-only filters. *Opt. Lett.*, 17, 1992. (to appear).

[71] J. Mait, D. Prather, and R. Athale. Crossed Bragg cell implementation of two-dimensional optical morphology. *Appl. Opt.*, accepted for publication, 1992.

[72] D. Casasent, R. Schaefer, and R. Sturgill. Optical hit-miss morphological transform. *Appl. Opt.*, 31:6255–6263, 1992.

[73] W. Ng, A.A. Walston, G.L. Tangonan, J.J. Lee, I.L. Newberg, and N. Bernstein. The first demonstration of an optically steered microwave phased array antenna using true-time-delay. *J Lightwave Technol.*, 9:1124–1131, 1991.

[74] C.A. Balanis. *Antenna Theory.* John Wiley and Sons, Inc., 1982.

[75] M.L. Skolnik. *Introduction to radar systems.* McGraw-Hill, Inc., second edition, 1980.

[76] D.D. Curtis. Frequency domain analysis and performance of a true time delay fiber optic beamforming network for array antennas. *Optoelectronic Signal Processing for Phased-Array Antennas II*, 1990.

[77] D. Dolfi, J.-P. Huignard, and M. Baril. Optically controlled true time delays for phased array antennas. *Proc. SPIE, Optical Technology for Microwave Applications*, 1102:152–161, 1992.

[78] D. Stilwell, M. Parent, and L. Goldberg. Microwave time delay beamforming using optics. *IEEE National Radar Conference*, pages 50–53, 1991.

[79] Leslie H. Gesell and Terry M. Turpin. True time delay beam forming using acousto-optics. *DARPA/Rome Laboratory Symposium on Photonics for Antenna Applications*, pages 196–201, 1991.

[80] P.R. Herczfeld, A. Paolella, A.S. Daryoush, W. Jemison, and A. Rosen. Optical control of mmic-based t/r modules. *Microwave J*, pages 309–322, May 1988.

[81] K.C. Gupta, R. Garg, and I.J. Bahl. *Microstrip lines and slotlines.* Artech House, Inc., 1979.

[82] J. Lee, H. Yen, R. Loo, A. Walston, and I. Newberg. System applications of photonics to phased arrays. *Proc. SPIE*, 1703, 1992.

[83] Nile F. Hartman and Larry E. Corey. Integrated optic time delay network for phased array antennas. *IEEE National Radar Conference*, pages 79–82, 1991.

[84] G. A. Vawter, V. M. Hietala, S. H. Kravitz, and W. J. Meyer. Direct optical phase shifter for phased-array systems. *Proc. SPIE, Optical Technology for Microwave Applications V*, 1476:102–105, 1991.

[85] I. C. Chang. Acousto-optic devices and applications. *IEEE Trans. Sonics Ultrason.*, SU-23, 1976.

[86] E. H. Young and S. K. Yao. Design considerations for acousto-optic devices. *Proc. IEEE*, 69, 1981.

[87] C. Camperi-Ginestet, M. Hargis, N. Jokerst, and M. Allen. Alignable epitaxial liftoff of GaAs materials with selective deposition using polyimide diaphragms. *IEEE Photon. Tech. Lett.*, 3:1123–1126, 1991.

[88] M. J. Wale and C. Edge. Self-aligned flip-chip assembly of photonic devices with electrical and optical connections. *IEEE Trans. Comp. Hybrids, Manuf. Technol.*, 13:780–786, 1990.

[89] D. G. Corondell and K. F. Jensen. Analysis of MOCVD of GaAs on patterned substrates. *J. Crystal Growth*, 114:581–592, 1991.

[90] M. C. Hamilton. RF frequency sorting in electronic warfare systems with integrated optical circuits. *Proc. SPIE*, 139:144–150, 1978.

[91] C. S. Tsai, L. T. Nguyen, and C. C. Lee. Optical pulse compression and very high-speed light beam scanning using guided-wave acoustooptic Bragg diffraction. *Topical Meeting on Integrated and Guided Wave Optics*, OSA 78CH1280-7 QEA:TuC3 1–4, 1978.

[92] W. S. C. Chang, C. S. Tsai, R. A Becker, and I. W. Yao. Convolution using guided acoustooptical interaction in thin-film waveguides. *IEEE J. Quantum Electron.*, QE-13:208–215, 1977.

[93] R. A. Sprague and K. L. Koliopaulos. Time integrating acoustooptic correlator. *Appl. Opt.*, 15:89, 1976.

[94] R. A. Sprague. A review of acousto-optical signal correlators. *Opt. Eng.*, 16:467–478, 1977.

[95] M. M. Minot and C. C. Lee. A new guided-wave lens structure. *J. Lightwave Tech.*, vol. 8:1856–1865, 1990.

[96] Y. Abdelrazek and C. S. Tsai. High-performance acoustooptic Bragg cells in ZnO-GaAs waveguide at GHz frequencies. *Optoelectronics*, vol. 4:33–37, 1989.

[97] T. R. Joseph, T. R. Ranganath, J. Y. Lee, and M. Pedinoff. Performance of the integrated optic spectrum analyzer. *Proc. SPIE*, 321:134–140, 1982.

[98] D. Mergerian, E. C. Malarkey, R. P. Pautienun, J. C. Bradley, M. Mill, C. W. Baugh, A. L. Kellner, and M. Mentzer. Advanced integrated optic rf spectrum analyzer. *Proc. SPIE*, 321:149–156, 1982.

[99] Y. Abdelrazek, C. S. Tsai, and T. Q. Vu. An integrated optic RF spectrum analyzer in a ZnO-GaAs-AlGaAs waveguide. *J. Lightwave Tech.*, 8:1833–1837, 1990.

[100] J. T. Boyd and D. B. Anderson. Effect of waveguide optical scattering on the integrated optical spectrum analyzer dynamic range. *IEEE J. Quantum Electron.*, QE-14:437–443, 1978.

[101] D. W. Vahey, N. F. Hartman, and R. C. Sherman. Optical waveguide scattering reduction II. Technical Report AFWAL-TR-80-1186, Battelle Columbus Laboratories, 1980.

[102] D. B. Anderson, R. L. Davis, J. T. Boyd, and R. R. August. Comparison of optical-waveguide lens technolgies. *IEEE J. Quantum Electron.*, QE-13:275–282, 1977.

[103] T. Suhara, K. Kobayashi, H. Nishihara, and J. Koyama. Graded-index Fresnel lenses for integrated optics. *Appl. Opt.*, 21:1966–1971, 1982.

[104] T. Q. Vu and C. S. Tsai. Ion-milled waveguide lenses and lens arrays in GaAs. *J. Lightwave Tech.*, vol. 7:1559–1566, 1989.

[105] M. C. Hamilton and A. E. Spezio. Spectrum analysis with integrated optics. In *Guided-Wave Acousto-Optics: Interactions, Devices, and Applications*, pages 242–245. Springer-Verlag, 1990.

[106] M. K. Barnoski, B. Chen, T. R. Joseph, J. Y. Lee, and O. G. Ramer. Integrated-optic spectrum analyzer. *IEEE Trans. Circuits Systems*, CAS-26:1113–1124, 1974.

[107] C. S. Tsai. Wideband acousto-optic Bragg diffraction in $LiNbO_3$ waveguide and applications. In *Guided-Wave Acousto-Optics: Interactions, Devices, and Applications*, pages 136–137. Springer-Verlag, 1990.

[108] T. Q. Vu and C. S. Tsai. GaAs-based integrated optic RF signal processor modules using InGaAs photoconductive receiver array. *OSA/LEOS Topical Meeting on Integrated Photonics Research*, 10:50–51, 1992.

[109] H. Matsueda and M. Nakamura. Monolithic integration of a laser diode, photo monitor, and electric circuits on a semi-insulating GaAs substrate. *Appl. Opt.*, 23:779–784, 1984.

[110] C. S. Tsai. Guided-wave acoustooptic Bragg modulators for wide-band integrated optic communications and signal processing. *IEEE Trans. Circuits and Systems*, CAS-26:1072–1098, 1979.

[111] W. H. Southwell. Planar optical waveguide lens design. *Appl. Opt.*, 21:1985–1988, 1982.

[112] W. H. Southwell. Geodesic optical waveguide lens analysis. *J. Opt. Soc. Am.*, 67:1293–1299, 1977.

[113] J. P. Delavaux and W. S. C. Chang. Fundamental limitations in the performance of chirped grating lenses on planar optical waveguides. *IEEE J. Quantum Electron.*, QE-21:86–94, 1985.

[114] G. C. Righini, G. Belli, M. Varasi, and A. Vannucci. Waveguide Fresnel lenses for integrated optical processors. *Proc. SPIE*, 1177:209–215, 1989.

[115] R. H. Weissman. Manufacturing low cost optoelectronic components. *LEOS Summer Topical Meeting on Integrated Optoelectronics*, IEEE Cat. # 92TH0421-8:6, 1992.

[116] T. Bicknell, D. Psaltis, and A. R. Tanguay, Jr. Integrated optical synthetic aperture radar processor. *J. Opt. Soc. Am. A*, 2:P8, 1985.

[117] M. N. Armenise, F. Impagnatiello, V. M. N. Passaro, and E. Pansini. Design of a GaAs acoustooptic correlator for real-time processing. *Proc. SPIE*, 1562:160–171, 1991.

[118] T. Tamir and S. T. Peng. Analysis and design of grating couplers. *Appl. Phys.*, 14:235–254, 1977.

[119] T. Suhara and H. Nishihara. Integrated optics components and devices using periodic structures. *IEEE J. Quantum Electron.*, QE-22:845–867, 1986.

[120] K. C. Chang and T. Tamir. Simplified approach to surface-wave scattering by blazed dielectric gratings. *Appl. Opt.*, 19:282–288, 1980.

[121] T. Tamir and S. T. Peng. Network methods for integrated optics devices. In *Applications of Holography and Optical Data Processing*, pages 437–446. Pergamon Press, 1976.

[122] N. N. Berg and J. N. Lee, editors. *Acousto-Optic Signal Processing: Theory and Implementation*. Marcel Decker, Inc., New York, 1983.

[123] A. Vander Lugt. Interferometric spectrum analyzer. *Applied Optics*, 20(16):2770–2779, 1981.

Sensing Applications

Chair

James B. Breckinridge
Jet Propulsion Laboratory

Optical Instruments for Remote Sensing from Space

Michael P. Chrisp

Jet Propulsion Laboratory, California Institute of Technology
4800 Oak Grove Drive, Pasadena, California 91109

ABSTRACT

This review covers the optical design of passive remote sensing optical instruments. The review concentrates on the design of spaceborne multispectral cameras and imaging spectrometers. The major designs that have been produced over the past ten years are discussed, and new designs for future imaging spectrometers are presented.

1. INTRODUCTION

The purpose of this review is to cover the optical designs that have been developed for passive remote sensing instruments over the last ten years. The review concentrates on the design of multispectral cameras and imaging spectrometers used on spacecraft, with some aircraft instruments included for completeness. A list of the instruments covered is given in Table 1, together with their acronyms and the references to the papers describing them. The earlier remote sensing instruments are well covered in Slater's book.[1]

Passive remote sensing instruments derive information from the scene by collecting the reflected solar radiation or emitted thermal radiation from each spatial area. This radiation is analyzed by the instrument for spectral content, and in some cases polarization properties.

Historically, remote sensing instruments have evolved from photographic film based systems. These were camera systems with filters, and scanner systems that wrote on film. The photographic film was then replaced with discrete electronic detectors, leading to whiskbroom scanning systems, such as Landsat, and the early infrared radiometers SMIRR and HCMR. Following on from discrete electronic detectors, silicon detector arrays were developed. These led to pushbroom systems such as SPOT and MISR.

The development of infrared detector arrays has led to the whiskbroom imaging spectrometer systems, AIS, AVIRIS and MODIS-T, and pushbroom multispectral cameras such as SPOT4.

Currently, pushbroom designs are common for spectral areas where detector arrays are available, and whiskbroom designs are used for the infrared regions where only discrete detectors are available.

	MULTISPECTRAL CAMERAS
MISR[2]	Multi-angle Imaging SpectroRadiometer
SPOT4[3]	HRVIR -High Resolution Visible and medium InfraRed Camera
SEVIRI[4]	Spinning Enhanced Visible-IR Imager
POLDER[5]	Polarization and Directionality of the Earth's Reflectances
VNIR/ASTER[6]	
SWIR/ASTER[7]	Advanced Spaceborne Thermal Emission and reflection Radiometer
TIR/ASTER[8]	
TM[1]	Thematic Mapper
MODIS-N[9]	Moderate Resolution Imaging Spectrometer - Nadir
	IMAGING SPECTROMETERS
MODIS-T[10]	Moderate Resolution Imaging Spectrometer - Tilt
MERIS[11]	Medium Resolution Imaging Spectrometer
SCIAMACHY[12]	SCanning Imaging Absorption SpectroMeter for Atmospheric CHartographY
ROSIS[13]	Reflective Optics System Imaging Spectrometer
AVIRIS[14]	Airborne Visible/Infrared Imaging Spectrometer
NIMS[15]	Near-Infrared Mapping Spectrometer
VIMS[16]	Visible and Infrared Mapping Spectrometer
SISEX[17]	Shuttle Imaging Spectrometer Experiment
HIRIS[18]	High Resolution Imaging Spectrometer
	GENERAL
MIPAS[19]	Michelson Interferometer for Passive Atmospheric Sounding
IMG[20]	Interferometric Monitor for Greenhouse Gases (ADEOS)
TES[21]	Tropospheric Emission Spectrometer
AOTF[22]	Acousto Optic Tunable Filter Spectrometer
MOES[23]	MultiOrder Etalon Sounder

Table I. Remote Sensing Instruments

2. DESIGN OF OPTICAL REMOTE SENSING INSTRUMENTS

This section is concerned with the optical designs used for remote sensing instruments. A number of current instruments are compared, and the correlation between their functional performance and their optical design type is discussed.

The type of optical design chosen for a remote sensing instrument depends upon various requirements: the number of spatial and spectral bands, the spatial resolution and field of view, the spectral resolution and wavelength range, the entrance pupil diameter and f-number at the focal plane. The optical design is also strongly driven by the availability of detector arrays in the spectral region of interest.

The optical design selected for a system is that which can best meet the above requirements together with the constraints. Spaceborne remote sensing systems are constrained by mass, volume, power consumption and thermal environment. Nowadays, limited financial resources also provide another strong constraint.

2.1. Optical Performance

A comparison between the designs for optical remote sensing instruments is given in figure 1, in terms of the number of spatial and spectral bands for each system. The number of spatial bands given are those that are intrinsic to the optical design, and does not take into account the use of a scan mirror in the whiskbroom systems. The graph provides an interesting view of the three main types of remote sensing instruments: multispectral cameras, imaging spectrometers, and Fourier transform spectrometers.

The multispectral cameras, which have a small number of spectral bands, are contained in a column on the left. Thematic mapper, a whiskbroom scanner, with 16 spatial bands and 7 spectral bands provides an interesting comparison to the most recent designs. These pushbroom designs, SPOT4 and VNIR/ASTER, provide a high number of spatial bands by taking advantage of the latest detector arrays.

The imaging spectrometer designs lie diagonally on the graph, with the latest designs such as ROSIS and HIRIS providing a large number of spatial and spectral bands.

The pure pushbroom designs with one spatial IFOV lie along the bottom of the graph. This is the domain of the Fourier transform spectrometers, which have the largest number of spectral bands and the highest spectral resolution. The only imaging Fourier transform spectrometer shown is TES, with 16 spatial bands, that is being built for EOS.

The spatial performance of the remote sensing instruments is shown in figure 1. This graph maps out the IFOV (Instantaneous Field of View) and total FOV for each instrument. Again, the FOV is for the intrinsic design, and does not take into account the use of the scan mirror in the whiskbroom systems.

For spaceborne remote sensing instruments the IFOV is driven by the requirement that enough photons be collected from the ground. Decreasing the IFOV directly reduces the solid angle of the light collected, which can only be compensated with a larger entrance pupil diameter. The IFOV of thematic mapper compares favorably with today's designs.

For imaging spectrometers the trend is for larger fields at lower spatial resolution, examples of this are MODIS-T and MERIS.

2.2. Multispectral Cameras

The basic components of a multispectral camera are shown in figure 3. It consists of a telescope followed by a filter and a linear or area array detector. Most systems have a number of spectral bands, obtained through the use of dichroic beamsplitters. In some designs the different field positions pass through different filters, and the spacecraft motion is used to superimpose the spectral bands for a given field point.

Figure 1. Spatial and Spectral Bands

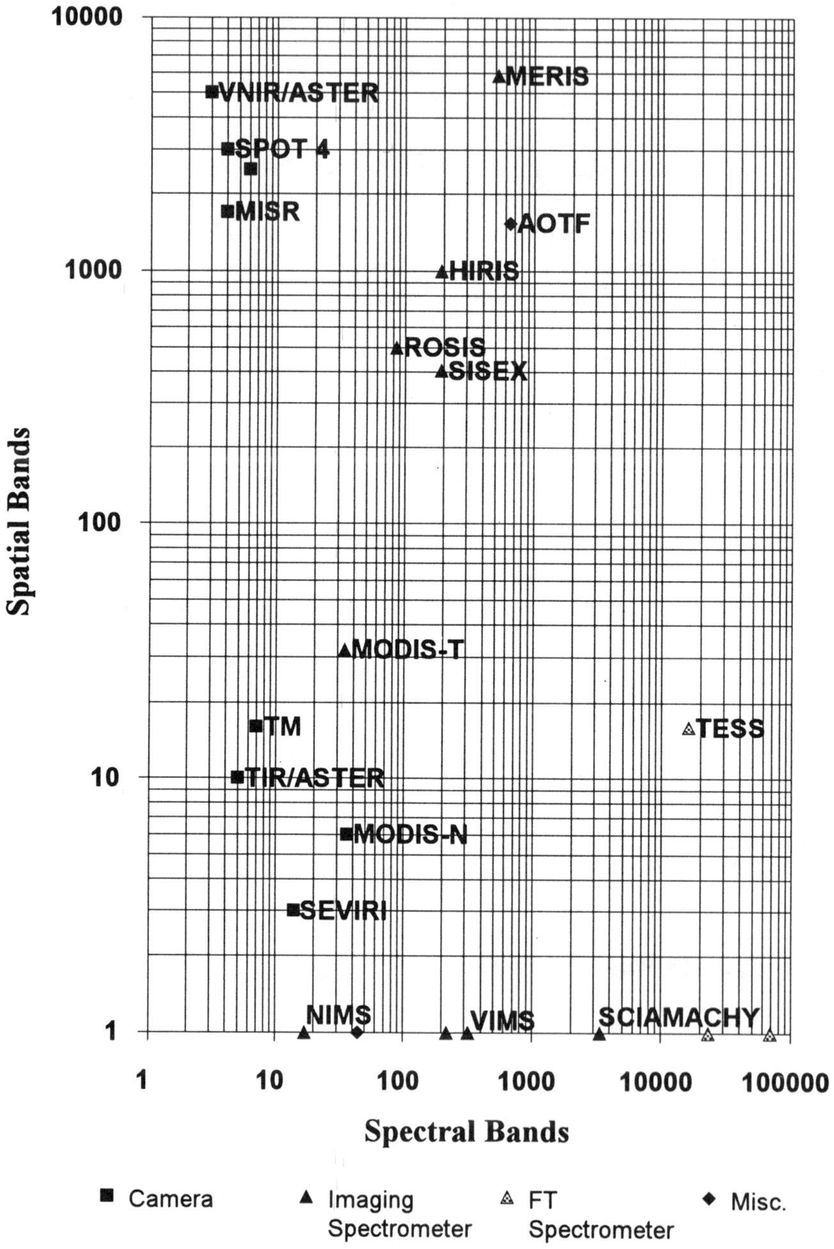

Figure 2. Spatial Performance

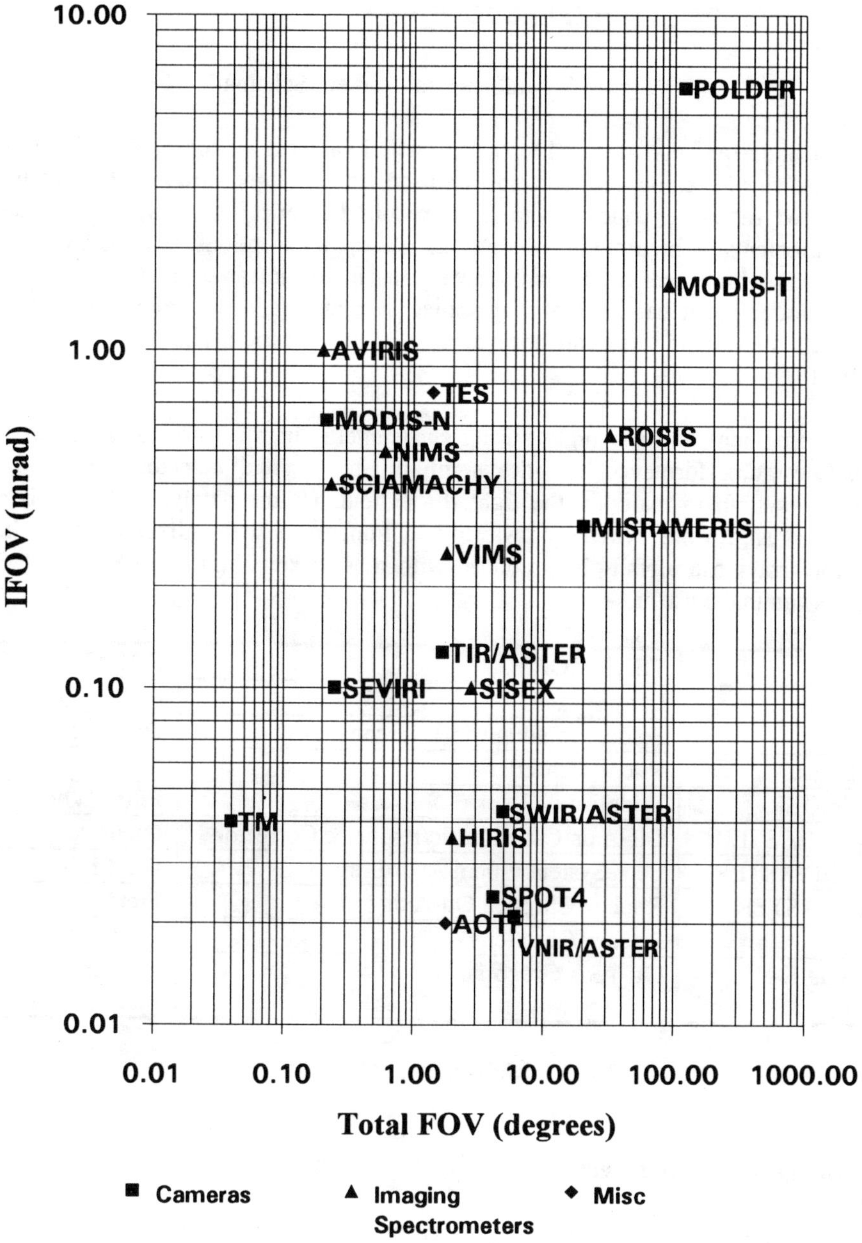

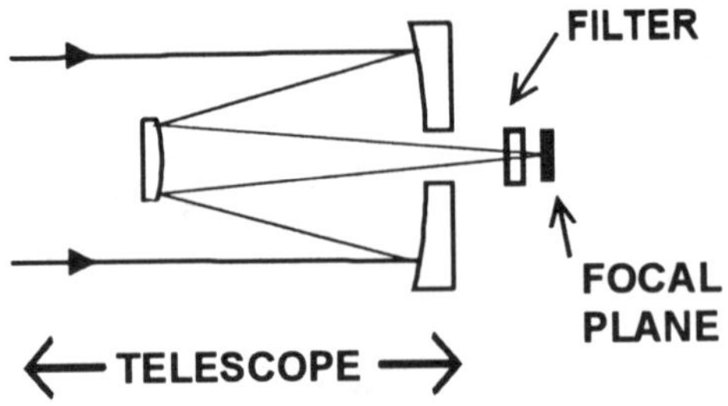

Figure 3. Multispectral Camera Schematic

The optical designs that are used for multispectral cameras are loosely classified in table 2. For the visible region, lens systems are used when the aperture is not excessive. For the infrared, the larger collecting apertures required favor reflecting systems, which also reduce problems with chromatic aberration. There are exceptions, such as SWIR system for ASTER which uses a 20 cm diameter lens in the infrared. The designers of this system claim that it has a weight advantage over an equivalent reflecting system. For large field angles and large collecting apertures, catadioptric designs tend to be used, as in SPOT4.

The development of wide-angle pushbroom cameras in the infrared is driven by the optical design performance. The ideal solution is to use small detector pixels with a fast f-number system to maintain the light throughput. Unfortunately, fast f-numbers and large field angles are extremely awkward to implement with mirror systems. In fact, detector arrays can already be produced with more pixels than remote sensing optical systems can make use of.

	TELESCOPE	FOCAL PLANE
MISR	Lens System	Field Filters
SPOT 4	Folded Catadioptric Telescope	Dichroic
SEVIRI	Ritchey Chretien	Field Filters
POLDER	Wide Angle Lens	Filter Wheel
VNIR/ASTER	Off-Axis Catadioptric/Aspheric Corrector	Dichroic
SWIR/ASTER	Lens System (20cm Aperture)	Field Filters
TIR/ASTER	Ritchey Chretien On-Axis	Field Filters
TM	Ritchey Chretien	Field Filters
MODIS-N	Off-Axis Gregorian	Dichroics Field Filters

Table 2. Multispectral Camera Designs

2.3. Imaging Spectrometers

The schematic design of an imaging spectrometer is shown in figure 4. The telescope system focuses the light onto the entrance slit of the spectrometer. The spectrometer portion then disperses the image from each point on the slit onto a two dimensional detector array On the detector array, the spatial dimension is in one dimension and the other dimension provides the spectral information. The motion of the spacecraft provides the pushbroom motion for the system.

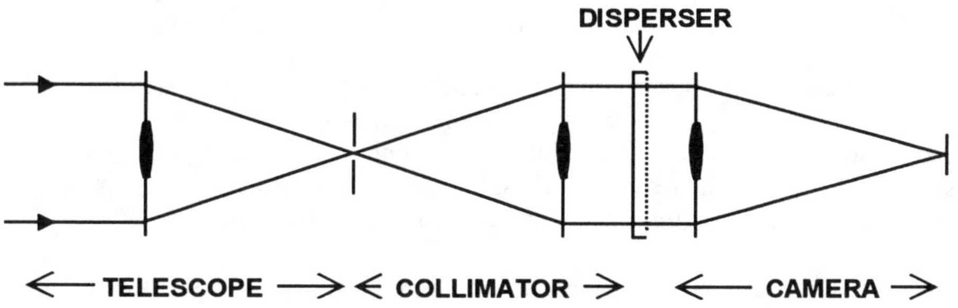

Figure 4. Imaging Spectrometer Schematic

The optical designs are loosely classified in table 3. The requirements for the angular spatial field of view and the spectral resolution tend to determine the design choice.

For increasing angular fields of view the telescope designs progress from simple paraboloids, to Ritchey Chretiens, to Schmidt type systems and catadioptric designs. Refractive designs are not used for two principal reasons. The first is because the narrow spectral bandwidths usually require a large collecting aperture to give an adequate signal to noise ratio. The second is to avoid problems with chromatic aberration due to the large spectral ranges.

	TELESCOPE	COLLIMATOR	DISPERSER	CAMERA
MODIS-T	Off-Axis Paraboloid	Off-Axis Paraboloid	Schmidt Grating	Spherical Mirror
MERIS	Off-Axis Catadioptric	Refractive Corrector	Concave Grating	Refractive Corrector
SCIAMACHY	Off-Axis Paraboloid	Off-Axis Paraboloid	Plane Grating	Lens System
ROSIS	Off-Axis Schwarzschild	Off-Axis Schwarzschild	Plane Grating	Off-Axis Schwarschild
AVIRIS	Paraboloid	Off-Axis Spherical Mirror	Schmidt Grating	Off-Axis Mirror
NIMS	Ritchey Chretien	Dall Kirkham	Plane Grating	Ritchey Chretien
VIMS	Ritchey Chretien	Dall Kirkham	Plane Grating	Ritchey Chretien
SISEX	Reflective Off-Axis Schmidt	Off-Axis Schmidt	Prism	Off-Axis Schmidt
HIRIS	Reflective Off-Axis Schmidt	Off-Axis Schmidt	Prism	Off-Axis Schmidt

Table 3. Imaging Spectrometer Designs

Most imaging spectrometer systems use gratings for dispersion because of their higher angular dispersion. The spectrometer camera focal length being too short to provide sufficient linear dispersion on the detector with a prism. The HIRIS and SISEX designs are exceptions to this rule and use multielement prisms to provide linear angular dispersions, but these designs are extremely large on the order of 1000kg.

The spectrometer portion for an imaging spectrometer provides and interesting design challenge. Its size is solely determined by the aberrations in the design. Basically, shrinking the spectrometer in size, while maintaining the area solid angle throughput, cause its aberrations to increase as the angles through it increase. Better spectrometer design enable smaller spectrometers to be used in imaging spectrometers.

For wide angle systems Schmidt type systems have been used. Two of the instrument designs reviewed have novel design solutions. The first is MERIS which uses an concave grating with a refractive corrector. The second is the design for ROSIS, which is based on an off-axis Schwarzschild type system. Although at first glance the mirrors in this system are quite large compared with the f-number and detector size in the spectrometer.

There will undoubtedly be more solutions for this problem. Variable groove space gratings and binary optical gratings provide interesting possibilities in the future. Most of the classical spectrometer design have been derived solely on the basis of obtaining good spectral resolution. More work needs to be done in the future to produce good designs which take the spatial imaging aspect into account.

3. TECHNOLOGY UTILIZATION

The optical systems used in remote sensing instrument have benefited most from improvements in optical surface fabrication. Aspheric surfaces are now commonly used because diamond turning has made their fabrication much easier. If necessary, then post-polishing can be used to reduce the surface roughness. The VNIR/ASTER system will utilize an aspheric lens, and MODIS-T and AVIRIS both utilize gratings ruled on aspheric Schmidt correctors.

Blazed holographic gratings are now produced with low scattered light, and are used in MODIS-T. None of the designs reviewed use aberration corrected holographic gratings, because their peak diffraction efficiency is only 33%. This compares with a peak blaze efficiency over 90% for a conventionally ruled grating. Increasing the entrance pupil diameter to overcome this loss in efficiency is usually to large a system impact.

Variable line spaced blazed gratings have been produced using numerically controlled ruling engines[24], and it is surprising that they are not utilized in any of the imaging spectrometer designs. They offer the possibility of reducing the size of the spectrometer system while maintaining a good diffraction efficiency. Binary optical gratings are likely to have an impact on spectrometer designs, and concepts based on their use are discussed in the next section.

In multispectral camera designs filter efficiencies are extremely important. The new optical coating techniques have helped this, and SPOT4 has an efficiency of 98% for its dichroic filters. Wide wavelength range high performance antireflection coatings will also improve the performance for multispectral camera optical systems.

Future multispectral camera designs are likely to incorporate active filter elements based on AOTF (acousto-optic tunable filters). An even more promising technology are liquid crystal based Lyot or Solc type filters. In comparison to AOTF based systems these have the advantage of low power consumption and possibility of larger throughput apertures, which is always of prime concern for a remote sensing instrument.

4. FUTURE CONCEPTS FOR IMAGING SPECTROMETERS

Future spaceborne systems will require lightweight imaging spectrometers. The next generation of miniature spacecraft will only have 6 kilograms available for the instrument payload. In the past the planetary remote sensing imaging spectrometers such as NIMS weighed 20kg.

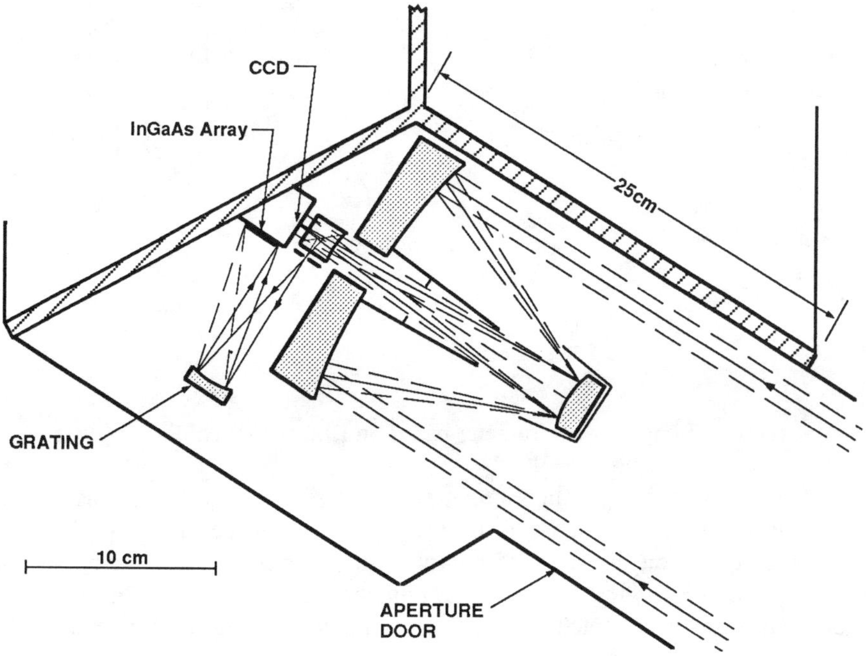

Figure 5. Microspacecraft Camera and Imaging Spectrometer

On microspacecraft for planetary remote sensing, future imaging spectrometers will be combined with the camera system to avoid the weight of an additional telescope system. Shown in figure 5 is one concept for an imaging spectrometer combined with a camera system. In this system a dichroic beamsplitter, separates out the light for the imaging spectrometer system. The single element spectrometer system is based on a variable groove spaced blazed grating which disperses and focuses the light onto an Indium Gallium Arsenide detector array.

Binary optical gratings are also undergoing rapid development, and offer the possibility of producing a grating with complete control over the blaze and the groove positions. Thus overcoming the diffraction efficiency problems associated with aberration corrected holographic gratings, and without the restriction on holographic construction geometry. Their principal problem at the moment is stray light, but lithography techniques are making rapid advances.

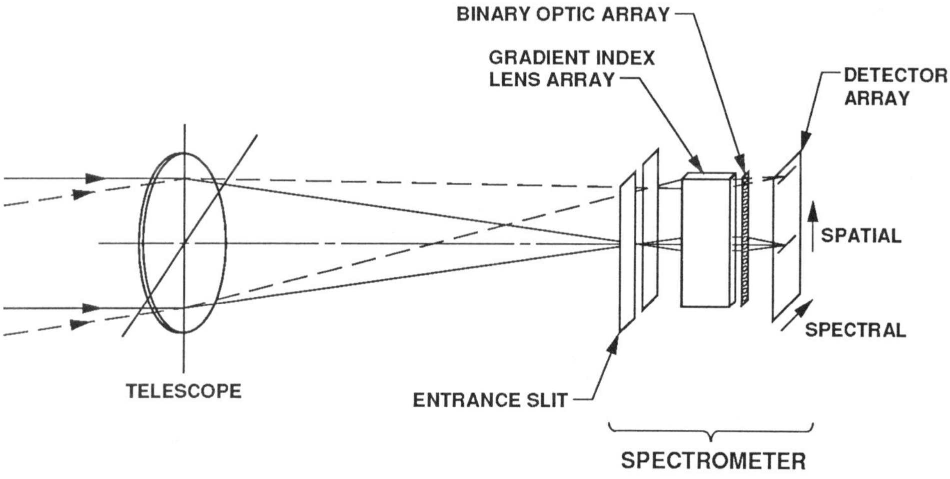

Figure 6. Imaging Spectrometer based on Binary Optical Grating Array

The use of binary optical gratings provides the possibility of new designs for imaging spectrometers. Shown in figure 6 is a design for a wide field of view imaging spectrometer based on an array of binary optical gratings. The use of an array of gratings reduces the angles through each individual grating, so enabling a reduction in size of the imaging spectrometer. With a conventional imaging spectrometer all the field angles must pass through the grating leading to severe aberration problems.

5. CONCLUSION

The optical design of imaging spectrometers is in its infancy. Unfortunately the very smallness of the field has limited its development. Hitachi has been using a numerically controlled ruling engine to produce variable line spaced gratings for the last decade[24], and yet there is still no facility with this capability in the United States.

Designing an imaging spectrometer with the current optical design programs is difficult. They are extremely weak at optimizing the off-axis spectrometer designs used nowadays. Any real design work usually requires constraints that are so awkward to implement that the optimization does not work properly. Unfortunately there are few users of the modules of the program associated with diffraction gratings, so there is no pressure on program vendors to improve them.

In the past, the performance of imaging spectrometers and multispectral cameras have been limited by the available detector array sizes. The advance in detector technology has led to detector array sizes which exceed the current optical designs. The performance is now limited by the optical design.

Future remote sensing instruments will be smaller and have more capabilities. The challenge of providing useful instruments for microspacecraft, where the instrument payload is on the order of 5 kg, will lead to new designs. The first approach to this problem will be the sharing of the optical telescope system by the camera and imaging spectrometer. The optical systems will also be made lighter by utilizing the new material technologies such as silicon carbide and metal matrix composites will help with these systems. Finally, optics technologies such as binary optical gratings, and gradient index optics will lead to new more compact designs.

6. ACKNOWLEDGEMENT

The preparation of this paper was carried out at the Jet Propulsion Laboratory, California Institute of Technology under a contract with the National Aeronautics and Space Administration.

7. REFERENCES

1. P. N. Slater, *Remote Sensing*, Addison-Wesley Publishing Company, Massachusetts, 1980.
2. D. J. Diner, "EOS Multi-Angle Imaging SpectroRadiometer (MISR)," *SPIE,* Vol. 1300, p. 163 (1990)
3. C. Fratter, J. F. Reulet, J. Jouan, "SPOT 4 HRVIR Instrument and Future High-Resolution Stereo Instruments," *SPIE,* Vol. 1490, p. 59 (1991)
4. P. Hollier, "The Imager of Meteosat Second Generation," *SPIE,* Vol. 1490, p. 74 (1991)
5. J. Lorsignol, P. Hollier, and J. Deshayes, "Polarisation and Directionality of Earth's Reflectances: the POLDER Instrument," *SPIE,* Vol. 1490, p. 155 (1991)
6. F. Takahashi, et. al., "Visible and Near Infrared (VNIR) Subsystem and Common Signal Processor Design Status of ASTER," *SPIE,* Vol. 1490, p. 255 (1991)
7. A. Akasaka, M. Ono, and Y. Sakurai, "Short Wavelength Infrared (SWIR) Subsystem Design Status of Aster," *SPIE,* Vol. 1490, p. 269 (1991)
8. Y. Aoki, H. Ohmae, and S. Kitamura, "Thermal Infrared Subsystem Design Status of ASTER," *SPIE,* Vol. 1490, p. 278 (1991)

9. *EOS Reference Handbook*, Nasa Godard Space Flight Center, Publication NP-144, May (1991)
10. T. Magner, "Moderate Resolution Imaging Spectrometer-Tilt (MODIS-T) Baseline Concept," *SPIE,* Vol 1492, p. 272 (1991)
11. G. Baudin, and R. Bessudo, "Medium Resolution Imaging Spectrometer (MERIS)," *SPIE,* Vol. 1490, p. 103 (1991)
12. J. P. Burrows and K. V. Chance, "Scanning Imaging Absorption Spectrometer for Atmospheric Chartography," *SPIE,* Vol 1490, p. 146 (1991)
13. R. Doerffer, et. al., "ROSIS - an Advanced Imaging Spectrometer for the Monitoring of Water Colour and Chlorophyll Fluorescence," *SPIE,* Vol. 1129, p. 117 (1989)
14. M. P. Chrisp, T. Chrien, and L. Steimle, "AVIRIS Foreoptics, Fiber Optics and On-Board Calibrator," *SPIE,* Vol. 834, p. 44 (1987)
15. I. M. Aptaker, "A Near-Infrared Mapping Spectrometer for Investigation of Jupiter and its Satellites," *SPIE,* Vol. 834, p. 196 (1987)
16. J. B. Wellman, J. Duval, D. Juergens, and J. Voss, "Visible Infrared Mapping Spectrometer (VIMS): a Facility Instrument for Planetary Missions," *SPIE,* Vol. 834, p. 213 (1987)
17. M. Herring, "The Shuttle Imaging Spectrometer Experiment (SISEX)," *SPIE,* Vol. 834, p. 181 (1987)
18. J. M. Conley, M. Herring, and D. Norris, "High Resolution Imaging Spectrometer (HIRIS)," *SPIE,* Vol. 834, p. 188 (1987)
19. W. Posselt, "Michelson Interferometer for Passive Atmosphere Sounding (MIPAS)," *SPIE,* Vol. 1490, p. 114 (1991)
20. K. Tsuno, Y. Kameda, and K. Kondoh, "Interferometric Monitor for Greenhouse Gases," *SPIE,* Vol. 1490, p. 222 (1991)
21. R. Beer, and T. Glavich, "Remote Sensing of the Troposphere by Infrared Emission Spectroscopy," *SPIE,* Vol. 1129, p. 42 (1989)
22. V. G. Duval, et. al., "Imaging Spectrometry in the Post-EOS Era," *SPIE,* Vol. 1129, p. 137 (1989)
23. J. Wang, P. B. Hays, and S. R. Drayson, "Multiorder Etalon Sounder for Vertical Temperature Profiling," *SPIE,* Vol. 1492, p. 391 (1991)
24. T. Harada, and T. Kita, "Mechanically Ruled Aberration-Corrected Concave Gratings," *Appl. Opt.*, Vol. 19, p. 3987 (1981)

MODIS: A Global Imaging Spectroradiometer for the Earth Observing System

William L. Barnes and Vincent V. Salomonson

Earth Sciences Directorate
NASA/Goddard Space Flight Center
Greenbelt, Maryland 20771

ABSTRACT

MODIS, a 36 band (0.42 - 14.24 micrometer) moderate resolution imaging spectroradiometer is the keystone sensor for the 15 year (1998-2013) Earth Observing System (EOS). MODIS sensors on both the AM and PM EOS spacecraft will each collect data from the entire globe every two days. This data will be used to generate numerous scientific products which will enhance knowledge of changes in the global climate system due to both natural and anthropogenic causes.

MODIS, under development by the Santa Barbara Research Center (SBRC), Goleta, California, and the EOS Project at the Goddard Space Flight Center (GSFC), Greenbelt, Maryland, will include bands with spatial resolutions of 250, 500 or 1,000 meters at nadir, SNR's as large as several thousand, onboard spatial, spectral and radiometric calibration, solar and lunar radiometric stability monitoring, knowledge of pixel location to less than 500 meters (including spacecraft errors) and radiometric accuracy of 5 percent or less for the 19 reflected solar bands and 1 percent or less for the 17 thermal emission bands. Polarization sensitivity will be limited to less than two percent for bands with wavelengths less than 2.2 micrometers.

2. INTRODUCTION

The National Aeronautics and Space Administration (NASA) has, as its input to the United States Global Change Research Program (USGCRP), formulated Mission to Planet Earth, a series of Earth observing spacecraft that will remotely sense global change over a variety of spatial, spectral and temporal scales. A major portion of Mission to Planet Earth will be accomplished via the Earth Observing System (EOS), a set of sun-synchronous polar-orbiting satellites that will generate a fifteen year global data set to enable detection and quantification of subtle changes in the Earth's climate. The heart of the EOS is the AM and PM series of satellites with local crossing times of 10:30 a.m. and 1:30 p.m. respectively. Both the AM and PM series will include three satellites, each designed for five years of on-orbit operation. AM-1 is slated for launch in mid-1998 with PM-1 planned for approximately 2 1/2 years later and the remainder of the AM and PM series being launched alternately on 2 1/2 year centers.

The key remote sensing system in both the AM and PM payloads is the Moderate Resolution Imaging Spectroradiometer (MODIS), a 36 spectral band calibrated radiometer each of which will view the entire globe once every 2 days for the 15 year mission life. MODIS is being developed for the EOS Project at the Goddard Space Flight Center (GSFC) by the Santa Barbara Research Center (SBRC) of Goleta, California, a subsidiary of the Hughes Aircraft Corporation. This paper will discuss the science and engineering requirements imposed on the MODIS design, the resulting system concept and modeled performance as viewed at the completion of the MODIS Preliminary Design Review in October 1992, approximately one year after start of development and two years prior to delivery of the Engineering Model.

3. SENSOR REQUIREMENTS

Following the launch of several satellites in the late 1970's and early 1980's including NASA's Nimbus-7 and the first of the NOAA series of meteorological satellites, it became apparent to members of the scientific community that advances in remote sensing radiometry coupled with the rapid decline in computational costs were enabling the possibility of quantitative global measurement of anthropogenic and natural changes in a variety of parameters. Consequently, NASA formed a series of science/instrument teams in early 1984 to examine which parameters could be exploited and what top-level sensor requirements would be required. The MODIS team examined a multi-band imaging radiometer with a spatial resolution (IFOV) of one kilometer that would build on the results coming from the Nimbus-7 Coastal Zone Color Scanner (CZCS) and the NOAA Advanced Very High Resolution Radiometer (AVHRR) and High resolution Infrared Sounder (HIRS). The recommendations[1] of this group called for two sensors; a nadir-viewing crosstrack scanner (MODIS-N) and a fore-aft tiltable imaging spectrometer (MODIS-T) in order to accommodate the wide variety of scientific requirements. Subsequent to this recommendation, NASA released the EOS Announcement of Opportunity and in 1989 formed the MODIS Science Team with Dr. Vincent Salomonson, selected as the Facility Science Team Leader. The Science Team organization, including members and support staff is shown in Figure 1.

In late 1990, the Science Team determined that a single system (MODIS-N) would suffice if it were to be included on both the AM and PM series. Also, in the interim since the original instrument team, several spectral bands were added and others deleted. Spatial resolution of two new bands was set at 250 meters and 5 bands at 500 meters with the remaining 29 bands having 1,000 meter nadir spatial resolutions. The final set of bands is shown in Table 1 together with an indication of their primary function.

The deliberations of the Science Team, together with the result[2,3] of formal design studies, were used to generate a set of performance specifications[4] which are the basis for the present contract with SBRC. A subset of requirements developed from these specifications is listed in Table 2. Although each of these requirements is important and each has an impact on system design, at least two can easily be characterized as system drivers, and therefore merit further discussion as to their origin. The spatial registration requirement, alternatively referred to as band-to-band registration, calls for all 36 spectral bands to be spatially registered to 0.1 pixel or better. This requirement arises due to the fact that most of remote sensing algorithms utilize two or more spectral bands and spatial misregistration can result in errors in the derived parameter, the magnitude of which is scene dependent[5]. The second driving requirement is that of radiometric calibration accuracy of 5 percent or better for the 19 reflected solar bands and 1 percent or better for the 17 thermal emission bands with neither of these values to be exceeded during the 5 year on-orbit life of the sensor. The requirement is a consequence of the EOS/Mission to Planet Earth requirement for a 15 year continuous data set for use in studying subtle changes in global climate. This data set must not be sensor dependent and will therefore need to be as well calibrated as the technology will allow. Mission success is directly dependent on data set calibration accuracy. A major portion of the MODIS system design complexity is due directly to the calibration requirements.

4. SYSTEM DESIGN

Based on the system requirements outlined in Tables 1 and 2, several top-level design choices are readily apparent. The wide spectral range, variable spectral bandpass and tight band shape and location requirements dictated the choice of a dichroic and bandpass filter approach. The wide spatial swath required for two day global coverage calls for a crosstrack scan mirror. Design limitations on dichroic beamsplitters forced the 36 channels to be allocated to 4 Focal Plane Assemblies (FPA's), 2 of which, due to detector technology, require cryogenic cooling and frequent in-flight calibration. The latter results in the need for a well-characterized in-flight blackbody and a view of space that will provide a two-point thermal calibration once every scan line. The need for in-flight stability monitoring of radiometric, spectral and spatial characteristics of the MODIS system led to the inclusion of a solar diffuser, a Solar Diffuser Stability Monitor (SDSM) and a Spectral Radiometric Calibration Assembly (SRCA) which performs a variety of tasks and will be discussed in detail in a later section. These choices have resulted in the design shown by subsystem in Figure 2.

In order to attain the required Signal-to-Noise Ratios (SNR) while minimizing the size of the collecting optics, the scan mirror is double-sided and the FPA's image 10 kilometers in the track direction at nadir. The latter means that each of the 1,000 meter spectral bands consists of a 10 element linear detector array while the 250 and 500 meter bands have 40 and 20 elements each. The focal plane layout is shown in Figure 3 which indicates the detector element sizes in micrometers and shows the outlines of the bandpass filter substrates. It should be noted that in seven cases two filters will be deposited on a single substrate. Also note that bands 13 and 14 which are ocean chlorophyll fluorescence bands, require 2 element Time Delay and Integration (TDI) to meet their required SNR's. The visible (VIS) and near infrared (NIR) FPA's use silicon photodiode arrays and are therefore uncooled. The short-wave/mid-wave and long wave infrared (SWIR/MWIR and LWIR) FPA's use HqCdTe detector arrays and are cooled by the passive radiative cooler to an 85k operating temperature. Bands 31-36 are photoconductive while the remainder of the HgCdTe arrays are photovoltaic.

The FPA's are superimposed optically via the dichroic filters and the bands within the FPA's are registered by careful sample timing and onboard processing. Sample timing for each FPA is on-orbit adjustable to facilitate crosstrack registration adjustments after launch.

The system parameters discussed above together with several others including data rates, size, weight, and power as developed by SBRC are summarized in Table 3. It should be noted that the data rates of 10.7 Mbit/sec in the day portion when combined with the night mode rate of 2.7 Mbit/sec and the 100 percent duty cycle results in a daily output of 6.2×10^{11} bits.

The MODIS design is highly modular as shown in Figure 4 including eight main instrument modules and four calibration modules. Instrument modules are (1) the mainframe assembly including covers and doors, (2) the scan mirror assembly, (3) the afocal telescope assembly, (4) the aft optics assembly including the dichroic subassemblies, four objective subassemblies and the visible and near infrared FPA's, (5) the radiative cooler assembly with the short-wave/mid-wave and longwave infrared FPA's, (6) the space-viewing analog electronics module, (7) the forward-viewing analog electronics module, and (8) the main electronics module. Calibration modules are (9) the solar diffuser assembly, (10) The Solar Diffuser Stability Monitor (SDSM), (11) the Spectro-Radiometric Calibration Assembly (SRCA), and (12) the blackbody assembly. An artist's concept of the fully assembled MODIS is shown in Figure 5. The direction of flight is into the page with crosstrack scanning along the 1.6 meter axis through the scan mirror and the radiative cooler. Details of the MODIS modules will be discussed below.

4.2 Mainframe assembly

The MODIS mainframe as seen in Figures 6 and 7 is a beryllium bonded and bolted rigid assembly consisting of three primary bulkheads and support panels to which the other modules, covers and mounting feet are attached. The assembly is lightweight, with high thermal conductivity, low thermal distortion, and high natural vibration frequencies.

4.3 Scan mirror assembly

The scan mirror assembly, which can be seen in Figures 4 and 6, includes the scan mirror, the mirror-to-motor adapter plate and the motor/encoder. The mirror is a double-sided continuously rotating single piece of beryllium that will be electro-discharge machined for light-weighting, nickel plated and silver coated. The mirror is 58 by 21 centimeters, roughly elliptical in shape and will weigh approximately 4.3 kilograms. The mirror is cantilever mounted to the motor/encoder. The motor is designed for 57 million revolutions with Δvelocity/velocity$\leq$ 1%. Velocity control is attained via a 14 bit redundant encoder. Bearing and lubricant design are being verified via life tests at normal and accelerated speeds.

4.4 Afocal telescope assembly

The energy coming off the scan mirror is collected and transferred to the aft optics assembly by means of an off-axis afocal Gregorian telescope. The telescope assembly as shown in Figure 8 consists of a fold mirror, which cancels the primary component of the scan mirror polarization, and the parabolic primary and secondary mirrors all of which are mounted on a graphite epoxy optical bench. The 17.8 centimeter unobscured aperture provides high throughput and the location of the secondary on the bottom of the bench results in excellent stray light control by inclusion of an intermediate field stop assembly at the point where the beam passes through the bench (Figure 8).

4.5 Aft optics assembly

The afocal telescope assembly (ATA) mounts directly to the Aft Optics Assembly (AOA) as indicated in Figure 9 which also shows how the output from the telescope secondary is passed through three dichroic beamsplitters to divide the radiant energy into the four spectral regions that are focused on the four FPA's via their respective refractive objective assemblies. It will be noted that fold mirrors are used in the SWIR/MWIR and LWIR arms to direct the beams through the graphite epoxy optical bench and onto the cold patch of the passive radiative cooler. The VIS and NIR arms incorporate polarization compensators in their design which, combined with the shallow (22.5°) angle of incidence on dichroics 1 and 2, will minimize system polarization. MODIS system assembly starts with aligning the dichroic assembly to the AOA optical bench and, subsequently, the preassembled VIS objective and FPA.

4.6 Radiative cooler assembly

The radiative cooler planned for MODIS is a minor modification of the design used for the Thematic Mapper (TM) and Enhanced Thematic Mapper (ETM) sensors that have flown on Landsats 4,5, and 6. An exploded view of the cooler is shown in Figure 10. The cooler is attached to the AOA optical bench on the surface opposite the afocal telescope assembly (Figure 4). The cooler is a three stage design with a 170° field of view to deep space. The 230 detector elements of the 20 spectral bands on the SWIR/MWIR and LWIR focal planes will be maintained at 85K by the radiative cooler. Details of the two Cold Focal Plane Assemblies (CFPA) (Figure 10) are shown in Figure 11 mounted on the cooler dewar stem. The LWIR CFPA requires two cables due to the numerous leads from the photo-conductive HqCdTe detectors, whereas the SWIR/MWIR CFPA, with all photo-voltaic detector elements, will require only a single flat cable. These assemblies include the spectral filters that are mounted in close proximity to the detector arrays. The VIS and NIR FPA's are almost identical to the SWIR/MWIR CFPA. This commonalty will greatly reduce fabrication, assembly and testing complexity.

4.7 Space-viewing analog electronics module

The Space-viewing Analog electronics Module (SAM) is mounted close to the radiative cooler and includes an oval hole through which the scan mirror views space for thermal calibration (Figure 4). SAM will include the photo-voltaic detector clock, gain and offset, DC restore and A/D conversion electronics.

4.8 Forward-viewing analog electronics module

The Forward-viewing Analog electronics Module (FAM) is placed on the upper portion of the MODIS front surface (Figure 7) and will contain the Photo-Conductive (PC) detector preamps, gain and offset, DC restore and A/D conversion electronics. The initial stage of amplification will be located at the radiative cooler. Locating both the SAM and FAM close to the radiative cooler will minimize the effects of Electro-Magnetic Interference (EMI).

4.9 Main electronics module

The Main Electronics Module (MEM) is positioned immediately above the radiative cooler (Figures 6 and 7) and contains the majority of the MODIS electronics including the power, command and control systems and the data buffer and formatting electronics.

4.10 Solar diffuser assembly

The solar diffuser assembly is a spectrally diffuse surface mounted near the top of the MODIS main frame (Figure 6) such that it is illuminated by the sun through an opening in the main frame when passing near the North pole, thereby enabling the scan mirror to pass solar radiance onto the detectors of the 19 reflective bands, which will provide a means of checking the stability of the sensor as often as once per orbit. The diffuse surface will have an effective albedo of 46 percent. There is a door on the front of the MODIS that will be commanded open only when a view of the sun is desired. Immediately behind the door is a screen that will block 90 percent of the solar energy thereby reducing the effective albedo to 4.6 percent. This screen can also be commanded open. This provides two levels of solar radiance at least one of which will be within the dynamic range of each of the solar reflectance bands.

4.11 Solar diffuser stability monitor

Tracking the stability and/or degradation of the solar reflectance bands via the solar diffuser views of the sun assumes that the sun and the diffuser are spectrally and radiometrically stable over a period of several years. The sun is stable to within the desired accuracy, but the diffuser is subject to contaminate deposition and solarization and must, therefore, be monitored to detect any post-launch change in its characteristics. This is the task of the Solar Diffuser Stability Monitor (SDSM), a device that alternately views the sun and the diffuser via a pointable mirror and measures the spectral output of each using a 10 centimeter diameter integrating sphere with 12 spectral filter/detector modules mounted on its inner surface (Figure 12). In order for the signals of the two views to be comparable, a two percent transmitting screen is placed between the mirror and the solar view. The spectral range of the 12 bands is from 0.4 to 2.3 micrometer using 9 silicon, 2 Ge and 1 InAs redundant detectors, with the InAs being cooled by thermoelectric coolers. Figure 13 is an artist's concept of the SDSM. The fold mirror immediately inside the entrance port has three viewing positions: 1) diffuser, 2) sun, and 3) black surround. The system will be capable of tracking changes in the solar diffuser reflectance to one percent accuracy over the VIS, NIR, and SWIR bands.

4.12 Sprectro-radiometric calibration assembly

The MODIS, via the Spectro-Radiometric Calibration Assembly (SRCA) (Figures 14 and 15), will be capable of (1) performing on-board radiometric calibration of the VIS, NIR, and SWIR bands to within five percent accuracy, (2) determining band-to-band registration to within 0.1 IFOV for all 36 spectral bands in both the crosstrack and along track directions, and (3) determining the spectral band shape and center of the VIS, NIR, and SWIR bands to an accuracy of 1.0 nm with a repeatability of 0.5 nm at 415 nm and scaled with wavelength. The SRCA is a self-contained system consisting of three major subsystems. The first of these is a source relay assembly consisting of a five centimeter diameter integrating sphere source with a redundant 1-watt and three redundant 10-watt lamps, an infrared resistive heater source and a six position filter wheel with three order filters, a short-wave/infrared beamcombiner, a 45 percent transmitting neutral density filter and an open position. The second subassembly is a monochrometer/relay that includes a Czerny-Turner monochrometer with a triangular rotatable mount with two gratings used over three diffration orders and a mirror for use in the relay mode, a grating motor with 40 arc- second steps, an entrance plane that can be commanded open or to the spectral slit position and an exit plane that can be open with an electronic phase delayed knife edge for crosstrack registration or a phased knife edge for in track registration or can be in the slit position which includes an unfiltered silicon photodiode and a didymium glass filtered self calibration silicon photodiode. The third subassembly is a collimator consisting of a two mirror 8.9 centimeter diameter cassegrain telescope with a fold mirror at the output that directs the energy to the MODIS scan mirror. The SRCA setup for the various calibration modes is shown schematically in Figure 14. A complete layout of the SRCA is shown in Figure 15.

4.13 Blackbody assembly

The blackbody assembly (Figure 6) is a V-groove design that fills the MODIS aperture and 30 IFOV's along-scan. The blackbody will operate at ambient and 315K with an effective radiometric temperature knowledge of $\leq$ 0.1K. The outer surface will be coated with specular black paint which results in a calculated effective emittance of 0.992.

5. SYSTEM PERFORMANCE

Extensive radiometric math models[6,7,8] have been developed by SBRC and GSFC personnel to predict MODIS system performance. These models are descendents of previous models that were used for system designs and iterated using the measured performance of these designs. Although numerous parameters have been modeled, only two will be discussed; radiometric precision and radiometric accuracy.

Radiometric precision is generally stated as a Signal-to-Noise Ratio (SNR) for reflectance bands (Bands 1-19) and as a Noise Equivalent Difference in Temperature (NEDT) for emissive bands (Bands 20-36). Predicted SNR values for MODIS Bands 1-19 are shown in Figure 16, where the dark bars are the required values and the diamonds are the modeled values with 3-sigma error bars. The NEDT's for Bands 20-36 are shown in Figure 17, where, again, the dark bars are the required values and the diamonds, with 3-sigma error bars, are the anticipated values for the MODIS design. NEDT is a noise figure and, therefore, smaller values are desirable. A more graphic representation of predicted precision is shown in Figure 18 which is a plot of percent margin when the predicted value is compared to the required value. A goal set forth in the MODIS specifications is to maintain a 36 percent margin. Figure 18 shows that, at the present design stage, only the last few bands fail to meet this goal and that all bands exceed the required values.

Radiometric accuracy requirements for MODIS are, as shown in Table 2, ≤ 5 percent for Bands 1-19 and ≤ 1 percent for Bands 20-36 The latter requirement has recently been modified for Bands 20, 31 and 32. Band 20 is now ≤ 0.75 percent with a 0.5 percent goal and Bands 31 and 32 have a 0.5 percent requirement and a 0.25 percent goal. The modeled accuracies for Bands 1-19 are shown in Figure 19. All bands are well within the 5 percent requirement. The predicted accuracy for Bands 20-36 are illustrated in Figure 20. In Bands 20, 31 and 32, an X shows the measured performance of the corresponding bands in the ERS-1 Along Track Scanning Radiometer (ATSR). The MODIS predicted values are similar to the ATSR values of all three bands.

The specifications[2] generated by the government contain numerous other requirements as outlined in Table 2. The SBRC design as discussed above and outlined in Table 3, meets the government's requirements and, in most cases, exceeds the specification with a margin that insures successful completion of the MODIS development.

The MODIS schedule calls for delivery of the Engineering Model in April 1995, delivery of the First Flight Model in June 1996, with subsequent flight models being completed every 18 months and launch of the EOS AM-1 spacecraft in June 1998.

6. ACKNOWLEDGMENTS

There are over 200 personnel from the Santa Barbara Research Center, the Goddard Space Flight Center, and numerous other science and engineering organizations across the United States and in several other countries who are involved in the development of MODIS. The present article only begins to describe the efforts of these talented professionals. The authors would like to express their appreciation of the continuing efforts of all concerned. In particular, the authors wish to acknowledge that many of the figures used in this paper are derived from recent reviews conducted by SBRC.

7. REFERENCES

1. W. Esaias and W. Barnes, "Moderate Resolution Imaging Spectrometer (MODIS) Instrument Panel Report," Earth Observing System Reports, Vol. IIb, 1986.

2. Goddard Space Flight Center, "Specification for the Moderate-Resolution Imaging Spectrometer-Nadir (MODIS-N)," GSFC Doc. No. 422-20-02, Goddard Space Flight Center, Greenbelt, Maryland, October 5, 1990.

3. ITT Aerospace, "Phase-A study report on the MODIS-N spectrometer for the Earth Observing System," ITT Aerospace/Optical Div., Ft. Wayne, IN, Contract No. NAS5-27145, 1985.

4. Santa Barbara Research Center, "Moderate Resolution Imaging Spectrometer-Nadir (MODIS-N), Phase-B Final Report," Santa Barbara Research Center, Goleta, CA, Contract No. NAS5-30149, July 1989.

5. V. V. Salomonson and J. L. Barker, "EOS moderate resolution imaging spectroradiometer: Phase C/D status and comments on calibration and georeferencing approaches," Proceedings of the 15th Annual AAS Guidance and Control Conference, Keystone, Colorado, February 8-12, 1992.

6. T. Pagano, "Radiometric Math Model," Santa Barbara Research Center, Goleta, CA, Contract No. NAS5-30800, CDRL No. 101, October 1992.

7. H. E. Montgomery, H. Ostrow, and G. M. Ressler, "Sensor Performance Analysis," NASA Reference Publication 1241, July 1990.

8. K. Forrest, "GSFC Radiometric Model," GSFC Internal Communication, October 16, 1992.

Table 1. MODIS SPECTRAL BANDS

BAND	λ	IFOV	Δλ	PURPOSE (Examples)
	LAND AND CLOUD BOUNDARIES BANDS			
1	659 nm	250 m	50 nm	Veg. Chlorophyll Absorption
2	865 nm	250 m	40 nm	Cloud and Veg. Land Cover Transformation
	LAND AND CLOUD PROPERTIES BANDS			
3	470 nm	500 m	20 nm	Soil, Vegetation Differences
4	555 nm	500 m	20 nm	Green Vegetation
5	1240 nm	500 m	20 nm	Leaf/Canopy Differences
6	1640 nm	500 m	20 nm	Snow/Cloud Differences
7	2130 nm	500 m	50 nm	Land and Cloud Properties
	OCEAN COLOR BANDS			
8	415 nm	1000 m	15 nm	Chlorophyll
9	443 nm	1000 m	10 nm	Chlorophyll
10	490 nm	1000 m	10 nm	Chlorophyll
11	531 nm	1000 m	10 nm	Chlorophyll
12	555 nm	1000 m	10 nm	Sediments
13	667 nm	1000 m	10 nm	Sediments, Atmosphere
14	681 nm	1000 m	10 nm	Chlorophyll Flourescence
15	750 nm	1000 m	10 nm	Aerosol Properties
16	865 nm	1000 m	15 nm	Aerosol/Atmospheric Properties
	ATMOSPHERE/CLOUD BANDS			
17	905 nm	1000 m	30 nm	Cloud/Atmospheric Properties
18	936 nm	1000 m	10 nm	Cloud/Atmospheric Properties
19	940 nm	1000 m	50 nm	Cloud/Atmospheric Properties
	THERMAL BANDS			
20	3.75 μm	1000 m	0.18 μm	Sea Surface Temperature
21	3.75 μm	1000 m	0.05 μm	Forest Fires/Volcanoes
22	3.96 μm	1000 m	0.05 μm	Cloud/Surface Temperature
23	4.05 μm	1000 m	0.05 μm	Cloud/Surface Temperature
24	4.47 μm	1000 m	0.05 μm	Tropospheric Temperature/Cloud Fraction
25	4.52 μm	1000 m	0.05 μm	Tropospheric Temperature/Cloud Fraction
26	4.57 μm	1000 m	0.05 μm	Tropospheric Temperature/Cloud Fraction
27	6.72 μm	1000 m	0.36 μm	Mid-Tropospheric Humidity
28	7.33 μm	1000 m	0.30 μm	Upper-Tropospheric Humidity
29	8.55 μm	1000 m	0.30 μm	Surface Temperature
30	9.73 μm	1000 m	0.30 μm	Total Ozone
31	11.03 μm	1000 m	0.50 μm	Cloud/Surface Temperature
	THERMAL BANDS			
32	12.02 μm	1000 m	0.50 μm	Cloud Height & Surface Temperature
33	13.34 μm	1000 m	0.30 μm	Cloud Height & Fraction
34	13.64 μm	1000 m	0.30 μm	Cloud Height & Fraction
35	13.94 μm	1000 m	0.30 μm	Cloud Height & Fraction
36	14.24 μm	1000 m	0.30 μm	Cloud Height & Fraction

Table 2. MODIS SYSTEM REQUIREMENTS

Orbit	705 km, 10:30 am descending
Swath	±55°, 2330 km
Spectral Coverage	0.407 μm - 14.385 μm, 36 Bands
Spectral Resolution	10 nm ≤ $\Delta\lambda$ ≤ 500 nm
Spectral Stability	λ & $\Delta\lambda$ Stable to ≤ 2 nm (VIS/NIR Bands) λ & $\Delta\lambda$ Stable to ≤ 1% (All other bands)
Spatial Resolution	250 m; 500 m; 1000 m at nadir (502 m; 1003 m; 2006 m; at ±55°)
Spatial Registration	≤ 0.1 IFOV
IFOV	0.354 mr, 0.709 mr, 1.418 mr (all ±6%)
Radiometric Range	0.002% ≤ρ≤ 100%, 3K≤T≤ 700K
Dynamic Range	NEΔL to Lmax (requires 12 bits)
Radiometric Performance	57 ≤ SNR ≤ 1087, @ 22° Solar Zenith, Typical L 0.05K NEDT @ 300K to 5.0K NEDT @ 700K
Polarization Insensitivity	≤ 2%, 0.43 μm ≤λ≤ 2.2 μm
Calibration Accuracy	Absolute Calibration: ±1%(λ >3 μm); ± 5%(λ <3 μm); ±2% reflectance (λ <3 μm)

Table 3. MODIS BASELINE PARAMETERS

Scanning	360° scan, double sided mirror, 20.3 rpm, 2.954 sec period
IFOV	0.354 mr (0.25 km), 0.709 mr (0.50 km), 1.418 mr (1.0 km)
Dwell Time	83.3 μs (0.25 km), 166.7 μs (0.50 km), 333.3 μs (1.0 km)
Telescope	2-mirror off axis afocal Gregorian, 4X magnification, EPD 17.8 cm
Reimaging Optics	Refractive, EFL 380.8 mm, 282.1 mm (LWIR only)
Spectral	36 bands from 0.41 μm to 14.3 μm, dichroic beamsplitters, bandpass filters
Focal Planes	Silicon (0.4 μm ≤λ≤ 1 μm), PV HgCdTe (1 μm ≤λ≤ 10 μm) PC HgCdTe (λ > 10 μm), CTIA readouts for PV, bipolar preamps for PC
Detector Sizes	Nominally square: 135 μm (0.25 km), 270 μm (0.5 km), 540 μm (400 μm LWIR) (1.0 km)
Detector Cooling	Thematic Mapper-type radiative, T~85K
Calibration	Laboratory, ground truth, on-board blackbody, solar diffuser, SRCA, and SDSM
Data Rates	10.7 MBPS (day mode), 2.7 MBPS (night mode) with cal and overhead
Size, weight & power	0.97 m x 1.59 m x 0.99 m, 238 kg, 192 w

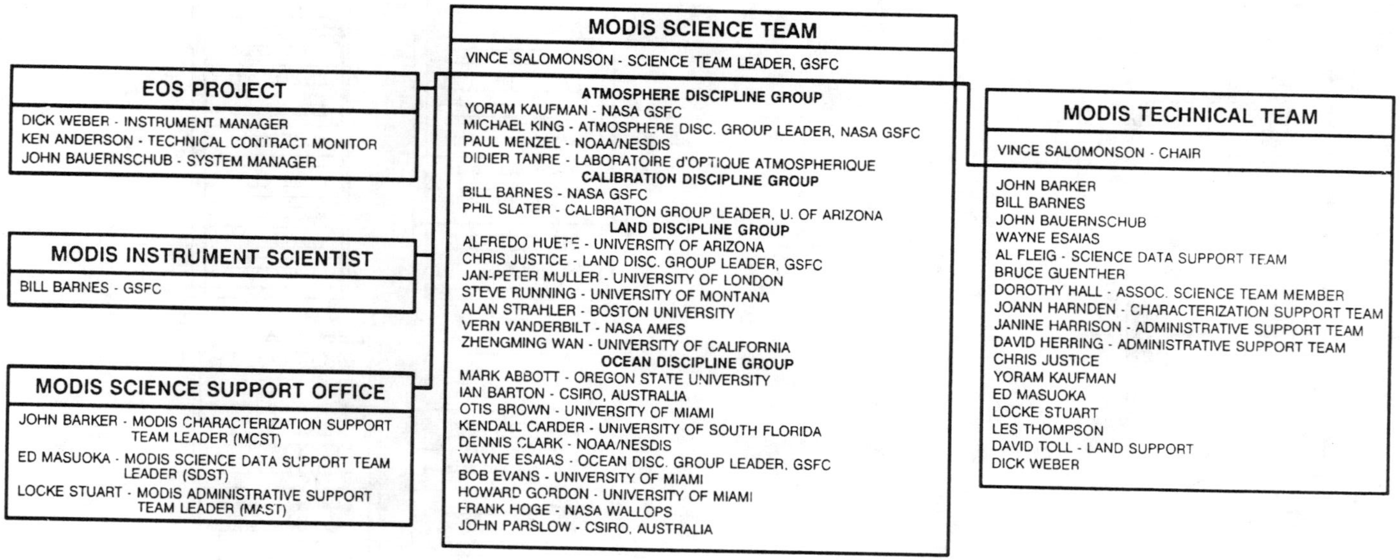

Figure 1. MODIS organization

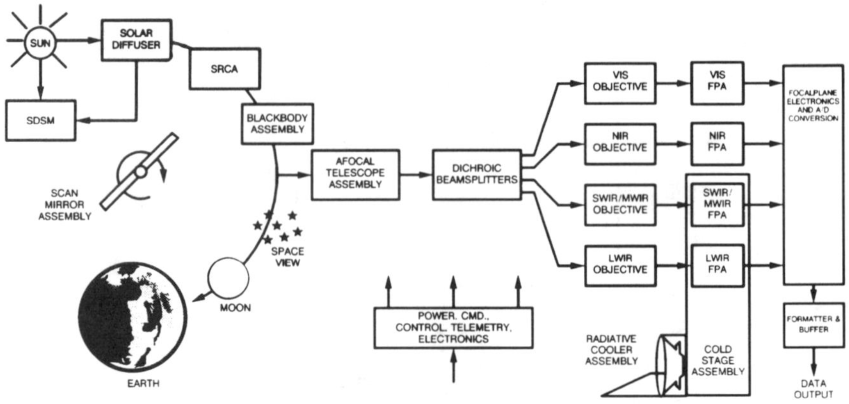

Figure 2. MODIS concept

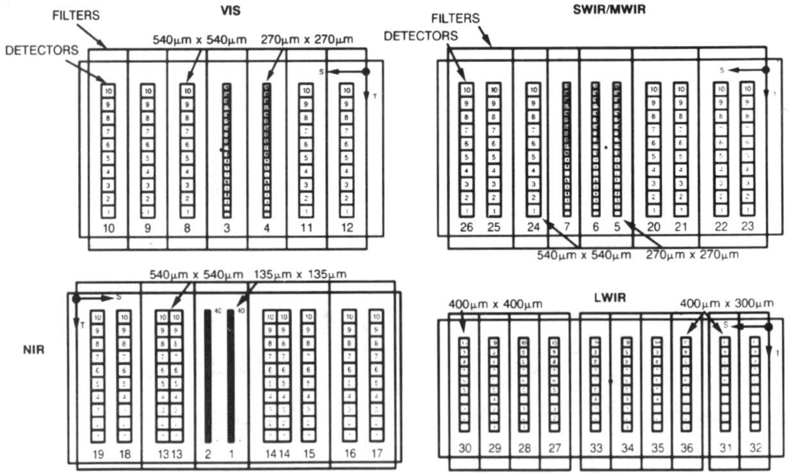

Figure 3. Focal plane layout

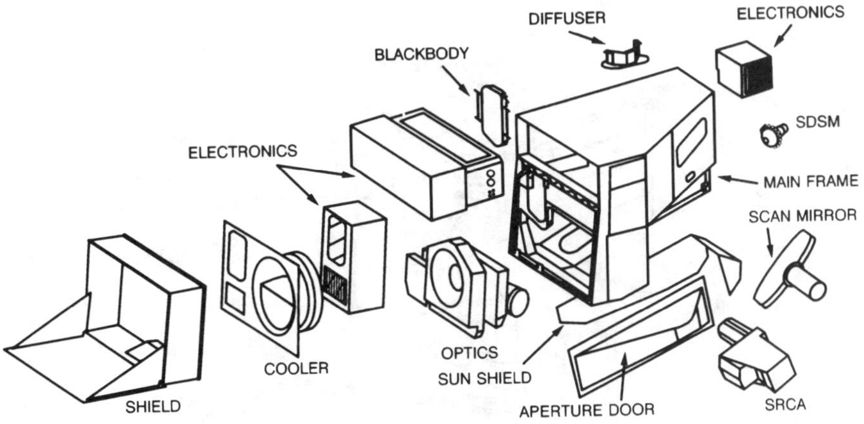

Figure 4. Subsystems

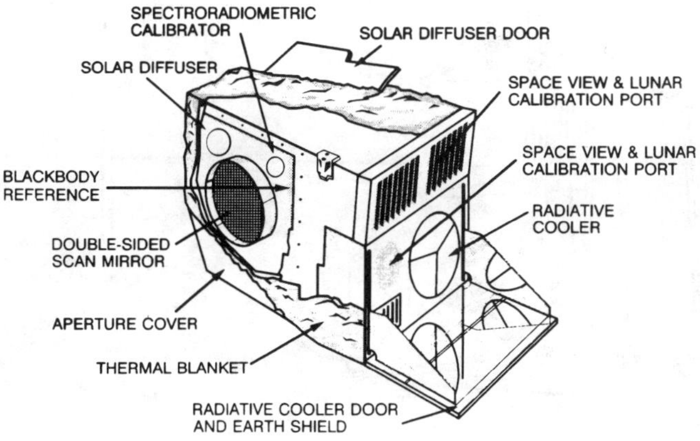

Figure 5. MODIS

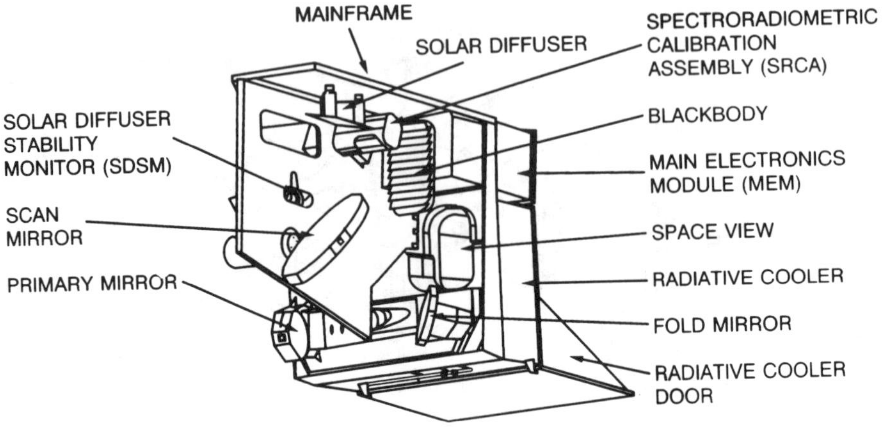

Figure 6. Scan cavity

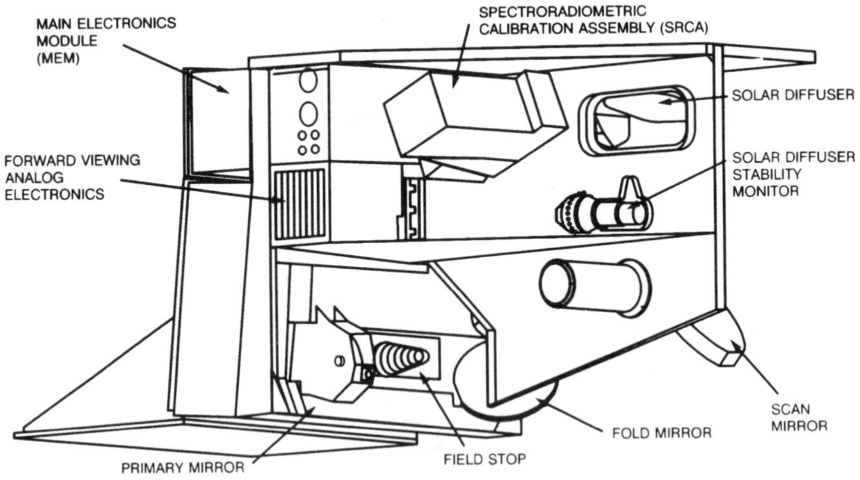

Figure 7. Forward side

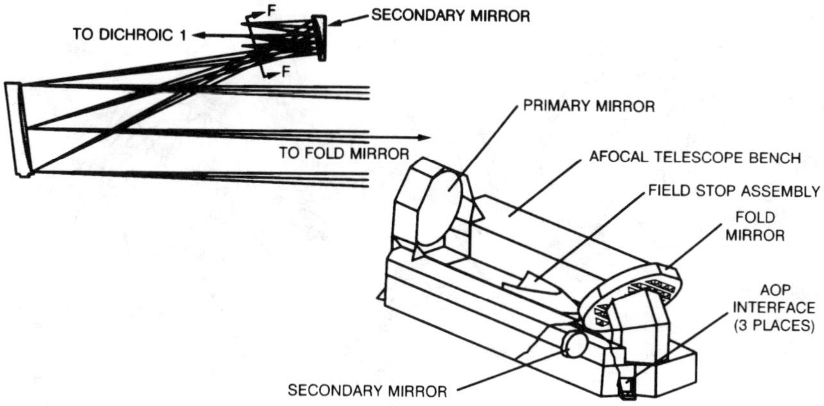

Figure 8. Afocal telescope assembly

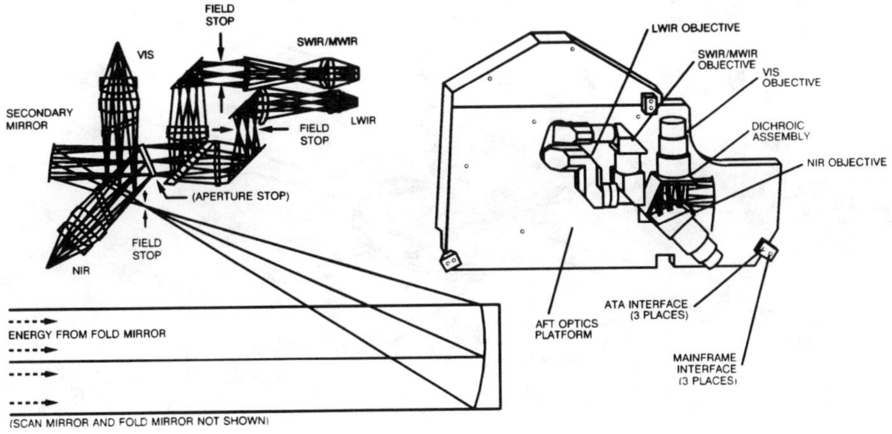

Figure 9. Aft-optics assembly

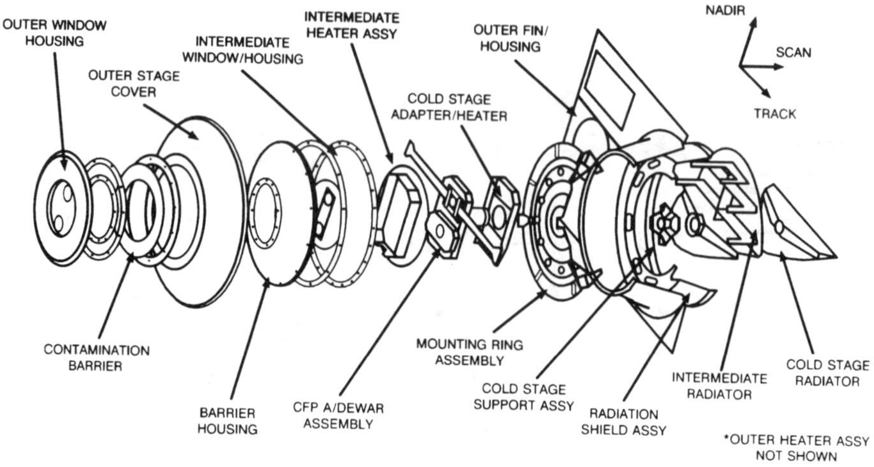

Figure 10. Passive radiative cooler

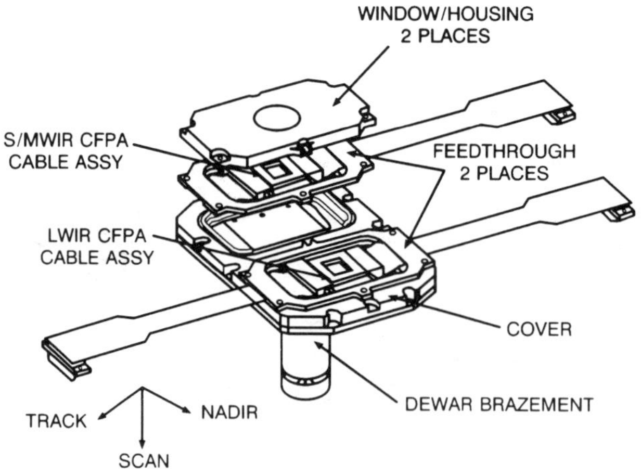

Figure 11. Cold focal plane and dewar assembly

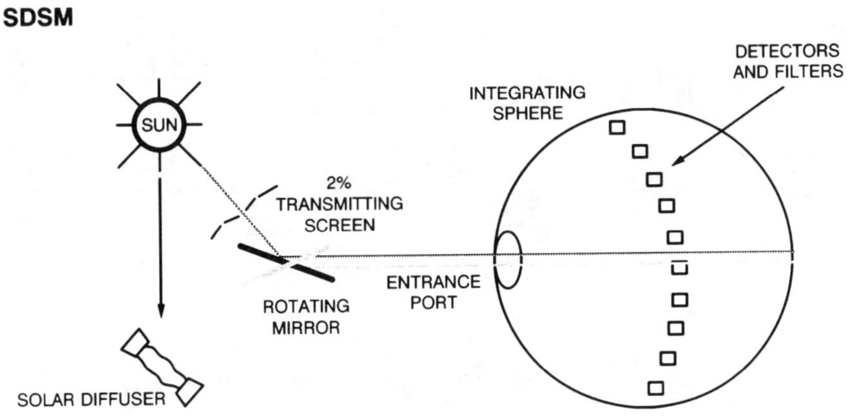

Figure 12. SDSM concept

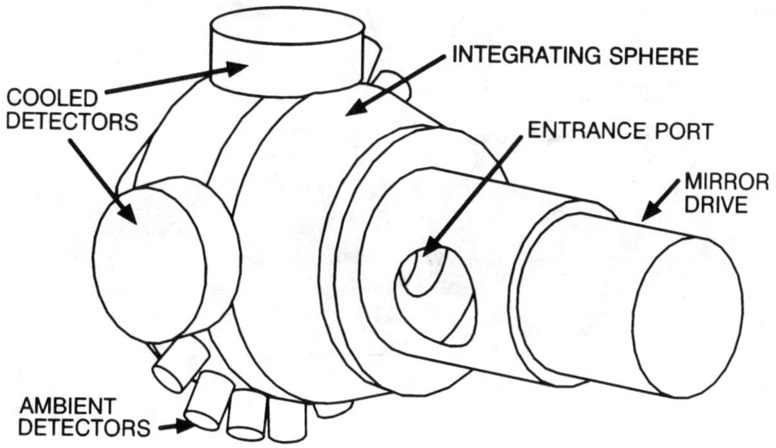

Figure 13. SDSM assembly

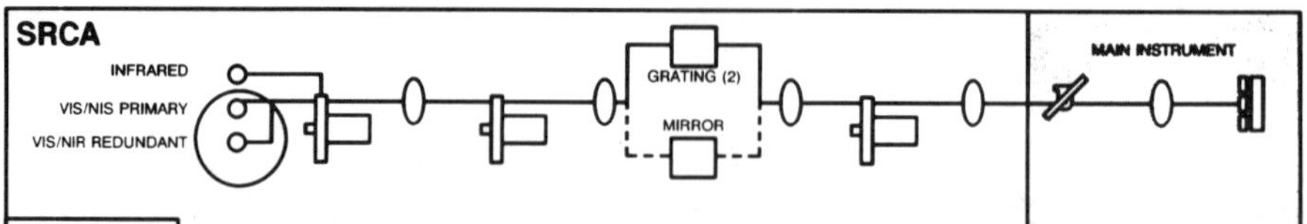

OPERATING MODE	SOURCES	FILTER WHEEL	RELAY OPTICS	ENTRANCE SLIT	COLLIMATING OPTICS	GRATING-MIRROR	FOCUSING OPTICS	EXIT SLIT	COLLIMATING OPTICS	SCAN MIRROR ASSEMBLY	TELESCOPE/ AFT-OPTICS	FILTERS/ DETECTORS
RADIOMETRIC CHECK	VIS/NIR	ORDER	X	OPEN	X	MIRROR	X	OPEN	X	X	X	X
SPECTRAL CALIBRATION	VIS/NIR	FILTERS	X	ENTRANCE SLIT	X	GRATING	X	EXIT SLIT/PHOTODIODE	X	X	X	X
SPATIAL REGISTRATION	VIS/NIR & IR	(TO DICHROIC	X	OPEN	X	MIRROR	X	RETICLE PATTERN	X	X	X	X
SELF CALIBRATION	VIS/NIR	X	X	ENTRANCE SLIT	X	GRATING	X	EXIT SLIT/DIDYMIUM PHOTODIODE				

Figure 14. SRCA operational modes

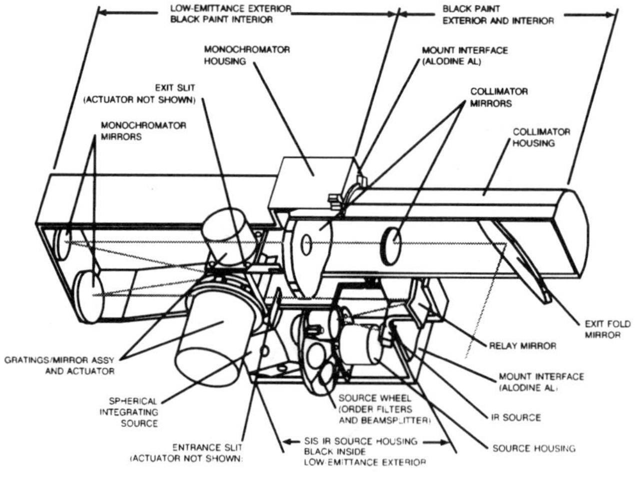

Figure 15. SRCA

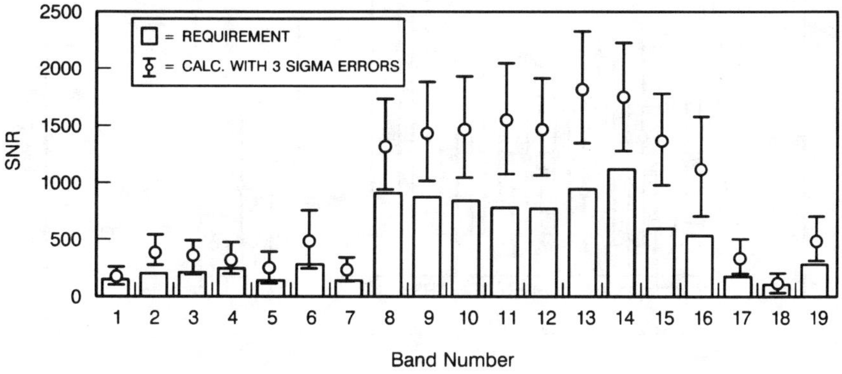

Figure 16. Modeled values of expected SNR compared to spec requirements

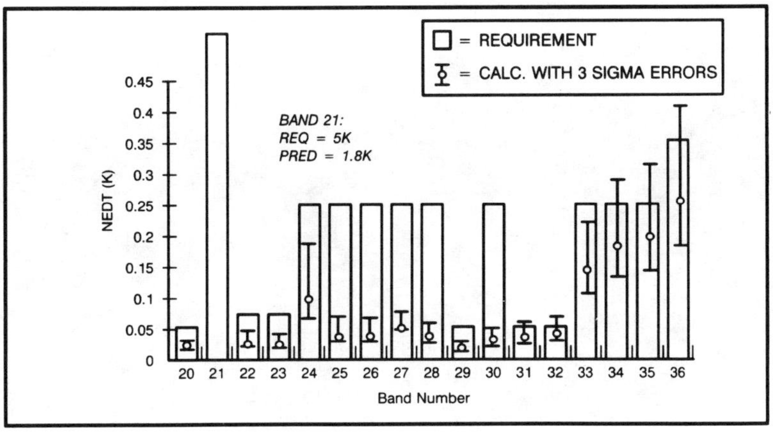

Figure 17. Modeled values of expected NEDT compared to spec requirements

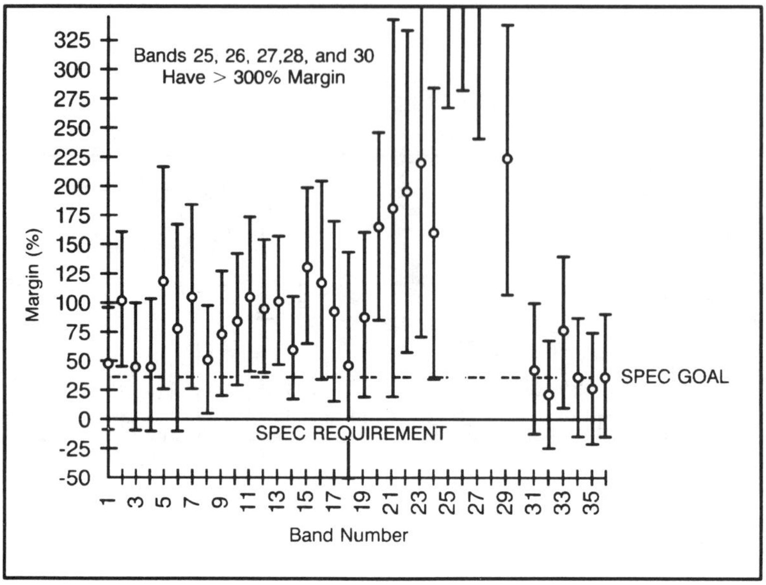

Figure 18. Modeled precision margins for MODIS
(SNR and NEDT)

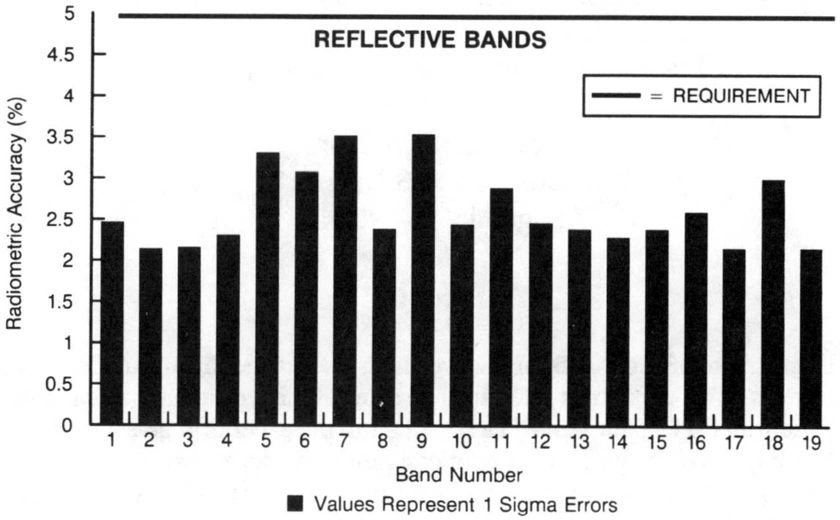

Figure 19. Modeled accuracy for the reflective bands

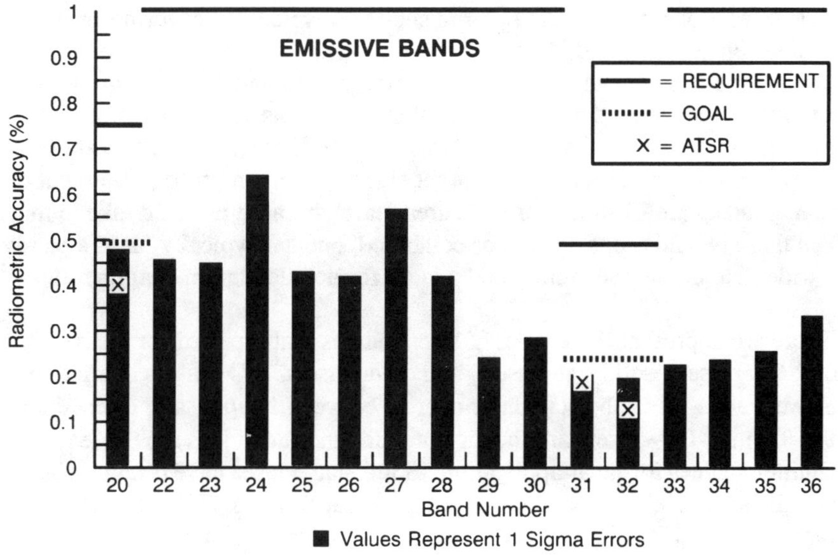

Figure 20 Modeled accuracy for the emissive bands

LASER RADAR SYSTEMS AND APPLICATIONS

Maurice J. Halmos and John H.S. Wang
Hughes Aircraft Co.
Electro-Optic Sector
P.O. Box 902 MS E1-B118
El Segundo, CA 90245

Abstract

Hughes Aircraft Co. has been involved with laser range finders and laser radars for over 20 years. This paper tries to summarize some of the typical laser radar sensors and applications in the last few years. The first half of the paper is to give a brief analytical description of the current issues in designing a modern ladar. The second half gives a review of laser sensors that have been developed by Hughes and by other companies highlighting some of the key features of those sensors.

1. Introduction

There have been numerous publications on laser radar sensors[1,2] over the last ten years. We can not and will not attempt to cover all of them in this review paper, however, we shall describe the theory and applications and to the best of our knowledge a few typical examples. Laser radars (or ladar - laser detection and range) are an extension of conventional microwave radar to shorter wavelengths covering from the ultraviolet to visible and infrared. Like microwave radar, ladar can measure simultaneously the range, velocity, reflectivity, and azimuth and elevation angle. Thus ladars operating in the infrared or shorter wavelength regions achieve high angular resolution, wide bandwidth, and high Doppler frequency sensitivity. Ladars are well suited for precise measurements useful in target classification and recognition, but on the other hand, ladars are ill suited for wide area search because it would take to much time. When these two missions need to be combined, one has typically fused a passive sensor for wide field of view searching and a ladar for detailed target examination.

There are at present three types of laser sources that are used for ladar applications; CO_2 lasers, solid state lasers, and diode lasers. CO_2 lasers operate in the infrared between 9-11 μm, solid state laser operate between 1-2 μm, and direct diodes between 0.7-0.9 μm. CO_2 lasers are the most mature and the diodes are the least mature laser transmitter sources for ladar application. Solid stated laser have become viable options with the advent of "low cost" diode pumping capabilities. The selection of a laser as transmitter is ultimately decided based on the ladar mission requirements.

In this paper we will discuss the theories of laser radar (section 2, 3, and 4), hardware requirements (section 5), and present and potential applications (section 6).

2. Signal to Noise Ratio (SNR) Equations

2.1. Receiver Detection Methods

There are two basic methods to detect transmitted and then reflected energy as shown in fig. 2-1. The first one, fig. 2-1a represents direct or incoherent detection where the total energy collected by the receiver aperture is converted to the signal current. The second method, shown in fig. 2-1b, is coherent or heterodyne detection receiver, where a local oscillator (LO) beam is mixed with the reflected beam from the target to generate an intermediate frequency (IF) signal current on the detector face. The advantage of coherent detection is that the signal phase is also detected along with the amplitude, thus allowing for Doppler shift estimation. In most cases, the coherent detection is more sensitive than direct detection, but it requires a more complicated receiver and is subject to more efficiency losses due to mismatch.

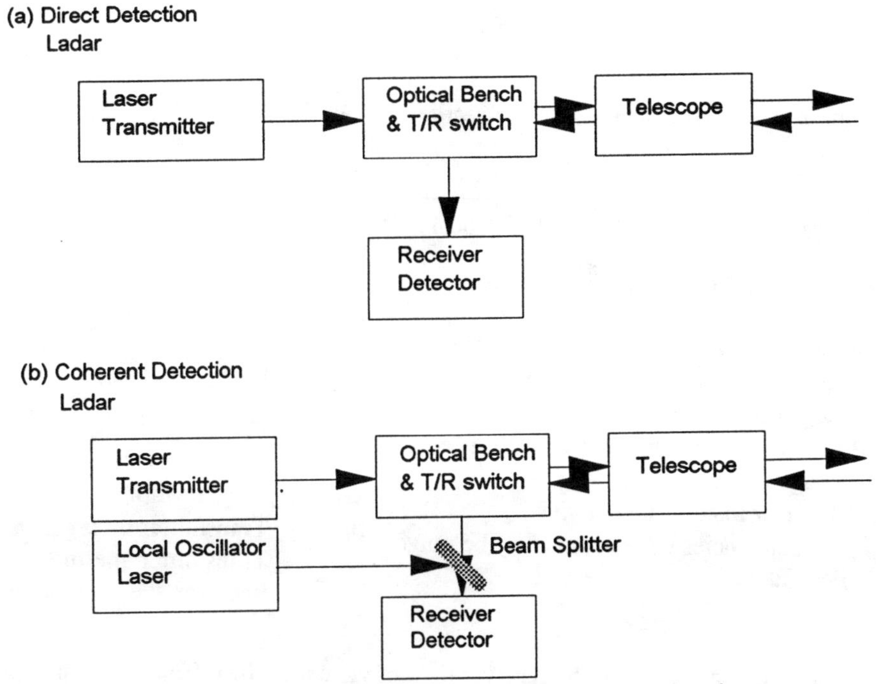

Figure 2-1. Laser radar detection methods, (a) direct or incoherent detection and (b) heterodyne or coherent detection.

2.2. Received or Signal Power

The received or signal power may be expressed as,

$$P_r = \frac{P_{laser}}{A_{beam}} \cdot \frac{\rho\, A_{eff}}{\pi} \cdot \Omega_r \cdot e^{-2\alpha R} \cdot \eta(R) \qquad (2\text{-}1)$$

where the first term is the power density at the target, second is the Lambretian (half plane) scattering cross-section of the target, third is the receiver solid angle, fourth is the atmospheric attenuation, and fifth is the total efficiency of the sensor. Figure 2-2 shows the physical configuration for Eqn.(2-1).

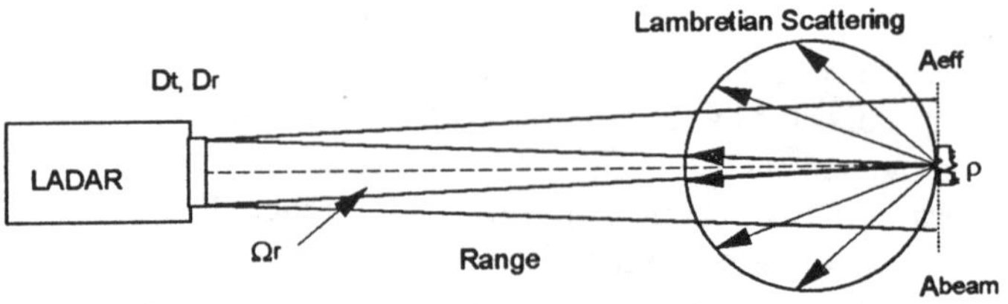

Figure 2-2. Physical representation for the main terms to calculated the power received in a ladar.

The receiver acceptance solid angle, Ω_r, is defined as

$$\Omega_r = \frac{\pi}{4} \cdot \min\left(\frac{D_{r^2}}{R^2} \, , \, (RFOV)^2 \right)$$

and the target effective geometrical cross-section , A_{eff},

$$A_{eff} = \min(\sqrt{A_{beam}}, H_t) \cdot \min(\sqrt{A_{beam}}, W_t) \, .$$

which cannot be larger than the beam size.

The terms used are:

P_r	Optical power signal received
ρ	Target reflectivity
α	Atmospheric attenuation per distance
R	Range
RFOV	Receiver instant filed of view
$\eta(R)$	All efficiencies where some may be range dependent
A_{beam}	Beam area size, $= \pi / 4 \cdot ((\theta_t R)^2 + D_t^2)$
H_t	Target height
W_t	Target width
θ_t	Transmitter divergence
D_t	Transmitter aperture
D_r	Receiver aperture diameter

For coherent detection, the receiver field of view must be diffraction limited in order to have maximum mixing efficiency, this means that typically, $RFOV = \lambda / D_r$.

2.3. SNR Expression

The signal to noise at the receiver is derived from the expression,

$$SNR = \frac{i_{sig}^2}{i_{noise}^2}$$ and the signal current is,

$$i_{sig}^2 = \left(\frac{q\eta_Q P_r M}{h\upsilon}\right)^2$$ for incoherent or direct detection and, (2-2a)

$$i_{sig}^2 = \left(\frac{q\eta_Q}{h\upsilon}\right)^2 2P_{LO}P_r$$ for coherent detection (2-2b)

where

P_r	Optical power signal received	q	Electron charge
P_{LO}	Optical power of LO laser	η_O	Quantum efficiency of detector
hν	Photon energy	M	Detector gain (i.e. APD)

Along with the signal detected currents there is the noise currents. The incoherent case may be expressed as,

$$i_{noise}^2 = I_d^2 + M^2F_m(I_b^2+I_s^2) \cdot \qquad (2\text{-}3)$$

where

I_d	Detector total noise (dark current and preamp noise) ~ 2 pA/$\sqrt{\text{Hz}}$	i_s	Signal current
I_b	Solar irradiant noise $\sqrt{2\cdot q\cdot i_b\cdot \text{BWn}}$	BW_n	Noise bandwidth $= \frac{\pi}{2}\cdot\frac{0.2}{\tau_{pulse}}$ for pulsed modulation
i_b	Solar dc irradiance dc current	M	Detector gain
I_s	Signal induced noise $\sqrt{2\cdot q\cdot i_s\cdot \text{BWn}}$	F_m	Detector excess noise = $M^{0.3}$ for Si-APD

and for the coherent case,

$$i_{noise}^2 = 2q\text{BW}_n\left(\frac{q\eta_Q}{h\upsilon}\right)(P_{LO}+P_r+P_{bk})+i_d^2 \qquad (2\text{-}4)$$

where

i_d	Detector noise (dark current and preamp noise)
P_{bk}	Solar irradiant noise and other background noise

The SNR expressions obtained by combining Eqn.(2-2), (2-3) and (2-4) are for the incoherent case,

$$\text{SNR} = \frac{(q\eta_Q P_r M / h\upsilon)^2}{I_d^2 + M^2F_m(I_b^2+I_s^2)} = \frac{M^2P_r^2}{(P_d + M^2F_m(P_b+P_s))(h\upsilon/\eta_Q)\text{BW}_n} \qquad (2\text{-}5)$$

and for the coherent case,

$$\text{SNR} = \frac{\left(\frac{q\eta_Q}{h\upsilon}\right)^2 2P_{LO}P_r}{2q\text{BW}_\text{n}\left(\frac{q\eta_Q}{h\upsilon}\right)(P_{LO}+P_r+P_{bk})+i_d^2} \Rightarrow \frac{P_r}{\text{BW}_\text{n}(h\upsilon/\eta_Q)} \tag{2-6}$$

where the approximation occurs when the LO power is larger than the other noise sources, in particular the detector noise power, which is the desired mode of operation.

The Noise Equivalent Power (NEP) is defined as the signal power that will make the SNR unity. We observe that for the two cases we have,

$$\text{NEP}^2 = (P_d / M^2 + F_m(P_b + P_s))\cdot \text{BW}_\text{n}(\text{h}\upsilon/\eta_Q) \quad \text{for incoherent case} \tag{2-7a}$$

$$\text{NEP} = \text{BW}_\text{n}(\text{h}\upsilon/\eta_Q) = \text{BW}_\text{n}\,\text{h}/(\lambda\eta_Q) \quad \text{for coherent case} \tag{2-7b}$$

We may use the results from Eqn. 2-7 to compare detection sensitivity between direct and coherent sensors. We observe that the direct detection NEP contains the coherent detection term multiplied by a systems noise term. When the coherent detection term is smaller than the system noise term, then coherent detection yields more sensitivity than direct detection. This is typically true with CO_2 wavelength of ~10 μm. As the wavelength shortens, however, the advantages of the added sensitivity of coherent detection begin to be canceled by the lower efficiency of coherent detection. This implies that one would typically not use coherent detection with a short wavelength laser (such as solid state) just to improved sensitivity, but only when phase and Doppler information is required.

Given a required SNR, one may calculate the laser transmitter energy required to meet this signal to noise ratio as a function of range by solving the S/N equation

$$P_{laser} \sim \text{NEP}\cdot\sqrt{\text{SNR}}\cdot\frac{\pi}{\rho\,\Omega_r\,e^{-2aR}}\cdot\eta(R) \quad \text{for incoherent case} \tag{2-8a}$$

$$P_{laser} \sim \text{NEP}\cdot\text{SNR}\cdot\frac{\pi}{\rho\,\Omega_r\,e^{-2aR}}\cdot\eta(R) \quad \text{for coherent case} \tag{2-8b}$$

3. Probability of Detection and Post Detection Integration

In the pervious section we discussed how to estimate the expected SNR from a sensor, however, it is the probability of detection, P_d, and probability of false alarm,

P_{fa}, that is usually specified for a given design. In this section we discuss the relationship between SNR and P_d. Coherent and incoherent laser sensors use post detection integration (PDI) in order to increase the detection probability of low cross-section targets. Figure 3-1 shows a functional diagram of how is PDI implemented, which is an averaging process after the second detector (i.e. at baseband). Our ability to model the performance a sensor using PDI requires an understanding of the statistics involved and how the probability of detection varies as function of signal to noise ratio for a variety of cases.

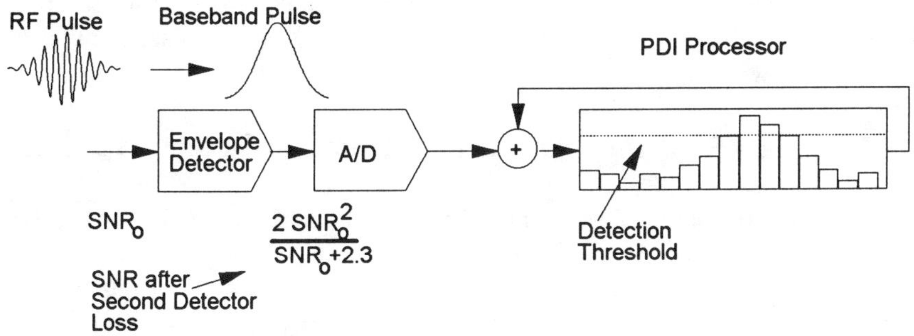

Figure 3-1. PDI functional diagram.

3.1. Second Detector Loss

The second detector loss occurs in coherent detection when the intermediate frequency (IF) signal is demodulated or brought down to zero frequency offset (baseband) as indicated in fig. 3-1. This loss is not present in direct detection because the signal of interest is already in baseband. There is signal loss during this process and an empirical relation is given by Barton,

$$SNR=2\ SNR_0^2/(SNR_0+2.3) \tag{3-1a}$$

where SNR_0 is the signal to noise ratio at the optical detector. Notice that for high SNR there is a doubling of the SNR when demodulating. In coherent detection one often refers to the SNR at the detector as the carrier to noise ratio, CNR, and is being used widely by other authors[3]. Using this notation one may write Eqn.(3-1a) as,

$$SNR = \frac{2 \cdot CNR^2}{CNR + 2.3} \tag{3-1b}$$

3.2. Swerling Cases

The statistics of the targets encountered with our laser sensors may be classified in 3 cases, as described by Swerling ,[4]

Case 0: Steady intensity return. This would include a "glint" or point source .

Case 1: Rayleigh distributed target with long correlation time (i.e. all signal samples will be correlated and have the same amplitude).

Case 2: Rayleigh distributed target with short correlation time (i.e. all signal samples will be decorrelated and have different amplitudes).

The latter two cases exist typically with heterodyne detection and also with direct detection when the receiver aperture is smaller or equal to the size of the average speckle. If the target is changing with respect to the laser source (moving or rotating) in time, there will be a decorrelation time for the return signal. If the sensor sample repetition interval is larger than the correlation time, then we have case 2, otherwise we will have case 1, or a combination of the two.

3.3. Required SNR for a Given Probability of Detection using PDI

We want to calculated the required SNR at the detector (SNR_0) to meet a given requirement of P_d and P_{fa}. Barton[5] calls this minimum SNR at the detector as the detectability factor D. We shall use Barton's notation and call the SNR_0 before the demodulator (at the optical detector), D, and the one after the demodulator, SNR.

The probability of false alarm caused by white noise has been derived[6] to be,

$$P_{fa} = e^{-\beta^2/2} \qquad \text{where } \beta^2 \text{ is the threshold to noise ratio.} \tag{3-2}$$

The probability of detection as function of SNR for all three Swerling cases may be expressed[7],

Case 0: Steady intensity returns

$$P_d = \frac{1}{2}\left[1 - \operatorname{erf}\left(\frac{\beta - \sqrt{\text{SNR}}}{\sqrt{2}}\right)\right] \tag{3-3}$$

Case 1& 2: Rayleigh distributed returns

$$P_d = P_{fa}^{\frac{1}{2.3+\text{SNR}/2}} \quad \text{or solving} \quad \text{SNR}/2 = -\left(\frac{\beta^2}{2\cdot\ln P_d} + 1\right) \tag{3-4}$$

Using Eqns. (3-3) and (3-4) we can solve for the signal to noise required at the optical detector, D. We shall use a subscript in the detectability factor, $D_x(n)$, indicates the Swerling case, and the argument n, indicates the number of pulses averaged. We find that for the single pulse case, the relation of the $D_x(1)$ factors to the SNR is the same for all three cases. However, the relation between the SNR and P_d is not the same for all three cases

For direct detection $D_{0,1,2}(1) = \mathrm{SNR}$, and for coherent detection we solve for Eqn.(3-1a),

$$D_{0,1,2}(1) = \frac{\mathrm{SNR}}{4} \cdot \left(1 + \sqrt{1 + \frac{8 \cdot 2.3}{\mathrm{SNR}}} \right) \tag{3-5}$$

Note that for a single pulse detection, the cases 1 & 2 are the same. For case 0, the SNR has to be numerically tabulated from equation (3-3) because there is no analytical expression for the inverse error function. As implied by the above equations, the D factor is the signal to noise ratio before the envelope detector (or second detector). This may increases the required signal to noise ratio, in order to compensate for this loss.

Using Barton's early work[8] we can now write the expressions for the required signal to noise ratios, D, when we PDI a number *n* pulses. For the glint type target, case 0, using PDI on *n* samples, we get[9]

For case 0,

$D_o(n) = D_0(1)/n$ for direct detection and,

$$D_o(n) = \frac{D_0^2(1)}{2n(D_o(1) + 2.3)} \cdot \left(1 + \sqrt{1 + \frac{4 \cdot 2.3n(D_o(1) + 2.3)}{D_0^2(1)}} \right) \tag{3-6}$$

for coherent detection.

Now we can write the expression for the PDI of the Rayleigh target in term of known quantities, for coherent case 1,

$D_1(n) = D_1(1)/n$ for direct detection and,

$$D_1(n) = \frac{D_1(1)}{n} \cdot \left(\frac{D_0(n) + 2.3}{D_0(n)} \cdot \frac{D_0(1)}{D_0(1) + 2.3} \right) \tag{3-7}$$

for the coherent case.

For case 2, we get

$$D_2(n) = \frac{D_1(1)}{n} \cdot \left(\frac{D_0(1)}{D_1(1)} \right)^{1-\frac{1}{n}}$$ for direct detection and,

$$D_2(n)=\frac{D_1(1)}{n}\cdot\left(\frac{D_0(1)}{D_1(1)}\right)^{1-\frac{1}{n}}\cdot\left(\frac{D_0(n)+2.3}{D_0(n)}\cdot\frac{D_0(1)}{D_0(1)+2.3}\right) \tag{3-8}$$

In the case 2 target, we observe that using PDI helps the detection in two ways. First, as with the glint case, the D is reduced by a factor n if we ignore the second detector loss, but also the statistics approach the case 0, which are typically much more favorable. If we ignore the second detector loss (let 2.3 factor be zero) and let n be large (1-1/n ~ 1) we see the $D_2(n)$ in equation (3-8) becomes $D_0(1)/n$. This occurs because the return signal Rayleigh fluctuations are averaged out. This gain is mostly felt when high P_d values are required (>90%).
Equation (3-7) assumes that all samples or pulses used for PDI are in correlated (same amplitude), and equation (3-8) assumes they are all independent. There may be situations that fall between case 1 and 2, where only n_e independent target samples are received during the observation time t_o.

$$n_e = 1+\frac{t_0}{t_c}\leq n \tag{3-9}$$

where t_c is the correlation time of the signal. The expression for the required SNR (coherent) in this case is given by,

$$D_2(n)=\frac{D_1(1)}{n}\cdot\left(\frac{D_0(1)}{D_1(1)}\right)^{1-\frac{1}{n_e}}$$ for the direct detection case and

$$D_2(n)=\frac{D_1(1)}{n}\cdot\left(\frac{D_0(1)}{D_1(1)}\right)^{1-\frac{1}{n_e}}\cdot\left(\frac{D_0(n)+2.3}{D_0(n)}\cdot\frac{D_0(1)}{D_0(1)+2.3}\right) \tag{3-10}$$

for coherent case.

We observe that this expression is very similar to equation (3-8) for case 2, however, if n_e is small (say unity), the average of the signal fluctuation does not occur. This is apparent by the fact the if we ignore the second detector loss, we get equation (3-10) becoming equal to D1(1)/n, which is the integration of case 1.

Using this equation we can calculate the reduction of the SNR per pulse as a function of integration number. Figure 2-1 shows that for a single shot detection (PDI of 1) the CNR required for a typical coherent sensor is ~200 or 23 dB. By performing PDI of 3, the required CNR per pulse drops by a factor of 20. This effect is described by the equation (3-10) where one not only improves by the fact that we use more energy for detection, but also because we average out the negative effect of speckle or Rayleigh fading. That is to say we move from Swerling case 2 towards case 1. If the average power out of the transmitter needs to be minimized, fig. 3-2 shows a second curve of the reduced SNR multiplied by the PDI number, thus being the total energy used on a given pixel. We see that the optimum PDI number is 7, and this will be true for most coherent systems. We note that the average energy required rises after this PDI number, and this is due to the fact that the CNR per pulse is approaching unity and the second detector loss (or efficiency) begins to have a considerable effect.

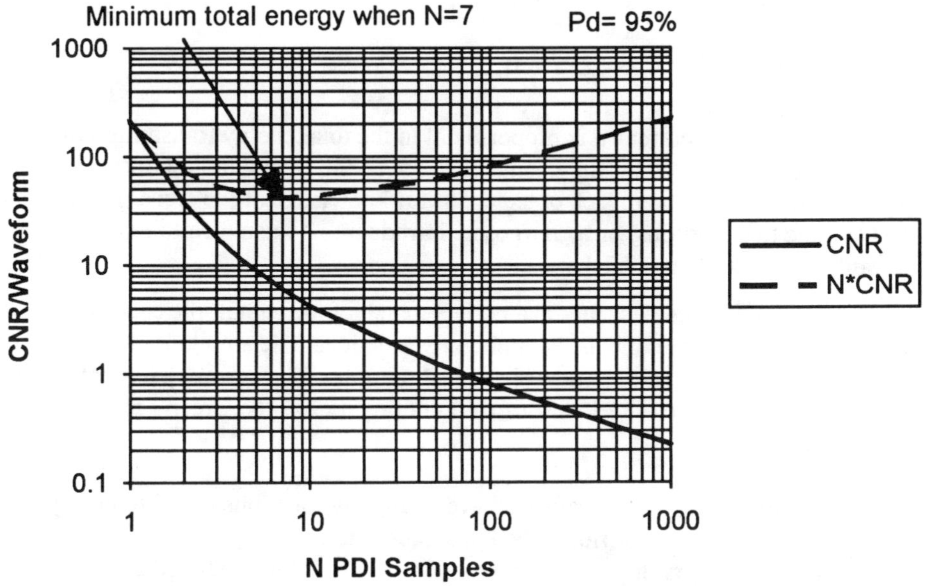

Figure 3-2. CNR required per waveform as function of number of PDI for a Pd of 95%. The minimum average power occurs when N=7.

4. Laser Radar Waveforms

Laser radar waveforms used to consist in simple AM sine, CW, and pulsed modulation. As laser sources became more mature and flexible, more complicated waveforms began to be used such as single linear chirps[10] and coherent train of chirps[11]. To understand and design these more complicated waveforms one needs to use the techniques developed in the conventional radar field such as the ambiguity function. The waveform design is concerned with specifying the radar signal that simultaneously provides; sufficient energy for detecting targets and estimating their parameters, sufficient range and velocity resolution, sufficient rejection of unwanted returns, clutter.

The effect of the transmitted waveform on both Doppler velocity and range is obtained by an extension of the correlation function of the received signal and the receiver response,

$$|\chi(t,f)| = \left|\int r(\tau)s^*(\tau - t)e^{-i2\pi f\tau}d\tau\right| \tag{4-1}$$

where t and f are time and frequency relating to the target's range and velocity. The receiver response, $r(t)$ is roughly a matched filter of the received signal, $s(t)$. The peak of the detected and filtered return occurs when the time delay (range) and the frequency shift (velocity) of the return are matched (or zeroed out). At that condition we have,

$$|\chi(t = 0, f = 0)| = 2 \cdot E \cdot \eta_{match} \tag{4-2}$$

The peak of the filtered return is proportional to the total energy transmitted in the signal waveform reduced by any mismatch between the signal and the receiver. A mismatch may occur due to lack of exact apriori knowledge of the signal parameters (lets say due to distortions) or by purposely introducing a mismatch to reduce some unwanted effect such as sidelobes.

When the received filter is a perfect match of the signal, the correlation function has been referred to as the ambiguity function. The ambiguity function yields the maximum for Eqn.(4-2) with η_{match} equal to unity. The correlation function may be thought as the response of the sensor to a point target or as the Green's function of the sensor response.

Most radar signal waveforms have ambiguity functions that fall loosely into three classes; knife edge (or ridge), bed of spikes, and thumbtack. For incoherent or direct detection sensors the ambiguity function is always knife-edge like because there is no sensitivity or dependence to frequency shift. Figure (4-1) shows a representation of these three classes.

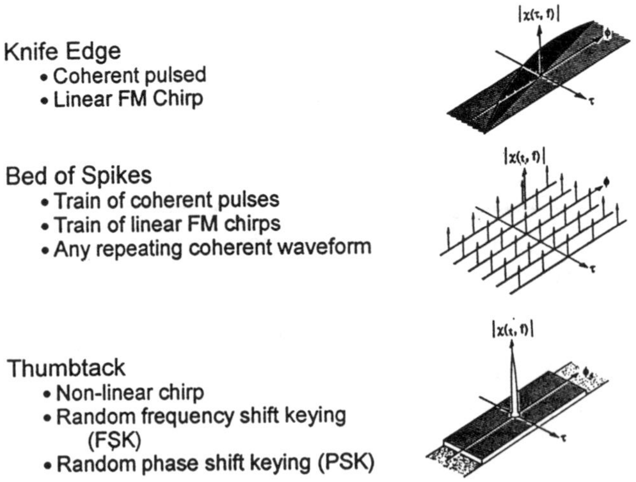

Figure 4-1. Ambiguity functions fall into three categories; (a) knife-edge, (b) bed of spikes, and (c) thumbtack.

4.1. Knife-Edge Response Function Waveform

Squared Pulse

In general these are single pulses with either no frequency modulation or a linearly changing frequency chirp. A rectangular pulse with no FM is described as,

$$s(t)=\frac{1}{\sqrt{\tau_p}}\mathrm{rect}\left(\frac{t}{\tau_p}\right)=\begin{cases}\dfrac{1}{\sqrt{\tau_p}} & |t|\le\tau_p/2\\ 0 & \text{Elsewhere}\end{cases}\tag{4-3}$$

The ambiguity function of this is given by,

$$\chi(t,f)=\mathrm{rect}\left(\frac{t}{2\tau_p}\right)\cdot e^{i\pi ft}\frac{\left(\tau_p-|t|\right)}{\tau_p}\cdot\frac{\sin\pi f\left(\tau_p-|t|\right)}{\pi f\left(\tau_p-|t|\right)}\tag{4-4}$$

Along the axis,

$$\chi(t,0)=\mathrm{rect}\left(\frac{t}{2\tau_p}\right)\cdot\frac{\left(\tau_p-|t|\right)}{\tau_p}\tag{4-5a}$$

$$\chi(0,f)=\frac{\sin\pi f\tau_p}{\pi f\tau_p}\tag{4-5b}$$

Figure (4-2) shows the plot of the ambiguity function cuts along the axis. We observe that the resolution in time and in frequency are approximately,

$$\delta t\approx\tau_p,\qquad \delta f\approx 1/\tau_p$$

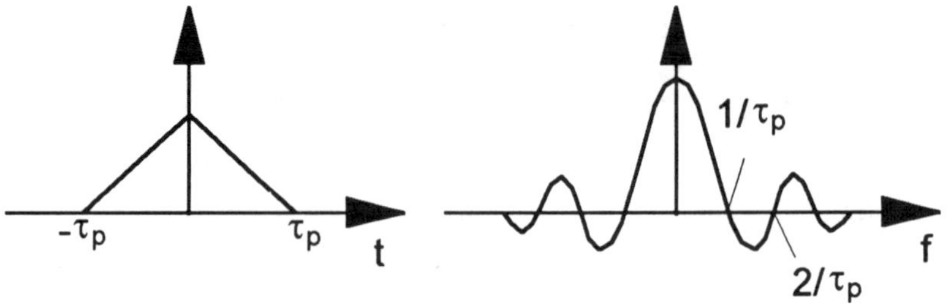

Figure 4-2. Ambiguity functions of rectangular pulse; (a) cut along the time axis, and (b) cut along the frequency axis.

Linear FM Chirp

A linear chirp is described by the following equation

$$S(t) = \frac{1}{\sqrt{\tau_p}} \operatorname{rect}\left(\frac{t}{\tau_p}\right) \cdot e^{i\pi\mu t^2} \tag{4-6}$$

where $f = \dfrac{\dot{\theta}}{2\pi} = \mu\, t$ and bandwidth $B = \mu\tau_p$

along the axis, the ambiguity function is when $B\tau_p \gg 1$

$$\chi(t,0) = \operatorname{rect}\left(\frac{t}{2\tau_p}\right) \cdot \frac{\left(\tau_p - |t|\right)}{\tau_p} \cdot \frac{\sin \pi Bt\left(1 - |t/\tau_p|\right)}{\pi Bt\left(1 - |t/\tau_p|\right)} \tag{4-7}$$

and

$$\chi(0,f) = \frac{\sin \pi f\tau_p}{\pi f\tau_p} \tag{4-8}$$

Figure (4-3) shows the plot of the ambiguity function cuts along the axis. We observe that the resolution in time and in frequency are approximately,

$$\delta t \approx 1 / B\,, \qquad \delta f \approx 1 / \tau_p$$

The product of the resolutions is now grater than unity,

$$\delta t \cdot \delta f \approx 1 / (B\tau_p)$$

which is the inverse of the bandwidth-time product, $B\tau_p$.

Large bandwidth-time product allow for large range-velocity resolution which is part of the waveform design.

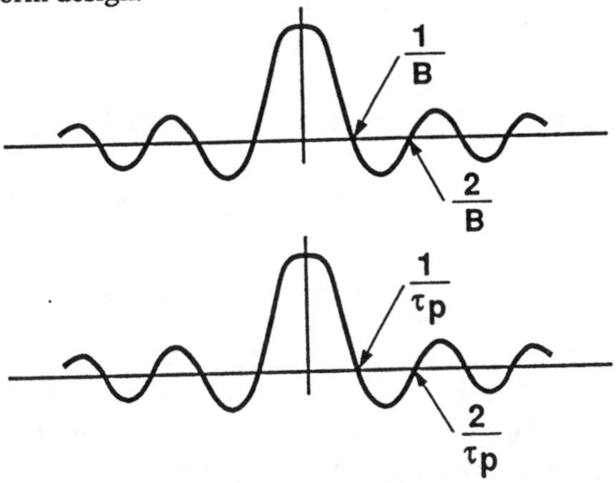

Figure 4-3. Ambiguity functions of FM Chirp pulse; (a) cut along the time axis, and (b) cut along the frequency axis.

4.2. Thumbtack Response Waveform

The ideal ambiguity function, which is usually not realizable and not processor efficient, has a single spike at the origin and is identically zero elsewhere. This means that a point target has a point response in the range-velocity plane and would be well resolved from a second point target. Typical waveforms to generate a thumbtack function are: (1) Nonlinear FM Chirp, (2) Random Frequency Shift Keying (FSK), and related (3) Random Phase Shift Keying (PSK).

Real waveforms, such as the ones mentioned, tend to have widely extending sidelobes that can be troublesome when multiple targets are to be resolved. Figure 4-4 shows the ambiguity function of an up and down chirp waveform. We observe that it is made up of two ridges that intercept at the origin and yield sort of a thumbtack peak. If a second and weaker point target is located at a range-velocity coordinate that falls on one of the ridges, that second target would be masked by the stronger one.

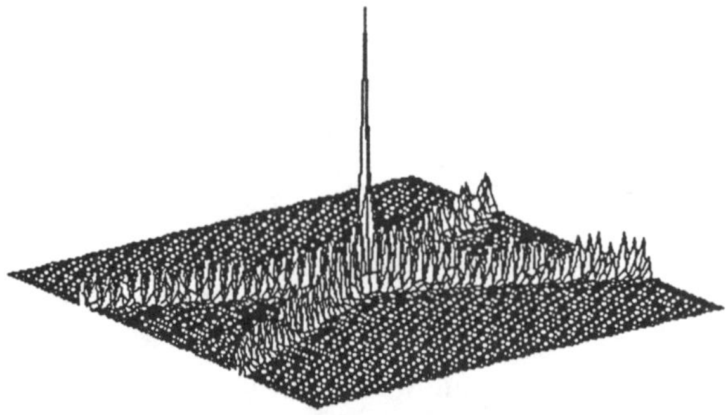

Figure 4-4. Ambiguity functions of an up then down LFM chirp waveform.

4.3. Bed-of-Spikes Response Waveform

This response function is derived from a coherent train of uniform spaced repeating sub-waveform such as a train of pulses or a train of FM chirps. The use and advantage of this response function is that may be a thumbtack like function but the total Doppler and time delay range is bounded. In other words there are repeating ambiguities in both the time delay and frequency directions. One would use the center spike as the point response and filter any responses past the first ambiguities. Figure 4-5 show a bed-of-spikes type response for a train of 4 square pulses and a train of 5 linear chirps respectively. The difference between the two waveforms is minimal from a processing point of view and one would choose between them based on the availability of a laser source for transmitter.

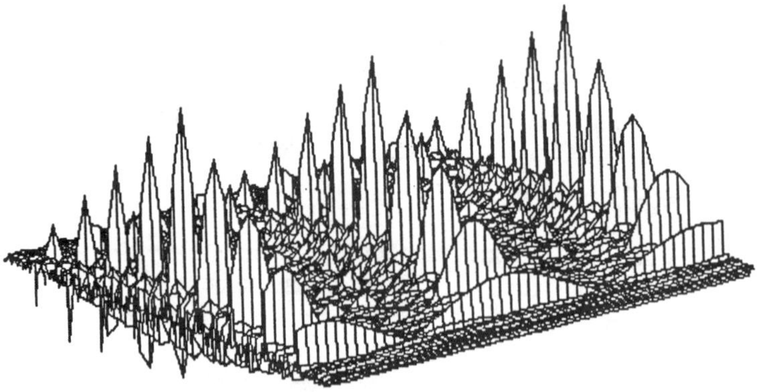

Figure 4-5. Bed of spikes ambiguity functions.

5. LADAR Transmitter Sources

Laser radar applications usually require transmitters that have: (1) high repetition rate (greater than 10 kHz), (2) good beam spatial coherence (better than 5 time diffraction limited, DL), (3) good modulation capabilities (AM, FM, Pulsed, etc...),.and (4) good energy or power capabilities (1-100 Watts average power output).

There are at present three types of laser sources that are used for radar applications; CO_2 lasers, solid state lasers, and diode lasers. CO_2 lasers operate in the infrared between 9-11 μm, solid state laser operate between 1-2 μm, and direct diodes between 0.7-0.9 μm. CO_2 lasers are the most mature and the diodes are the least mature laser transmitter sources for radar application. Solid stated laser have become viable options with the advent of "low cost" diode pumping capabilities. Table 5-1 is a summary of these laser sources.

Table 4-1. Laser transmitter sources.

Laser	**λ μm**	**Features**	**Drawbacks**
CO_2	9-11	Efficient source. Highly coherent. Supports AM, FM, and pulsed operation. Eye-safe wavelength. Dominant in the ladar efforts.	Requires a high power rf or high voltage supply source. Needs long gain lengths for high energy output. Need high vacuum technique, need cryo-cooled detectors.
Ho-Tm	2	Eye safe laser	Low repetition rates, low energy. Low efficiency except for diode pumping.
Erbium	1.54	Typical fiber based. Technology development is pushed by the fiber communication area.	Low power, low efficiency, low repetition rate.
Nd:XXX	1	With the development of diode pumping, efficiency, modulation and coherence have vastly improved. Rivals CO_2 in many applications, and is finding a niche in small light weight applications.	Not eye-safe. Pumping diodes are still relatively expensive.
Laser Diodes	0.7-0.8	Very compact and efficient. Array configurations are possible. Infant technology.	Single element average power is low. Not very good spatial or temporal coherence. Limited modulation capabilities to usually AM (sine or chirp).

Prior to the development of high power diode lasers, CO_2 lasers have been widely used for ladar because of their high electric efficiency (5-10%), good beam quality (near diffraction limited) and good frequency stability for coherent detection. Flash lamp pumped solid state lasers have low electric efficiency (0.1-1%) and poor spatial and temporal coherence hence mainly used in the laser range finder and designator area. Since the emergence of high efficiency and power diode lasers (peak power out of diode bars of 60 Watts and eff. >40% reported) in the early 1980's, it has been used to pump many solid state lasers thus increasing the efficiency and also the spatial and temporal coherence of the solid state lasers. Solid state laser radar is now a fast developing area eating away at the CO_2 dominance.

Another over-the-horizon technology is the use of direct diodes as the laser source. At present only incoherent sensors have been built and the maximum detection range has been limited due to the power limitations of the diodes themselves. Coherent diodes exist but only at very low power. Direct diodes for ladar are an exciting technology, because it allows for very small and efficient ladars to be built. Diodes lend themselves to be fabricated in one and two dimensional arrays, thus multiple beam ladars would be viable.

Selection of the proper laser for the ladar depends critically on the mission and application. From the mission requirements, one needs to know the target information, atmospheric transmission, the effective range, the accuracy and precision of the measured parameters, probability of detection and false alarm, and optical restrictions and requirements. In the next section we discuss the various ladar applications and the accompanying laser transmitters.

6. Laser Radar Systems and Applications

We shall describe to the best of our knowledge a few typical examples together with their critical components, as illustrated in Table 6-1 . The laser radar types that we include in our discussion are limited to portable units. Several outstanding large laser radar systems with ground based facility, such as MIT's Lincoln Lab, will not be included in this report. We classify the ladars into three groups based on the type of transmitter used. The first group consists of sensors with CO_2 transmitter. The second group are solid state laser sensors pumped with either a flash lamp or a diode array laser. The third group of sensors uses a direct diode array laser as the transmitter.

Table 6-1. Summary of laser sensors and associated laser sources.

Systems	Transmitter	Wave-form	Detector	Det. Type	Imag. Type	Potential Application	Developer
Imaging CO_2 Laser Radar	Waveguide CO_2 Laser	AM, FM, CW, Pulsed	HgCdTe	Cohe rent	AAR, AAD, RD, RRDI	Multi-functional	HAC, MM, Raytheon
Remote Spectrometer	TE CO_2 Laser λ agile 9-11 μm	Pulsed	HgCdTe	In-cohe rent	I-R, DISC, DIAL	Chemical Detection Air Pollution Detection	HAC, LSI
Microburst Detector	Waveguide CO_2	Pulsed	HgCdTe	Cohe rent	RD	Wind Sensing Microburst Detection	Lockheed UTOS, NASA
Wind Sensor	Flash lamp and Diode Pumped Coherent Nd:XXX (1μm) Ho-Tm (2 μm) Er (1.5 μm)	Long Pulse	InGaAs	Cohe rent	RD	Wind Sensing Microburst Detection	CTI, Light-wave
Imaging Solid-State Laser Radar	Incoherent Nd Doped SS Laser (1 μ)	Pulsed	Si APD	In-cohe rent	AAR AAI	Multi-functional	Spectra Physics HAC Hercules LTV
Imaging Diode Laser Radar	GaAs Diodes 0.7- 0.9 μm	AM Sine AM Chirp	Si APD	In-cohe rent	AAR	Multi-functional	HDOS SEO

The imaging type functions are:

- AAR for angle-angle-range requires scanning
- AAD for angle-angle-Doppler, requires scanning
- RD for range-Doppler, no scanning is necessary
- RRDI for range resolved Doppler imagery (synonymous with ISAR, inverse synthetic aperture radar)
- I-R for intensity and range
- V Vibration spectrum

Laser radar has many potential applications. They can be classified into three categories: (1) multi-functional radar for target detection and recognition, (2) aircraft avionics, and (3) chemical agent and pollution detection. For target detection and recognition applications, the three-dimensional laser radar imaging can perform many functions such as wire detection and obstacle avoidance, terrain and feature following, three-dimensional target identification, and precision targeting, as illustrated in Figure 6-1. Potential applications includes striking aircraft, helicopter, and missiles for airborne applications. It also can be used on ground-to air and ground-to ground applications.

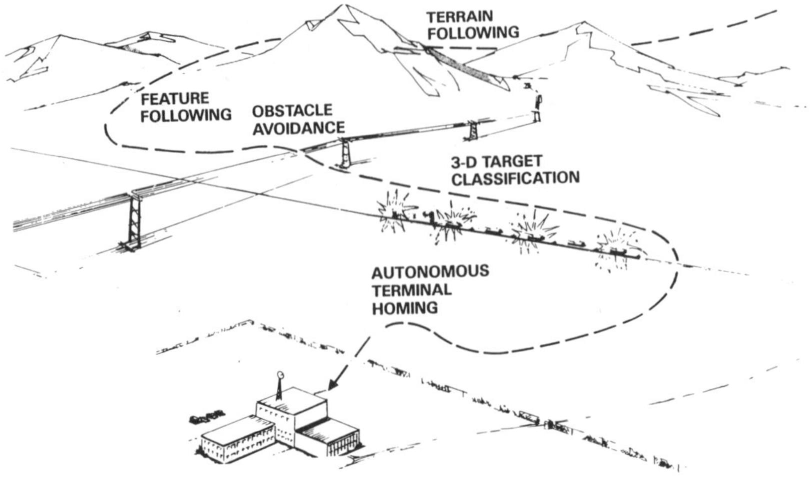

Figure 6-1. Multifunction ladar allows for automatic guidance of air vehicle.

In the case for the wide area target search, laser radar has some limitation because it takes too much time due to its small beam divergence. One will use a passive sensor for searching for this function and then hand-over to laser radar for target identification, as illustrated in fig. 6-2. Notice that laser radar is only used with a very small field of view for final target identification.

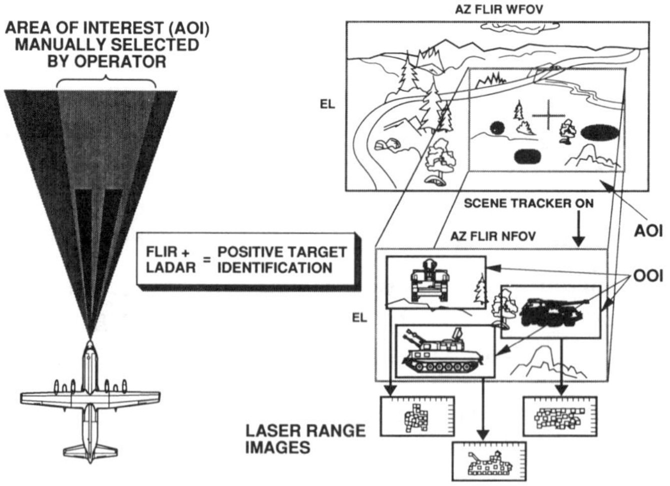

Figure 6-2. Wide area search target identification using a ladar assisted FLIR.

For the commercial avionics aircraft industry , there is a safety requirement to develop a multi-functional sensor suite on board for the detection and avoidance of hazards during all phases of flight. Major functions for the sensor include obstacle avoidance in the fog-in condition during taxi, takeoff and landing, also wind shear and wake vortex detections during approaching and landing; aircraft velocity vector measurement, and ash, hail, and clear air turbulence detection during the mid-flight as illustrated fig. 6-3. A laser radar can perform these functions with various laser power requirements for different functions. The sensor has to be eye-safe, lightweight, compact, low maintenance and low cost. Time line and conditions for which each function will be used are shown in fig. 6-4.

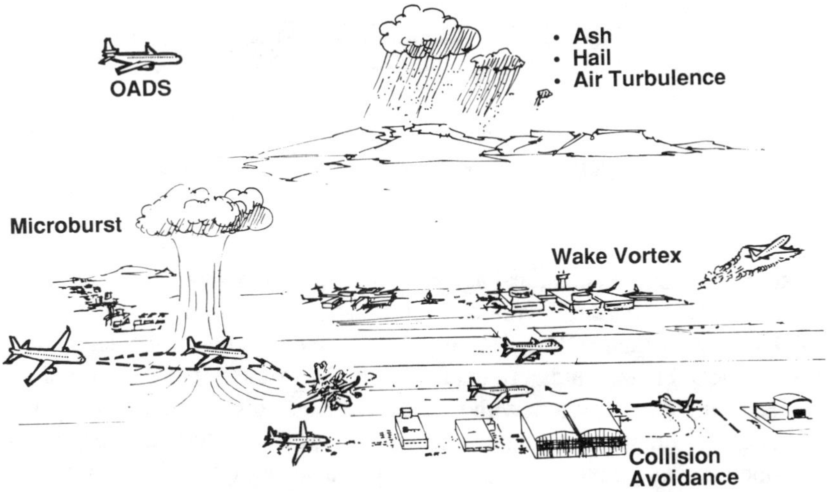

Figure 6-3. Pictorial representation of the aircraft threats that the ladar must address.

Low altitude wind shear is recognized by the commercial industry as a major hazard. FAA has mandated that all commercial aircraft must be equipped with a wind shear detection sensor in the near future. The phenomenon of wind shear/microburst hazard is shown in fig. 6-3. When an approaching and landing aircraft enters a microburst cloud, it will sense an airlift force which will force the pilot to adjust the aircraft downward. As soon as the aircraft enters the core of the microburst a downward force will accelerate the nose causing the aircraft to dive and thus possibly causing it to crash to the ground. If the wind velocity profile can be measured 5-8 km ahead, the microburst will be detected and the pilot will have sufficient time to avoid the hazard. Lockheed[12] is conducting a flight test with a coherent CO_2 transceiver developed by UTOS under a contract by NASA. The results indicate that microburst indeed can be detected. A 2μm diode-pumped coherent sensor is being developed by Coherent Technologies, Inc.(CTI) for Lockheed in the parallel effort for comparison.

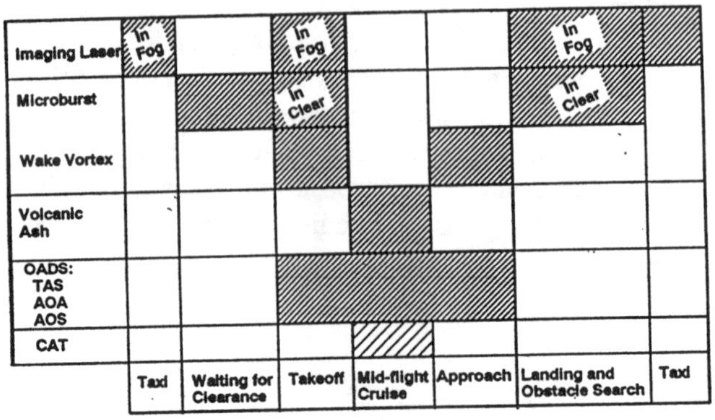

Figure 6-4. Timeline of functions required of the commercial avionics ladar during phases of a flight.

Remote sensing of chemical agents and air pollutants has been actively investigated with systems employing differential absorption lidar (DIAL) and differential scattering lidar (DISC) techniques. Since most chemical agents have absorption in the 9-11 μm band, a frequency-agile CO_2 transmitter is required for these applications. Using a lidar system with multi-wavelength laser, one can not only detect but also identify and map the contaminated area. Figure 6-5 illustrates the general concept, functions, and the first remote active spectrometer (RAS) that Hughes developed. This work is supported in part by U.S. Army CRDEC (Chemical Research and Development Center).

DIAL is usually used for chemical detection in the vapor phase. There are two ways to process the information: the column content DIAL and the range-resolved DIAL. In the first method, laser pulses are transmitted through the chemical cloud, reflected off a terrain background, and detected by the sensor receiver. Because absorption of the chemicals is wavelength dependent, the difference in amplitude of pulses at different wavelengths can be used to detect and characterize chemical clouds. In the range-resolved DIAL, one measures the return signal from the atmospheric aerosol scattering instead of the signal from the terrain scattering. By using the time-gated signal processing, one can derive the range-resolved chemical concentration. It requires, however, a transmitter with much larger output energy because the aerosol has a much smaller scattering cross-section than that of terrain. For the detection of aerosol chemicals and ground contamination, one has to use the DISC technique instead of DIAL.

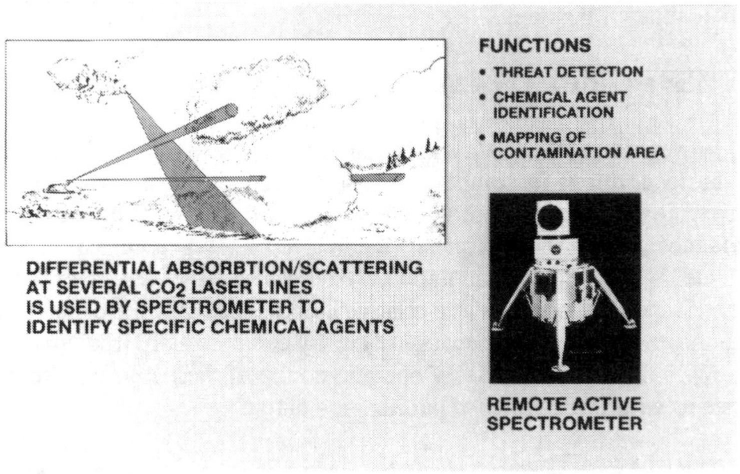

Figure 6-5. Remote chemical sensing with a CO_2 active spectrometer.

This type of remote sensor can also be used to monitor air pollutants from such industries as oil refineries, off-shore platforms, chemical plants, paint industry, hazardous waste sites and toxic waste treatment plants. Figure 6-6 compares a remote air monitor with several current techniques. Notice that gas chromatography, mass spectrometer and Fourier transform Infrared (FTIR) methods either use labor-intensive local sampling or have limited range capability. The remote sensing based on the frequency-agile laser can have the effective range of several kilometers and it does not require a reflector to enhance the return signal like what FTIR needs for 500 m range or less.

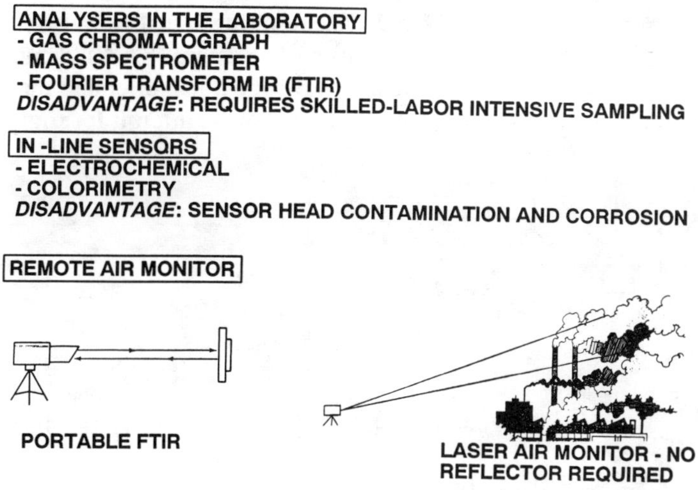

Figure 6-6. Laser based pollution sensor may detected pollutants at kilometers range compared with current instruments which operate only locally.

In the following subsections, we will describe briefly each laser radar system and its potential applications.

6.1. Waveguide CO_2 Laser Radar

Waveguide CW CO_2 lasers have in the past been one of the preferred laser sources for ladar due to its compactness and high coherence and ease of modulation. There are typically four different waveforms attained; CW with no modulation, amplitude modulated (AM), frequency modulated (FM), and pulsed. The modulation is accomplished with a proper modulation crystal placed either within or outside the laser resonator. Since CO_2 operates at a relative long wavelength, the coherent detection approach increases the sensor sensitivity greatly compared with the direct detection method. With the exception of CW operation, most of these sensors are imaging laser radar systems with three dimensional imaging output.

The CW CO_2 laser radar can be used for extremely accurate Doppler velocity measurement, which requires a very stable or actively frequency-stabilized transmitter. The return signal scattered from the target has a frequency shift due to the radial relative velocity of the sensor and the target. By mixing the return signal with a fraction of the transmitter beam in a homodyne mode, one can derive the relative velocity from the frequency shift of the mixing signal. Hughes has developed such a CO_2 velocimeter.

For the modulated waveguide laser, an additional CO_2 laser is used as the local oscillator, and hence operates in the heterodyne mode (as opposed to homodyne which uses only one laser). The output is usually an angle-angle-range-Doppler (AARD) image. A block diagram of a typical FM chirp laser radar is shown in fig. 6-7. The transmitter is linearly chirped up and down by applying a linear voltage ramp on a CdTe crystal. The return signal is mixed with a local oscillator signal and then its frequency is converted into proper IF band. The resulting signal is pulse compressed using a matched filter made of a surface acoustic wave device (SAW), demodulated , and coupled into a range counter. Figure 6-8 shows the transmitter and the local oscillator and Figure 6-9 shows the integrated sensor on an aircraft for flight test.

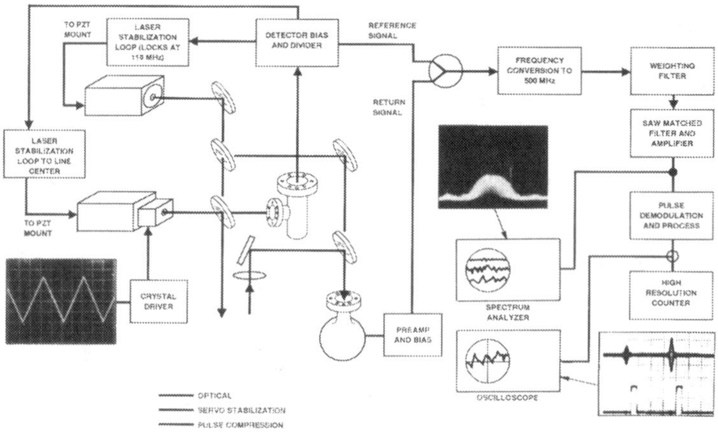

Figure 6-7. Block diagram of a linear FM chirped laser radar sensor.

Figure 6-8. Photo of the transmitter and LO lasers, both are rf exited waveguide CO_2.

Figure 6-9. FM Chirp multifunction sensor integrated on airplane nose.

We have flown sensors with AM, Linear FM Chirp, and pulsed waveforms. Figure 6-10 illustrates the AAR 3-D image of a flight tested laser radar using an AM modulated format, and fig. 6-11 illustrates the same type of image using a pulsed ladar from a roof-house test. Notice that one may detect the wires, obstacles such as the power tower, and use the 3-D imaging for target identification. This sensor may be developed into the multi-function ladar for targeting applications as well as commercial avionics applications.

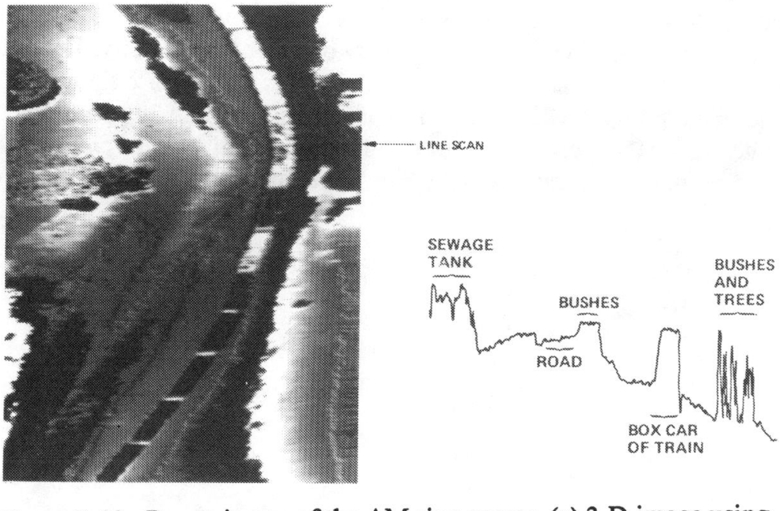

Figure 6-10. Range image of the AM sine sensor, (a) 3-D image using false colors, (b) line scan cross-section plotting range vs. cross-range.

Figure 6-11. Pixel-registered range image and passive IR of power lines from roof top test.

6.2. Frequency-Agile TE CO_2 Laser Sensor

Hughes has been developing frequency agile CO_2 laser for chemical agent detection over the last 10 years. The first prototype, Remote Active Spectrometer (RAS), was developed in 1986[13]. It demonstrated successfully the concept of remote chemical detection. In the RAS system, four TE CO_2 lasers are used. Each laser has 50 mJ output with 10 Hz repetition and wavelength tuning rates. Currently, we are developing a frequency-agile transmitter to replace this four laser system. The transmitter and the breadboard sensor with their parameters are shown in fig. 6-12. This transmitter can generate over 100 mJ pulse over 60 different lines with the pulse repetition rate of 200 Hz in the burst mode of 16 scanning lines with 40% duty cycle. We have demonstrated over 53 million shots over a period of 27 days and the laser experiment was terminated by time considerations. The sensor has been successfully tested in the field and shown to have a sensitivity of better than 1% signal stability. This level of sensitivity will allow us to detect chemicals with a noise equivalent concentration-path product (NECL) of 5-10 ppm-m for freon and 30 ppm-m for 1,3,butadiene. Since the sensor is designed to have an effective range of 2 km, the detection concentration is calculated to be 2 to 5 ppb for freon and 15 ppb for 1,3,butadiene.

Laser	
Size	4 cu ft
Weight	100 lbs
Lifetime	10^8 shots
Shelflife	1 year
Output energy	> 100 mJ, all lines
Pulselength	120 nsec
Transverse modes	10
PRF	200 Hz, 40% duty cycle
Temp range	0-40 °C
Shock/vibration	3g/2g, all axes

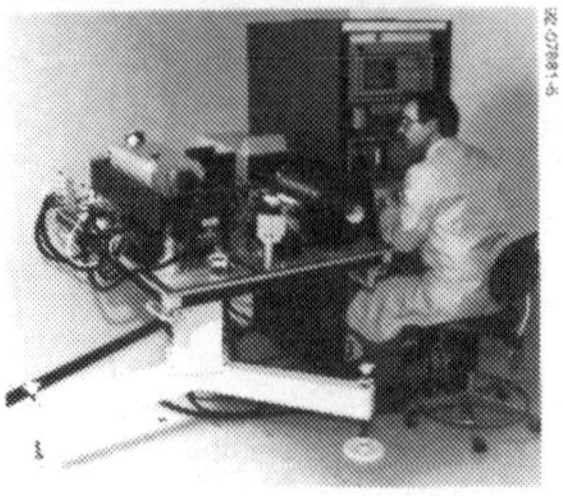

Sensors	
Telescope aperture	25 cm
Telescope FOV	4 mrad
Detector, D*	PV HgCdTe, 6×10^{10}
Bandwidth	5 MHz
Gimbal Az/El, accuracy	90°/20° ± 87 μrad
Data acquisition	Digital scope
W x L x H	40 x 44 x 36 in^3
Weight	250 lbs
Computer control for wavelength selection	Automatic

Figure 6-12. Frequency agile laser transmitter and the breadboard sensor listing key parameters.

6.3. Coherent Solid State Laser Radar

Many coherent solid state laser radars have been reported[14,15,16,17,18,19]. The transmitter can be either flash lamp pumped or diode pumped.(see discussion in the transmitter section). The emerging technology development of low cost and high efficiency diode array laser makes the diode-pumped laser radar very attractive. We will review a few sensors developed by Coherent Technologies, Inc. The first example is an eye-safe 2 μm coherent laser radar[16], the block diagram is shown in fig. 6-13. This laser radar uses a frequency-stabilized master oscillator to slave drive a Tm, Ho-doped solid state laser to provide the coherent source. It produces an output of 22 mJ with a 200 Hz repetition rate. It has been demonstrated to detect wind velocity at ranges greater than 20 km in the horizontal direction, greater than 5 km in the vertical direction, and to detect the range over hard target greater than 145 km on the mountain side. Using the similar coherent technology with a 1 Joule Nd:YAG laser[19], Lockheed and CTI has demonstrated a measurement of the crosswind velocity with a vertical range up to 25 km, as shown in fig. 6-14. This can be used to avoid launch vehicle hazards due to excessive crosswind. A diode pumped eye-safe laser radar offers many advantages because it has all solid state components, long lifetime, and is light weight and compact. It can be developed into a multi-functional sensor for commercial avionics applications.

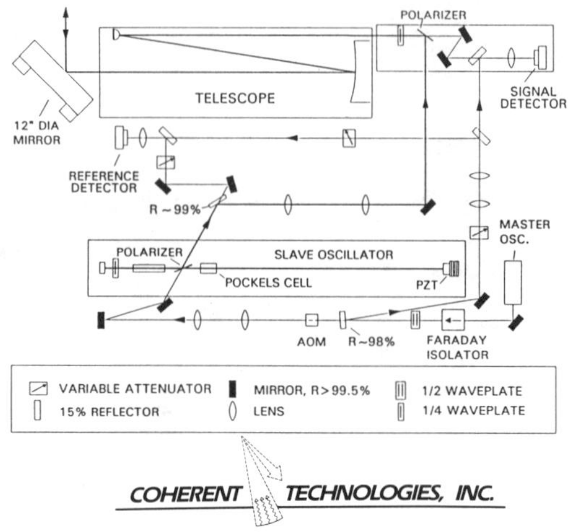

Figure 6-13. Solid state coherent transmitter for wind sensing application.

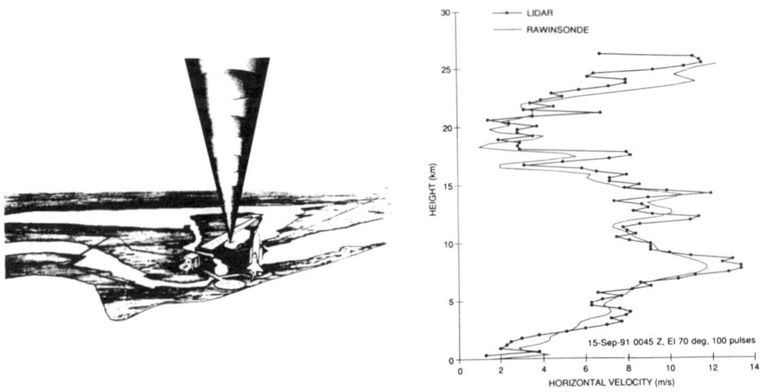

LOCKHEED/COHERENT TECHNOLOGIES, INC.

(a) (b)

Figure 6-14. (a) Lockheed and CTI wind sensor demonstration at the Kennedy Space Center, and (b) wind measurement data for heights up to 25 km.

6.4. Incoherent Diode-Pumped Solid State Laser Radar

Coherent sensors allow for complex waveform generation for high degree of fidelity and precision. However, coherent detection sensors are inherently more complicated (i.e. LO laser required). One of the present trends in laser sensor technology is to develop inexpensive, light, and compact systems. This trend makes the use of diode-pumped incoherent solid-state laser sensors very attractive. Diode pumping is a basically simple process and the expectation of cheap diodes in the future are well founded based on the semi-conductor history. A high repetition rate Q-switched diode-pumped solid-state lasers are now commercially available from companies such as Spectral Physics Lab[20]. As described in the wavelength trade section, there is no significant SNR improvement from incoherent to coherent detection systems at short wavelength, therefore, the high repetition rate Nd:XXX transmitter is suitable for the development of an angle-angle range imaging sensor. It is potentially useful for the tactical multi-functional laser radar applications.

6.5. Direct Diode Array Laser Radar

The direct diode laser is better than the diode-pumped laser for solid state laser radar because there is no conversion loss due to the optical pumping process. The diode laser, however, usually operates in the CW mode and it is difficult to make it coherent and reasonable large average power(greater than a few watts). An example of a GaAs laser radar with direct detection scheme, developed by HDOS (Hughes Danbury Optical System), is shown in fig. 6-15. The AM transmitter signal passes through a rapid scanner and exits through a window to the target. The return signal is collected by a receiver telescope and detected by a silicon avalanche photo-diode. The modulated phase delay provides the relative range information. The laser radar imaging is provided by the unidirectional scan in one axis and the airplane motion in the second axis. An image obtained from a flight test is shown in fig. 6-16. This laser radar can be developed into a multi-functional laser radar. It is, however, limited to short range applications at present because the power of available diode laser is quite low and some effort is required to scale up the power.

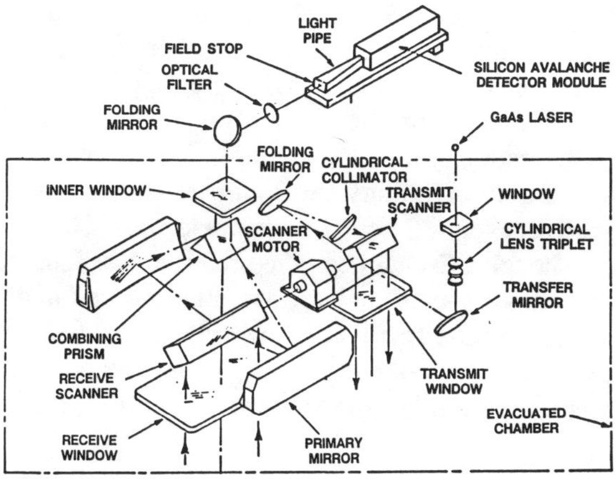

Figure 6-15. Diode bases laser radar schematic.

Figure 6-16. Angle-Angle-Range image obtained using the HUGHES Danbury diode laser radar..

7. Conclusions

Laser radar started as a modest enhancement to FLIR (Forward Looking Infra-Red) systems, however, now it has developed into an entity of its own. We derived or highlighted some of the basic equations that describe the performance of a ladar. These analytical descriptions help in designing the specific parameters of a sensor such as aperture, laser parameters, coherent versus direct detection, etc... We touched on the many uses and applications such as multi-functional sensor to perform 3-D images (AAR and AAD). This type of sensor we showed lends itself to be used for altitude detection, hazards detection, feature or target recognition for automatic guidance or taxing. These uses are found both in military and commercial avionics world. Other applications discussed were for chemical and pollutant detection, where CO_2 laser based ladars are typically used. Though CO_2 has been the preferred laser source in the past, we pointed out that diode-pumped solid state laser based ladars are rapidly gaining ground. This is particularly true were small size, low weight, high efficiency, and low maintenance are the drivers. Direct diodes as the ladar transmitters were also discussed as possibly the coming technology and one diode based sensor was presented.

8. Acknowledgments

The authors acknowledge technical contributions and useful conversations with their Hughes associates, D. Michael Henderson, Robert L. Duvall, David B. Cohn, Henry W. Steinkamp, Jack Finzi and Kuei-Ru Chien. They also thank Sammy Henderson from Coherent Technologies, Inc., Tom Danckworth from Hughes Danbury Optical Systems, Inc. and Russell Targ from Lockheed to provide information on their systems. The frequency agile CO_2 laser has been developed in part with the support of Chemical Research and Development Engineering Center (CRDEC).

References

[1]A.V. Jelalian, "Laser Radar Systems," Artech House, 1992

[2]C.G. Bachman, "Laser Radar Systems and Techniques," Artech House, INC., 1979

[3]J.H. Shapiro, B.A. Capron, and R.C. Harney "Imaging and target detection with a heterodyne-reception optical radar," APPLIED OPTICS, Vol. 20, No. 19, October 1981.

[4]P. Swerling, "Probability of Detection for Fluctuating Targets", RAND Corp. Research Memo, RM-1217, March 17, 1957, Reprinted in Trans. IRE, IT-6, No. 2, April 1960.

[5]David K. Barton, "Modern Radar System Analysis," Artech House Inc., MA 02062

[6]A. D. Whalen, "Detection of Signals in Noise," Academic Press,1971

[7]M.J. Halmos "Comparison of Glint Heterodyne Detection and Direct Detection," HUGHES IDC Ref#:72561.12, May 23, 1990

[8]David K. Barton, "Simple Procedures for Radar Detection Calculations," IEEE Transactions on Aerospace and Electronic Systems, Vol. AES-5, No. 5, p-837, September 1969.

[9]M.J. Halmos, "Calculations of the Probability for Post Detection Integration (PDI) Process Using Swerling Cases 0, 1, and 2," HUGHES IDC Ref# 72671.13, 7 June 1990.

[10]M.J. Halmos, D.M. Henderson, and R.L. Duvall III, "Pulse compression of an FM chirped CO_2 laser," Applied Optics, Vol.28, No.17, 1 September 1989.

[11]A.L. Kachelmyer, "Range-Doppler Imaging with Laser Radar," The Lincoln Lab. Journal, Vol. 3, No. 1, 1990.

[12]R. Targ, M. Karaya, R.M. Huffaker and R. Bowles, Applied Optics, 15, 2013 (1991)

[13]J.V. Cernius, D.A. Elser and J. Fox, "Remote Active Spectrometer," IRIS Specialty Group on Active Systems, Oct. 1986, Monterey, CA.

[14]T.J. Kane, W.J. Kozlovsky, R.L. Byer, and C. Byvik,"Coherent laser ladar at 1.06 μm using Nd:YAG lasers,"Optics Letters, Vol.12, 239-241, 1987.

[15]K.P. Chan, D.K. Killinger, and N. Sugimoto, "Heterodyne Doppler 1-μm lidar measurement of reduced effective telescope aperture due to atmospheric turbulence," Applied Optics, Vol. 30, No. 18, 1991

[16]S. Henderson, C. Hale, J. Magee, M. Karaya and A.V. Huffaker," Eye-safe coherent laser radar system at 2.1 μm using Tm,Ho:YAG lasers," Optical letters, 16, 773 (1991)

[17]S. Henderson, P. Suni, C. Hale, S. Harmon, J. Magee, D. Bruins and E. Yuen, " Coherent Laser Radar at 2 μm Using Solid-State Lasers," IEEE Trans. Geoscience and Remote Sensing, Jan 1993.

[18]R. Targ, M. Karaya, R.M. Huffaker and R. Bowles, "Coherent lidar airborne widshear sensor: performance evaluation," Applied Optics, Vol. 30 No. 15, p.2013 ,1991.

[19]J. Hawley, R. Targ, R. Bruner, S. Henderson, C. Hale, S. Vetorino, R. Lee, S. Harper and T. Khan, "Performance of a 1 μm 1 Joule coherent launch site atmospheric wind sounder," Proceeding of SPIE Symposium on Optical Engineering and Photonics in Aerospace Sensing, April 1-5 1991, Orlando, FL.

[20]G.J. Kintz and T. Baer, "Single-Frequency Operation in Solid-State Laser Material with Short Absorption Depths," IEEE J. QE, Vol. 26, No. 9, September 1990.